AF552618

National Skill Development Programme

Machine Shop Technology Practices

M. Adithan

Published by

ATLANTIC

PUBLISHERS & DISTRIBUTORS (P) LTD

7/22, Ansari Road, Darya Ganj, New Delhi-110002
Phones : +91-11-40775252, 23273880, 23275880, 23280451
Fax : +91-11-23285873
Web : www.atlanticbooks.com
E-mail : orders@atlanticbooks.com

Branch Office
5, Nallathambi Street, Wallajah Road, Chennai-600002
Phones : +91-44-48531784, 28411383
E-mail : chennai@atlanticbooks.com

Printed in India at Nice Printing Press, A-33/3A, Site-IV, Industrial Area, Sahibabad, Ghaziabad, U.P.

Preface

The Government of India is giving a lot of impetus to skill development amongst the youth in the country. The National Skill Development Corporation (NSDC) launched by the Government is a unique Public-Private Partnership (PPP) Enterprise which has been mandated to skill 30% of the overall target of 500 million people by the year 2022. NSDC has been incorporated as a not-for-profit organization under Section 25 of the Companies Act, 1956 by the Ministry of Finance. It has been formed as 51:49 shareholding of Private Sector and the Government. Besides the Ministry of Finance, all prominent industry bodies have contributed to the initial capital of this venture to address the need for training and skill development.

Skill development programmes envisaged under the scheme are being run by 17 Ministries/Departments in the Government of India. Large budgetary outlays have been earmarked for these programmes. Public interest demands that duplication is avoided between the work areas of the relevant sectoral entities and NSDC. There is need to develop synergy between the efforts of the Government and NSDC, which catalyses private sector interest in skill development, to optimally employ underutilized assets. For this, training, curriculum and maintaining standards will play a crucial role. Expertise needs to be developed amongst the faculty and the students of engineering colleges, institutes of technology, polytechnics and ITIs, so that the students can be trained to meet the requirements of various types of industries.

Machine Shop Technology uses three basic tools which are the lathes, drill and mill. Under these are used different types of tools in a machine shop and its processes. Boring on the lathe, centre drilling, counter sinking, reaming, collets, four jaw chuck, grooving, operations and control, and three jaw chuck are all associated with engine lathes. The computer numerical control (CNC) is a type of drilling machine that is automated and utilizes electronic controls to operate.

The micro, small and medium scale industries which use machine shop technology will be greatly benefitted from the initiative taken by the Government under the National Skill Development Programme since industries will get qualified and well-trained technical personnel for various industrial, production and manufacturing processes.

In this context, National Skill Development Programme as well as instructional sources, materials and books especially prepared for use by the technical personnel assume great significance. This book explains various machine shop technology processes, including safety in machine shop; lathe operations; drilling, reaming and tapping; milling and milling machines; boring, broaching and grinding; jigs and fixtures; microfinishing processes; machine shop measuring devices; robotics and computer-aided processes; planning, product development cycle; and the impact of innovations on production, among others.

Various processes have been explained in a lucid manner for easy understanding of the readers. The book contains a large number of labelled diagrams of machines, tools and parts. Besides explanations of operations, there are objective-type questions in each chapter with Answer Keys to help the students. A glossary of technical terms has been provided at the end.

The book will be useful for machine operators, technical trainees, students of engineering, shop floor supervisors, and engineers engaged in industrial operations.

Dr. M. Adithan
Senior Professor in Mechanical
Engineering, VIT University,
Vellore, Tamil Nadu

Contents

Preface .... *iii*

1. Safety in Machine Shop **1**
- Safety Precautions 1
- Objective Type Questions 2

2. Ferrous Materials and Non-Ferrous Materials **6**
- Basic Properties of Materials 6
- Ferrous Materials 10
- Alloying Elements in Steels 14
- Non-Ferrous Alloys 20
- Aluminium Alloys 21
- Brass 24
- Bronze 25
- Magnesium Alloys 26
- Objective Type Questions 28

3. Shaping and Planing **50**
- Machine Tools and Machining Processes 50
- Shaping Operation 53
- Planing Operation 53
- Slotting Machines 54
- Geer Shaping Machines 54
- Revision Questions 55
- Objective Type Questions 55

4. Turning (Lathe Operations) **60**
- Constructional Details of the Lathe 60
- Turning between Centres 64
- Fixed Rest (Steady) 66
- Types of Turning Tools 66
- Tool Materials used for Making of Turning Tools 66
- Tool Grinding Gauge 67

– Clamping and Adjusting of Turning Tools 67
– Right Hand and Left Hand Turning Tool..................... 68
– Precautions While Working with Lathes....................... 69
– Revision Questions 69
– Objective Type Questions.. 70

5. Drilling, Reaming and Tapping **83**
– Advantages of Twist Drills.. 83
– Parts of a Twist Drill 83
– Errors in Holes due to Improper Drill Grinding............ 87
– Drill Breakages 88
– Countersinking and Counterboring 89
– Drilling Machines 90
– Precautions to be Observed in Drilling 91
– Reaming.. 91
– Types of Reamers 92
– Tapping 93
– Revision Questions 95
– Objective Type Questions.. 95

6. Milling and Milling Machines **105**
– Cylindrical Milling.. 105
– Face Milling... 106
– Milling Cutters 106
– Milling Machines.. 109
– Precautions to be Observed while Working on Milling Machine 110
– Evaluative Questions 111
– Objective Type Questions.. 111

7. Boring **118**
– Classification of Boring Machines 118
– Specifications of a Boring Machine.............................. 124
– Boring Tools 124
– Boring Bars 126
– Boring Head 128
– Precision Boring.. 130
– Boring Defects and Causes.. 131

– Review Questions 132
– Objective Type Questions 132

8. Broaching **133**
– Principal Parts of Broach 134
– Broaching Machines 137
– Applications of Broaching 141
– Advantages of Broaching 141
– Limitations of Broaching 141
– Specifications of Broaching Machine 142
– Broaching Tools 143
– Review Questions 146
– Objective Type Questions 147

9. Grinding **152**
– Types of Grinding Machine 153
– Various Terms Related to Grinding Wheel 168
– Factors in the Selection of Grinding Wheels 175
– Standard Codification of a Grinding Wheel 176
– Wheel Dressing 177
– Wheel Truing 179
– Balancing of Grinding Wheel 179
– Mounting a Grinding Wheel on the Spindle 180
– Wheel Speed and its Impact on Grinding 181
– Work Speed and its Effect on Grinding 182
– Common Troubles, Causes and Their Remedies in Grinding Process 183
– Safety in Grinding Machines 184
– Questions 184
– Revision Questions 185
– Objective Type Questions 185

10. Jigs and Fixtures **195**
– Uses of Jigs and Fixtures 196
– Difference Between a Jig and Fixture 196
– Principles of Location 197
– Locating Devices 198
– Principles of Jig and Fixture Design 206
– Purpose of Clamping Elements 207

– Types of Clamp 208
– Support (Rest) of the Workpiece 212
– Drill Bushes and Jigs 212
– Fixtures for Milling Operations 217
– Welding Fixtures 221
– Materials for Jigs and Fixtures 225
– Economics of Jigs and Fixtures 225
– Practical Case Studies 227
– Review Questions 229
– Objective Type Questions 229

11. Micro-Finishing Processes **233**
– Honing 234
– Honing Tools 236
– Lapping 241
– Polishing 247
– Buffing 248
– Superfinishing 249
– Superfinishing Attachment on a Lathe 253
– How Micro-Finishing Processes Differ from Grinding? 254
– Review Questions 255
– Revision Questions 256
– Objective Type Questions 256

12. Machine Shop Measuring Devices (Metrology) **259**
– Classification of Measuring Devices 259
– Linear Measurements 261
– Angular Measurement 274
– Checking of Surfaces 282
– Surface Plate 285
– Limit Gauges 287
– Gear Measurements 291
– Screw Thread Measurement 295
– Tool Maker's Microscope 301
– Profile Projector 306
– Autocollimator 310
– Coordinate Measuring Machine (CMM) 311

– Review Questions 313
– Objective Type Questions 314

13. Surface Roughness and Its Measurement **328**
– Factors Contributing to Quality of Surface Finish 328
– Some Important Surface Characteristics 329
– Purpose of Finishing Surfaces with Respect to Their Function 332
– Analysis of Surface Profiles 332
– Symbols for Designating Surface Roughness 334
– Surface Roughness Obtained from Various Processes 334
– Methods of Measuring Surface Roughness 335
– Review Questions 337
– Objective Type Questions 338

14. CNC Machines **339**
– Introduction to Computer Numerical Control 339
– Definition of NC 339
– Comparison of NC and Conventional Machines 341
– Features of a CNC Machine Tool 344
– Electrical and Electronic Hardware 346
– Programming CNC Machines 348
– Objective Type Questions 350

15. Robotics **364**
– Introduction 364
– Parts of a Robot 364
– Robot Kinematics 367
– Classification of Robots 368
– Programming of Robots 372
– Sensors 374
– Applications of Robots 377

16. Computer Aided Process Planning (CAPP) **386**
– Introduction 386
– Methodology in Conventional Process Planning 388
– Manual Process Planning 390
– Evolution of CAPP 391

– Two Important Types of CAPP Systems 393
– CAD/CAM Integration and CAPP Features 400

17. Product Development Cycle **402**
– Product Development Cycle 402
– Definitions of Design 402
– Rapid Prototyping 407
– Rapid Tooling 407
– Objective Type Questions 409

18. Impact of Innovations on Product Development **416**
– Impact of Innovations on National Economy 416
– Trade Policy: Outward Oriented and Inward Oriented 417
– Export Competitiveness 418
– Product Design as a Strategy to Win International Market 418
– Innovation Process 419
– Mental Blocks for Creativity 420
– Ideation Process 422

19. Computer Integrated Manufacturing (CIM) **424**
– Definitions of CIM 425
– Components of CIM 430
– Barriers to CIM Implementation 431
– Move to CIM 432
– Tangible and Intangible Results (Benefits) 434
– Developments in CIM 435
– Future Direction of CIM 437

Glossary of Terms 442

1

Safety in Machine Shop

Safety Precautions

Safety precautions and general instructions for operation of machine tools and equipment in machine shop:

1. The machine tools/equipment must be installed and levelled properly on the foundations as per the manufacturers' drawings and instructions.
2. Electrical installation, electrical connections and earthing should be undertaken by qualified persons/electricians.
3. Familiarize yourself with the controls, switches, knobs, levers and hand wheels before you work on it.
4. Do not operate any switches, levers or knobs before you know their functions and how they work.
5. Know the proper sequence of operation of the machine tool.
6. Know the location of any emergency switch, if it is installed on the machine.
7. Satisfy yourself that the tool and the workpiece are properly secured and clamped.
8. Stop the machine before you leave the workplace even if it is for a short time.
9. Know the basic maintenance procedures like lubrication, oiling and removing chips/swarf daily.
10. Check the bearings (for any peculiar noise) and guide ways periodically and adjust them, if required.

11. Clean the machine regularly. Remove swarf/chips daily. Do not remove them by hand/fingers. They are quite sharp. Use always a brush.
12. Do not touch the chips while they are being produced. They will be quite hot.
13. Lubricate the machine tool bearings as per the lubrication chart and use proper grade of oil/grease as given by the machine manufacturer.
14. Do not remove/disturb any protective safety devices installed on the machine, such as guards, switches, shields.
15. Never take any measurements or use instruments when the job is in motion. An accident may be caused and/or measuring devices will get damaged.
16. Wear protective clothing as appropriate. Do not use full sleeve or loose shirts. Do not wear any ring, wrist watch or bangle while operating the machine tool.
17. Familiarize yourself with the accident risks and potential hazards associated with the operation of the machine tool.

OBJECTIVE TYPE QUESTIONS

Choose the correct alternative:

1. In isometric projection, two of the axes are
   (a) 15 degrees with the horizontal
   (b) 30 degrees with the horizontal
   (c) 60 degrees with the horizontal
   (d) 120 degrees with the horizontal
2. The device/instrument most often used in measuring cross-sectional areas shown on an engineering drawing is a
   (a) Pantograph (b) French curve
   (c) Geodimeter (d) Planimeter

3. For a sphere
   (a) any section will be an ellipse.
   (b) auxiliary views are necessary to fully describe the figure.
   (c) a bottom view will be necessary to adequately represent the object.
   (d) all views of the object are similar.
4. The system of presenting views of component in engineering drafting is
   (a) Trimetric projection (b) Oblique projection
   (c) Isometric projection (d) Orthographic projection
5. Which one of the following will not be put in the title block of an engineering drawing?
   (a) number of hours required to make the drawing
   (b) name of the designer
   (c) scale of the drawing
   (d) date of completion of the drawing.
6. On a drawing of a part, the heaviest lines would usually be
   (a) invisible lines (b) main dimension lines
   (c) the outline (d) the centre lines
7. If the front and top views of a solid object are circles of the same diameter, then the object would be a
   (a) Torus (b) Ellipsoid
   (c) Parallelopiped (d) Sphere
8. A pictorial drawing, two of the axes of which are 30 degrees with the horizontal, is known as
   (a) a diametric drawing (b) an isometric drawing
   (c) a perspective drawing (d) an oblique drawing
9. In the part drawing of a component, the symbol ∇ refers to the
   (a) class of fit
   (b) finish on that surface

(c) type of knurl to be machined on that surface
(d) class of steel to be used for the component

10. A detailed drawing of component from which the component is fabricated is known as a
(a) perspective drawing (b) contour drawing
(c) trilinear drawing (d) shop drawing

11. The name of the instrument used to find the areas of irregular shapes is
(a) Planimeter (b) Micrometer
(c) Perthometer (d) Vernier caliper
(e) Indicator

12. An instrument consisting of four jointed pins forming a parallelogram used for copying drawings or maps to any desired scale is called a
(a) Proportional divider (b) Beam compass
(c) Planimeter (d) Pantograph

13. On a drawing showing front, rear and side elevations and plan, the projected views are most likely to be
(a) Isogonic (b) Orthographic
(c) Isographic (d) Isometric

14. Of the following, the most important reason for checking an engineering drawing is to
(a) check accuracy of the scaling
(b) eliminate unnecessary sections
(c) rectify the errors or mistakes
(d) check for the time taken to complete the drawing

15. An approved method of obtaining the area of an irregular figure is by means of a
(a) Slide caliper (b) Micrometer
(c) Planimeter (d) Pantograph

16. On the part drawings of components which require machining, symbol ∇ denotes
(a) type of steel to be used for the component
(b) finish required on that surface

(c) class of fit
(d) tolerance value on the dimension
(e) welding is to be done

17. On a production drawing, the details of a Metric thread are specified as: M20 × 1.0. In this 1.0 represents
(a) the depth of the thread in mm
(b) the pitch of the thread in mm
(c) the diameter of the wire (in mm) used to measure the thread characteristics
(d) the diameter of the thread in mm

18. Surface roughness on a drawing is represented by
(a) Triangle (b) Rectangle
(c) Square (d) Parallelogram

19. RMS value stands for
(a) Root minimum square value
(b) Root mean square value
(c) Root maximum square value

20. CLA (Centre Line Average) value is used to represent
(a) Surface dimensions
(b) Surface hardness
(c) Surface roughness

21. How many roughness grade numbers have BIS (Bureau of Indian Standard) specified?
(a) 4 (b) 8 (c) 12 (d) 16

ANSWER KEY

1. (b)	2. (d)	3. (d)	4. (d)	5. (a)
6. (c)	7. (d)	8. (b)	9. (b)	10. (d)
11. (a)	12. (d)	13. (b)	14. (c)	15. (c)
16. (b)	17. (b)	18. (a)	19. (b)	20. (c)
21. (c)				

2

Ferrous Materials and Non-Ferrous Materials

1. Basic Properties of Materials

For a material to be suitable for the function it has to perform, it must have certain properties.

For example, a milling cutter must be harder than the materials to be cut; a material to be drawn into wire must have the property to flow into the required shape without breaking or cracks appearing in the material. A steel hook used on cranes must have enough strength to carry the load.

Some basic properties of materials required are ductility, malleability, elasticity, toughness, hardness and strength (Tensile strength, compressive strength or shear strength).

Ductility

A ductile material is one which can be drawn into wires by a tensile force. The metal from which the finest wire is produced is the most ductile material. Wires are produced by pulling rods of a specified size through holes in dies.

Ductility can be measured by a standard bend test. A strip is alternately bent and straightened through a prescribed angle, usually 90° or 180°. The number of bends required to cause fracture is taken as a measure of ductility.

Metal	Ductility
Gold	Highest
Silver	Ductility Increases ↑
Aluminium	
Iron	
Copper	
Zinc	
Tin	
Lead	Lowest

Malleability

Malleability of a material is defined as the ability to be hammered or rolled out without cracking. Under cold conditions malleability is less, but most are malleable when heated to a suitable temperature. A standard bend test is used to measure malleability.

Metal	Malleability
Gold	Highest
Silver	Malleability Increases ↑
Aluminium	
Copper	
Tin	
Lead	
Zinc	
Iron	Lowest

Elasticity

Elasticity is the property of the material to ragain its original shape after deformation. When a material is subjected to a force, a change of shape can be observed. If the force is tensile, the material is elongated. If the force is compressive the material get compressed. The elasticity of a material is very important in structures such as in bridges and buildings, aircraft and vehicles. The loads applied are absorbed by the elasticity of the materials.

Toughness

Toughness is the resistance offered by the material to fracture when subjected to impact loads, or shocks loads. Toughness is indicated by shock resistance of material.

An Izod impact testing machine is used for testing the toughness of the material. Notched specimens are used for measuring toughness. Toughness is a measure of the energy observed by the material when it fractures or breaks.

Hardness

Hardness is the property of a material which offers resistance against wear and surface indentation. Hardness is measured on different scales, (e.g. Rockwell Hardness, Brinell Hardness and Vickers Pyramid Hardness). High numbers of hardness indicate hard materials; lower numbers indicate soft materials. At higher temperatures hardness gets reduced.

Strength

The strength of a material is its ability to resist the load applied without failure. Depending on the type of forces acting on the material, strength may be defined as Tensile strength, Compressive strength or shear strength.

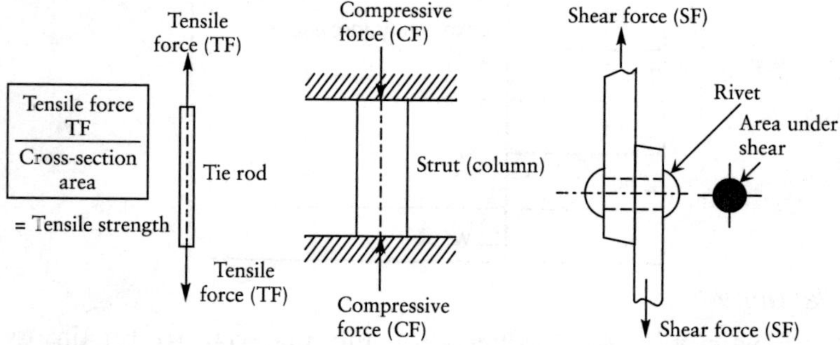

Fig. 1: Tensile Stress, Compressive Stress and Shear Stress

1. Tensile strength is the maximum force per unit area the material will withstand if loaded in tension.
2. Compressive strength is the maximum force per unit area the material will withstand if subjected to a compressive load.
3. Shear strength is the maximum force per unit area the material will withstand if a shear force is applied.

Ultimate Strength of Materials

Material	Approximate Ultimate strength in N/mm²		
	In Tension	In Compression	In Shear
Cast iron	150-180	620	90
Wrought iron	360	360	280
Mild steel	430	430	340
Copper	180	180	110
Aluminium	110	110	90
Brass	180	150	120

(N: Newton, unit of Force)

Brittleness

A brittle material breaks easily under a sharp blow, or impact force (although it may resist a steady load). Glass is a brittle material.

Machinability

It is the properly of a material which permits the material removal easily in a machining operation (i.e. by using a cutting tool), economically (less tool wear) and by providing a reasonably good surface finish.

Examples in Strength Calculations

1. A solid bar of diameter 28 mm was subjected to a tensile force of 44000 N. Calculate the tensile stress in the bar. Cross-sectional area of the bar

$$= \frac{22}{7} \times \frac{d^2}{4} \text{mm}^2, (d = 28 \text{ mm})$$

$= 616 \text{ mm}^2$

Tensile stress

$$= \frac{\text{tensile force}}{\text{cross-sectional area}}$$

$$= \frac{44000\ N}{616\,\text{mm}^2}$$

$= 71.4\ N/\text{mm}^2$

2. What shear force will be required to shear off a bar 38 mm × 9 mm if the ultimate shear stress of the material is 340 N/mm²?

   The area being sheared is

   $38 \times 9 = 342 \text{ mm}^2$

   $$\text{and stress} = \frac{\text{force}}{\text{area}}$$

   $$340 \text{ N/mm}^2 = \frac{\text{shear force N}}{342 \text{ mm}^2}$$

   $\therefore$ Shear force = 340 × 342 N = 116280 N

   = 116.28 kN

Materials can be Classified into

- Ferrous materials (such as Iron and Steel),
- Non-ferrous materials (such as Copper, Lead, Zinc, Tin, Nickel, Aluminium and their alloys),
- Synthetic materials (such as resins, fibers, plastics, rubber, wood, cotton, etc.).

2. Ferrous Materials

In Steel, there are categories such as common structural steel (e.g. m.s., mild steel), tool steel, chorium steel, etc. and cast steel.

In Iron, there are varieties such as CI (Cast Iron), chilled CI and malleable iron. Plain steel and Alloy steel are distinguished mainly by the % composition of alloying elements such as Si, Mn, Al or Ti, or Cu.

$$\text{Plain Steel} -- \begin{bmatrix} <0.5\%\ \text{Si} > 0.5\% \\ <0.8\%\ \text{Mn} > 0.8\% \\ <0.1\%\ \text{Al or Ti} > 0.1\% \\ <0.25\%\ \text{Cu} > 0.25\% \end{bmatrix} ----- \text{Alloy Steel}$$

C.I. (Cast Iron)/ Gray C.I.

C.I. contains Iron and Carbon, Carbon % ranging from 2-4% of the total material. Also, small % of Si, Mn, Phosphorus, Sulphur, Ni, Cr and Mo are present.

It is called gray C.I. because the structure of the material seen under a microscope will appear gray (grey).

Cylinder parts, machine housings, engine blocks, large gears, piston and piston rings are made of gray C.I.

Gray C.I. can be easily machined (free machining). Physically, it is relatively brittle, has little ductility and cannot be shaped by hammering or forging.

Ordinary gray C.I. is soft due to the presence of free Carbon (graphite) in the form of graphite flakes interspersed throughout its structure.

Chilled C.I./ White C.I.

When C.I. is chilled, it is called Chilled C.I. It is extremely hard, has high resistance to wear and difficult to machine. However, it can be machined by grinding or by using hard tool materials like Ceramic tools (Aluminium Oxide). When C.I. is chilled, its metallurgical structure changes to Iron Carbide (Fe_3C), called combined Carbon.

The hardened guide ways of lathe beds are examples of chilled C.I. Rail road wheels are made of chilled C.I.

Malleable Iron

In malleable iron all the Combined Carbon (viz. Iron Carbide), Fe_3C in the white C.I. is transformed to free or temper carbon or temper graphite in a nodular form instead of in the form of flakes.

Thin walled castings, wheels and keys which are subjected to high strain rates and impact loads are made of malleable Iron.

Other uses of malleable iron are automobile parts such as axles, brakes and steering gears, boiler tanks, fittings, household appliances and rail road components to withstand repeated impact loads.

Wrought Iron

Wrought Iron is another member of the iron group of materials. It has high purity containing iron silicate, a form of slag. The slag constituent is distributed as a fibrous micro-structure in the iron matrix. Rolling process is used in the manufacturing of wrought iron products such as plates, sheets, bars, rods, tubes, pipes and different types of structural steels (angles, channel and I sections). The fibrous micro-structure is in the direction of rolling.

Wrought products has high resistance to corrosion. They are extensively used in marine installations and in petroleum industries. They can withstand fatigue stresses. They can be easily machined and holds paint-protective coatings.

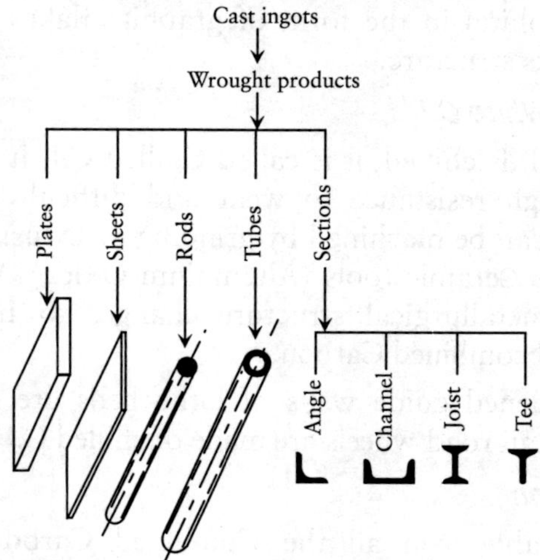

Fig. 2: Metal working processes (Forming Processes)

Ductile Iron

Ductile Iron has high ductility. Small amounts of Mg added to ordinary C.I. during the melting operation changes the microstructure of the material from graphite flakes to spheroids. This results in stronger, tougher and more ductile type of iron. It also has outstanding wear resistance.

Steels

Steel is an alloy of iron and carbon. The carbon content ranges from 0.06% to 1.7%. Within this range of carbon content there are 3 different types of steel.

Low Carbon Steel

Carbon content varies from 0.06% to 0.15%. These steels have low tensile strength, but good formability at room temperature. They are not hard steels, but their machinability is poor. Weldability is good.

There are low carbon steels with carbon varying from 0.15% to 0.30%. This type of steel has high strength, but less ductility. These steels are suitable for carburizing, where outer portion is harder with a tough core. Structural steels are of this type.

Medium Carbon Steel

Carbon content varies from 0.30% to 0.52%. This is called machinery steel. Many machine components are made of this steel. They are harder and stronger but less ductile than low carbon steels.

High Carbon Steel

Carbon content ranges from 0.55% to 0.95%. Steels in this category are readily hardened and are heat treated. Many hand tools are made from this steel.

There are many alloy steels produced for specific purposes such as higher strengths, deeper hardening, better wear resistance, ability to withstand high temperature, corrosion resistance and rust resistant (stainless steel) and for tool steels (high speed steel, etc.)

All these steels vary in their composition, micro-structure, strength and hardness, machinability characteristics, weldability, castability, and foregability. Their prices are also different.

Cast steel

If greater strength and tenacity are required cast steels are preferred. For parts subjected to severe strains cast steel can be used.

Free Cutting Steel

Sulphur is added to improve the machinability of the steel. They are called free cutting steels. Screw parts and fastener parts from free cutting steels are produced by automated machines at high speeds.

Procedure in the Selection of Steel

First establish the characteristics/requirements of the product to function properly. Then select the steel that fulfills the requirements of physical strength, resistance to corrosion, heat resistant, machinability and price (economical). Alloy steels are expensive.

3. Alloying Elements in Steels

Carbon steel is an alloy of carbon and iron. The presence of carbon, usually in excess of 0.60 per cent is essential for assuring the hardenability of steels to the levels needed for tools. Raising the carbon content by different amounts up to a maximum of 1.3 per cent increases the hardness slightly and the wear resistance considerably.

Hardenability is the property of the material which determines depth to which steel can be hardened to martensite under stated conditions of cooling.

Alloy steels have been developed to a great extent and may be regarded as steels which contain specified amounts of other elements which are added in addition to iron and carbon. The carbon is necessary for hardening and tempering; the alloying elements modify the effect of the treatment.

Alloying elements in steels are nickel, chromium, vanadium, molybdenum, tungsten, manganese and silicon.

Examples of the use of alloy steels are components for vehicle engines, modern heavy guns, tanks, cutting tools, crushing machinery, dredger parts, oil drilling equipment, railway crossings, dies, helmets, turbines, gear wheels, jet engines, chemical vessels etc.

Alloy steels are used for the following reasons:

1. To enable the heat treatment to take place uniformly in large masses of steel without the outer skin behaving differently from the inside, as would be the case if bulky sections of 'straight' carbon steel were treated.
2. To obtain a combination of mechanical properties by less drastic heat treatment than would be necessary in a plain carbon steel.
3. To enable special qualities to be imparted to the metal, such as high strength, hardness, resistance to wear, springiness, or resistance to corrosion.

Nickel (Ni)

A nickel steel usually contains from 3.0 to 3.5 per cent nickel and from 0.2 to 0.4 per cent carbon. It possesses toughness and good wear-resistance properties. It is used in the manufacture of armour plate, gun manufacture, bridge construction, in railway lines and in the automobile and aircraft industries. One of the properties required for armour plate is that it will not crack when perforated by a high velocity projectile.

A steel containing 36 per cent nickel and 0.2 per cent carbon is called invar. It is non-magnetic and stainless and has a low coefficient of thermal expansion (length is invariable). It is used in clocks and measuring instruments.

"Case-hardening nickel steel" contains 3 per cent nickel and 0.15 per cent carbon.

Chromium (Cr)

(a) High alloy tool steel

Chromium element is added in amounts up to 12 per cent for tool steels. It improves hardenability and, together with high carbon, provides both wear resistance and toughness, a combination valuable in certain tool applications. However, high chromium raises the hardening temperature of the tool steel and thus can lead to hardening deformations. A high per centage of chromium also affects the grindability of the tool steel. The chromium reduces the grain size.

(b) Stainless steel

A minimum of 13 per cent chromium is added to steel, which increases the corrosion resistance. Stainless steel with 13 per cent chromium and less than 0.1 per cent carbon is used for turbine blades. The effect of carbon in raising the hardness is very rapid. Consequently, the stainless steel used in cutlery contains 13 per cent chromium and only 0.3 per cent carbon and can be hardened and then ground to a good finish.

(c) Chromium carbon steels

Carbon steels containing chromium characterized by their high hardness, mostly contain 0.5 to 2.25 per cent chromium and 0.8 to 1.2 per cent carbon. They also have the properties of toughness and stiffness. Typical applications of this steel are in the manufacture of files, saws, ball-bearings, safes, armour plates, crushing and grinding machinery parts and permanent magnets for which purposes they are used in their hardened condition.

(d) Chrome and nickel steels

Termed as 18/8 steel, this important stainless steel contains 18 per cent chrome and 8 per cent nickel. If heated and quenched it is quite soft and is very suitable for all cold working applications. The cold working leads to considerable work hardening, but annealing can be repeated by heating and quenching. Many food vessels and tools for chemical plants are made of 18/8 steel and can be cold worked.

Vanadium (V)

This element contributes to the refinement of the carbide structure and thus improves the forget ability of alloy tool steels. Vanadium has a very strong tendency to form a hard carbide (Vanadium carbide) which improves both the hardness and the wear properties of tool steels; however, a large amount of vanadium carbide makes the grinding of the tool very difficult. Vanadium increases the temperature at which the grains begin to enlarge. Other effects are a greater retention of hardness in the tempered steel, a secondary hardness is shown by the High Speed Steel (HSS) and a superior and finer microstructure. Vanadium steels are used for high speed tool steels, spring steels, heat resistant and nitriding steels, etc.

Molybdenum (Mo)

Molybdenum has a similar effect to that of tungsten on steel, but in a more marked degree, 1 per cent of the element being as effective as about 2.5 per cent tungsten. The steels containing nickel and chromium sometimes suffer from temper-brittleness. This is a defect which arises if the steel is held at a temperature between 250°C and 500°C. The tensile strength remains high, but the shock resistance becomes dangerously low. This temperature range is often found in steam power plant. This trouble can be overcome by the introduction of 0.5 per cent molybdenum. A typical alloy contains 3.0 per cent nickel, 0.8 per cent chromium, 0.5 per cent molybdenum and 0.25 per cent carbon.

Small quantities of molybdenum yield a finer 'grain' and improve the properties of many of the heat-resistant alloys. Nickel-chrome steels containing small amounts of molybdenum, used for such purposes as dies in drop-forging, do not wear so readily and have a longer life than those free of molybdenum. This element is also a constituent of some of the 'super high speed' steels of composition of about 0.9 per cent carbon, 3.0 to 3.75 per cent chromium and 3.0 to 3.75 per cent molybdenum.

Tungsten (W)

This is one of the important alloying elements of tool steels, particularly because of two valuable properties: it improves 'hot-hardness', that is the resistance of the steel to the softening effect at elevated temperatures and it also forms hard, abrasion-resistant carbides, thus improving the wear properties of tool steels.

The property that tungsten imparts to steel is that of hardening in the air, after heating to the required temperature. These alloys are termed self-hardening steels. The composition of tungsten steels is usually 5.0 to 15.0 per cent tungsten and from 0.4 to 2.0 per cent carbon.

Air hardening steels

These are steels adequate in carbon content, completely hardened by cooling in air or a suitable gas when at a temperature above the critical or transformation point at which the pearlitic or austenitic microstructure changes to martenstic form, i.e. the steel is above its transformation temperature range before cooling.

Classification of tungsten steels are as follows

(a) Chisel and punch steels for cold work

These are medium deep-hardening steels, harden in oil and have reasonable resistance to wear, together with good resistance against impact. They are used for shear blades and can be used for dies.

(b) Tungsten die steels for hot working

These have a fine grain microstructure but are liable to crack if hardened in water, so they are better hardened in air or oil. They have a reasonable wear resistance, low distortion and high red-hardness. They readily carburize and are extensively used for hot-shaping of metals, dies for forging brass and punches and shears for hot work. They can also be used for die-casting aluminum and permanent moulds for brass and bronze.

Various hot working and cold working processes are shown in Fig. 3.

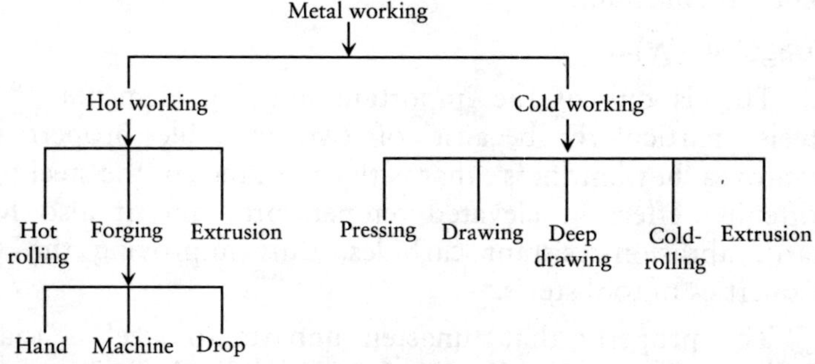

Fig. 3: Manipulation of metals through hot working and cold working

(c) Tungsten chromium die steels for hot working

They harden well in either air or oil, have reasonable resistance to wear and fairly high strength together with red-hardness. They have greater strength than the tungsten die steels and are much more resistant to impact, but readily carburise. Applications are dies for die-casting aluminum and aluminium alloys where impact is severe on occasions.

Manganese (Mn)

Manganese carbon steel can be classified into three common compositions:

1. Upto 1.5 per cent manganese, but not exceeding 2.5 per cent has higher forging properties than ordinary carbon steel.
2. Upto 1.5 per cent manganese with not more than 0.4 per cent molybdenum for automobile crankshafts, medium tensile bolts, shafts and spindles in machinery.
3. 11 to 14 per cent manganese and from 0.8 to 1.25 per cent carbon is known as Hadfield manganese steel. With the high manganese content the steel becomes ductile and very hard. The ductility of the steel is brought out by sudden cooling, the process being opposite to that employed for carbon steel. This alloy has very high work hardening properties. It can be hot forged, but is very difficult to machine as cutting quickly leads to a hardned surface. It is machined with high cobalt, high-speed steel tools. The steel is wear-resistance to a high degree, particularly when under severe pressure (load) and abrasion. With the work-hardening surface, still retaining the toughness and ductility of the core, this steel is used in railway switches and crossings, dredger parts, such as pins, links, bucket lips and bushes, steam shovel parts, parts of crushing and grinding machines, pulverisers and conveying plant.

The steel can be cast, rolled into bars and rails or forged. Its strength in the rolled condition can vary between 900 MN/m^2 and 1080 MN/m^2.

Silicon (Si)

The element silicon is a non-metal and occurs widely in combination with other elements in sands, clays and other minerals. For most engineering purposes, the silicon should be kept below 0.5 per cent, since, although higher proportions cause slightly better resistance to corrosion, particularly as regards attack by dilute sulphuric acid, they result in drastic lowering of the tensile strength, ductility and resistance to shock. Silicon steels for corrosion resistance have up to 15 per cent silicon. They have reasonable strength but cannot resist severe shock, nor corrosion by acids. With up to 4 per cent of silicon, the hot-forming properties of the steel is improved. In combination with other certain alloying elements the silicon content is sometimes raised to 2 per cent for increasing the strength and toughness of steels used for shock resisting tools.

4. Non-Ferrous Alloys

Classification of various non-ferrous metals are shown in Fig. 4.

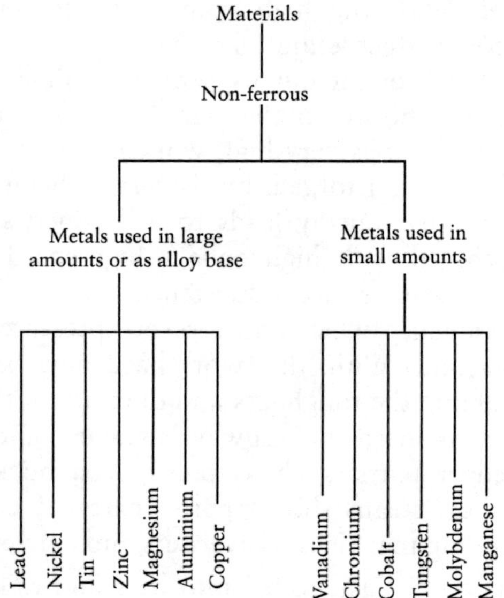

Fig. 4: Classification of metal (non-ferrous metals)

5. Aluminium Alloys

Wrought Aluminum Alloy: Non-heat Treatable

This alloy contains about 1.25 per cent manganese as a means of improving the resistance to corrosion and making the alloy easier to weld. This is available in the form of strip, sheet, etc. Various degrees of hardness or temper can be obtained, but the alloys cannot be heat treated. Mechanical properties of the order 100 to 200 MN/m^2 are possible according to the temper. The hardness values vary between Brinell hardness number 30 and 60. The alloy is used for roofing sheets and for canning of fruit and vegetables.

Other types of wrought alloy contain about 5 per cent magnesium which are widely used on account of their combination of high strength and corrosion resistance. These aluminium-magnesium alloys, develop further strength as a result of cold working, and are used in the construction of marine parts and in ship building.

Duralumin

Wrought Aluminum Alloy; (4 per cent copper): Heat Treatable

Duralumin is an alloy of copper 4 per cent, 0.4 to 0.7 per cent manganese, 0.4 to 0.7 per cent magnesium, the remainder being aluminium. It is widely used for forgings, stampings, bars, sheets, tubes and rivets in the aeronautical, automobile, shipbuilding and civil construction industries. The alloy is readily hot worked by heating to a temperature of about 500°C. This alloy is heat-treatable. After forging on annealing, duralumin may be cold-worked to an appreciable degree, and a combination of both hot and cold working is generally employed for the production of sheets. The alloy is obtained in the fully softened state by annealing at about 400°C.

Duralumin exhibits the phenomenon of age hardening to a large extent. If the alloy is heated to about 480°C and quenched in water, its hardness increases rapidly for the first day and then slowly up to a maximum value after 4 or 5 days. When it has aged it is too hard for work to be done on it, and must be annealed. The heat treated and aged duralumin has a tensile strength of about 300 MN/m^2.

The alloy has good machining properties. In its annealed condition the Brinell Hardness No. is about 60; in the aged condition it is about 100. Its density is little more than that of pure aluminium.

Cast Aluminium Alloy: Non-heat Treatable

Contains about 10 to 12 per cent silicon, the remainder being aluminium. These alloys have a close-grained structure, are ductile, can be mechanically worked and have a silvery white tint. This Alloy can be sand-cast and also die-cast, being ductile and of medium strength with excellent corrosion resistance. It is extremely castable and can be used to produce thin cross-sections of complicated form. The castings are dense, non-porous and will withstand high fluid pressure without leakage. Since silicon is lighter than aluminium, alloys of the two (i.e. silicon and aluminium) are lighter than most other aluminium alloys and it is this property resulting in applications where lightness is of primary importance. For this reason, together with its excellent corrosion properties, this alloy finds considerable applications for marine purposes. The alloys retain their strengths very well at moderately high temperatures, a property which renders them suitable for moving parts of internal combustion engines.

Metal	**Melting temperature °C**
Tin	230
Lead	330
Zinc	420
Aluminium	660
Magnesium	650
Silver	950
Copper	1060
Gold	1060
Cast Iron	1200-1350
Nickel	1450
Iron	1500
Tungsten	3260

Cast heat-treatable Aluminium Alloy

This alloy is used for aircraft castings and for pistons of diesel engines and for other types of pistons. The composition is typically 2.5 to 4.5 per cent copper, 1.8 to 2.3 per cent nickel and 1.2 to 1.7 per cent magnesium, with a little iron and silicon as impurities, the balance aluminium. A small titanium content is sometimes added.

An improvement in the strength of castings is observed if the alloys are cast in cast iron moulds, rather than in sand mouldings, by the gravity die-casting process. The alloy can be forged.

The strength of this alloy can be increased 50 per cent by age-hardening and it can be heat treated either in the cast or wrought conditions. This alloy provided the basis of a number of special aluminium alloys developed by automobile and aircraft engine manufactures.

The tensile strength of the alloy is about 340 MN/m^2.

Solution heat-treatment and ageing are required for the best properties. The hardness obtained by ageing may be accompanied by lower ductility.

(a) Solution heat-treatment

The first step consists of heating the alloy to a temperature below the melting temperature, usually in the range 510°C to 520°C for aluminium alloys, to put as much as possible of the alloying constituent into solid solution. The temperature is maintained for about 20 minutes to 60 minutes, according to the thickness of the part. This permits the entire material to reach a uniform temperature and the dissolving of the alloying elements in the solid solution to take place throughout. In effect, the alloying constituent has been dissolved in the aluminium and dispersed as completely as possible. The alloy is quenched and after quenching, it is in a relatively soft condition.

(b) Ageing (Age hardening)

After the solution heat-treatment the hardness of the alloy increases, rapidly at first, and then more slowly, up to a maximum value after a period of about four to five days. This phenomenon is called ageing. The age hardening process may be considerably accelerated and completed in a few hours by reheating the quenched alloy in boiling water, or some other medium, so long as the temperature does not exceed 150°C.

6. Brass

The term "brass" is applicable only to alloys of copper in which zinc is the main alloying element. A brass made of 70 per cent copper and 30 per cent zinc is referred to as cartridge brass. This alloy can be cold-rolled into sheets, drawn into wires, formed into tubes, pressed and cold-worked.

To manufacture a cartridge case, a pre-determined size of a circular blank of a certain thickness is used. The blank is pressed into the shape of a shallow cup. By a series of further pressing operations the cup is pushed and squeezed through successively smaller and smaller holes in a steel die, resulting in the walls of the cup being elongated until eventually a tube with a relatively thick base and thin walls is obtained. Further operations are carried out to the base to form the recess for the detonating cap and ensure that the cartridge fits only one type of breech (i.e. gun barrel). The rim of the cartridge is also softened by annealing so that it may be bent to clip the bullet.

Machining of 70/30 Brass

During the machining of 70/30 brass the chip or swarf comes in long ribbon form at a fast rate. It has sharp, jagged edges and the machined surface is usually torn and rough.

Cracking of Brass

When 70/30 brass has been severely cold-worked, cracking may occur. Cracking also occurs when the part is brought into contact with mild corrosive agents, such as salty water or air containing traces of acid. Cracks suddenly appear, sometimes after years of service. The remedy is to anneal at 280°C after cold-working; this treatment relieves internal stresses, but has little effect on the hardness.

Another important brass in engineering applications is Muntz metal or red brass of composition 60 per cent copper and 40 per cent zinc. This alloy can be particularly adaptable to casting and hot working by rolling, extrusion and hot stamping. Muntz metal can be machined readily. Castings are very suitable for pump parts. Machined components manufactured are bolts, pins, spindles, etc.

7. Bronze

(a) Tin bronze

An alloy of copper and 15 per cent tin produces a castable tin bronze. It is used for the manufacture of heavily loaded bearings, pumps, gears, nuts, etc., which have good mechanical strength. It also resists acid corrosion very well.

(b) Phosphor bronze

This alloy is of 5 per cent tin, about 0.1 per cent phosphorus, the remainder being copper; it has a close-grained microstructure. Phosphor bronze can be mechanically worked without difficulty, but is liable to become brittle if heated to high temperatures. It has considerable toughness, tensile strength and has yellowish tint.

The alloy's tensile strength ranges from 340 MN/m^2 to 740 MN/m^2 and the Rockwell hardness B numbers between 30 and 100.

It is readily cold-worked into rods, wires and sheets. Typical applications include bellows, Bourdon pressure gauge tubes, clutch discs, diaphragms, fuse springs, switch parts, perforated sheets, chemical equipment and welding rods. In wire form it can be bent cold through an angle of 120° without fracture.

Metal	Specific gravity
Magnesium	1.74
Aluminium	2.56
Zinc	7.1
Tin	7.2
Iron	7.8
Copper	8.6
Nickel	8.8
Silver	10.5
Lead	11.36
Mercury	13.6
Gold	19.3
Water	1.0

(c) Aluminium bronze

Two copper-base alloys containing aluminium are important.

1. Copper-base alloy containing 5 to 7 per cent aluminium which has a fine golden colour. It is highly ductile and may be cold-worked into tubing, sheets, strips and wires. It is made into parts that require strength at elevated temperatures, a low coefficient of friction against steel, or where a combination of strength and corrosion resistance is required. Aluminium bronze is used for bushings, gears, valve parts, bearings, sleeves, screws, pins and for fabricated sections.
2. Copper-base alloy containing 10 per cent aluminium and a castable alloy. It cannot be cold-worked to the same extent as the previous aluminium bronze, but can be mechanically worked in the hot state. This aluminium bronze or aluminium copper alloy is the most widely used in engineering applications for diecast parts. The castings produced possess high yield strength and good ductility, combining their hardness and wear-resistance properties, making them useful for pump castings, rods and shafts, valve fittings, nuts and bolts, propellers for small vessels, for electrical work and for heavy duty bearings.

   This alloy can be heat treated in a manner similar to a medium-carbon steel.

8. Magnesium Alloys

When alloyed with aluminium, manganese and zinc, the magnesium alloy requires sufficient mechanical strength and corrosion resistance to produce useful alloys for construction works. Magnesium alloys can be subjected to all the usual metallurgical treatments, including casting, diecasting, rolling, forging, extruding and pressing. A number of magnesium alloys are used as castings, but some alloys are available in wrought form. Certain precautions have to be taken when magnesium alloys are machined owing to the possibility of frictional heat at the tool nose generating high heat to ignite the powdery chips produced. Tools are therefore kept as sharp as possible and chips are not stored (exposed to air), but kept in a container with a lid. Powdery chips are likely to get ignited leading to fire hezard.

Wrought Magnesium Alloys

These alloys contain about 1.5 per cent manganese and are called manganese-aluminium alloys. The manganese improves the corrosion resistance properties and facilitates welding.

Usually the alloys are manufactured in five different tempers, ranging from soft to hard. The tensile strengths vary from 90 MN/m^2 to 200 MN/m^2. The Brinell hardness number varies from 27 to 54.

The magnesium alloys have a specific gravity about two-thirds of that of an aluminium alloy. Due to its lightness, the alloy is used a great deal in the construction of aircraft, space vehicles and automobiles.

There alloys are available as sheets, extruded rods, angles, sections and tubes.

Cast Magnesium Alloy

The cast magnesium alloy, has a composition of 9.5 per cent aluminium, 1.5 per cent zinc and 1.0 per cent manganese and the remainder magnesium.

The aluminium and zinc act as hardeners to raise the mechanical property of the alloy. Permanent-mould and die-casting produces large complicated thin-walled cast parts used in aircraft and space vehicles. Other castings are made for trucks, power drill parts, chain saw parts, textile and printing machinery parts, crank cases and gearboxes for vehicles, cameras etc. The lightness and high mechanical strength make it an ideal alloy for many applications. Strength varies from 140 MN/m^2 to 220 MN/m^2.

OBJECTIVE TYPE QUESTIONS

Choose the correct alternative

1. A ferrous metal is one which
   (a) does not contain iron
   (b) is an alloy
   (c) is a mixture of copper and brass
   (d) contains iron
   (e) contains aluminium
2. An alloy is
   (a) any article made of pure metal
   (b) a metal which has not yet been refined
   (c) chemically combined mixture of two or more metals
   (d) a mixture of two metals one of which must be aluminium
   (e) a mixture of two metals one of which must be iron
3. Mild steel belongs to the category of
   (a) Low-carbon steel
   (b) Medium-carbon steel
   (c) High-carbon steel
   (d) Tool steel
4. Steels whose Carbon content ranges from 0.05 to 0.30% are classified as
   (a) Low-carbon steels (b) Medium-carbon steels
   (c) High-carbon steels (d) Tool steels
5. In white cast iron, the Carbon is present in the form of
   (a) Austenite (b) Cementite
   (c) Eutectic (d) Ferrite
   (e) Graphite
6. The important property of high Silicon (12-18%) cast iron is its
   (a) high brittleness (b) high ductility
   (c) high hardness (d) high malleability
   (e) high corrosion resistance

7. Of the following types of pipes of the same diameter, the one that is *least* corrosion resistant is
   (a) Copper (b) Steel
   (c) Wrought iron (d) Brass
8. Mild steel has got the following metallic structure
   (a) Face Centred Cubic structure
   (b) Body Centred Cubic structure
   (c) Close packed hexagonal structure
   (d) Orthorhombic crystalline structure
9. Alpha Iron has got
   (a) B.C.C. crystalline structure
   (b) F.C.C. crystal structure
   (c) H.C.P. structure
   (d) Orthorhombic crystalline structure.
10. Carbon present in the chemically combined form in cast iron is known as
    (a) Austenite (b) Cementite
    (c) Ferrite (d) Martensite
    (e) Pearlite
11. The temperature and Carbon content at which eutectic reaction occurs in Fe-C equilibrium diagram are
    (a) 723°C and 0.02%C (b) 723°C and 0.80%C
    (c) 738°C and 0.69%C (d) 1130°C and 2.00%C
    (e) 1130°C and 4.30%C
12. The temperature and Carbon content at which eutectoid reaction occurs in Fe-C equilibrium diagram are
    (a) 723°C and 0.02%C (b) 723°C and 0.80%C
    (c) 1130°C and 2.00%C (d) 1130°C and 4.30%C
13. In grey cast iron the Carbon is in the form of
    (a) Cementite (b) Flakes
    (c) Nodular aggregates of graphite (d) Spheroids
    (e) Austenite

14. The property of a high Silicon cast iron is
    (a) its high brittleness
    (b) its high freezing temperature
    (c) its high hardness
    (d) its malleability
15. Malleable cast iron possesses ______ than ordinary grey cast iron.
    (a) higher compressive strength
    (b) higher % C
    (c) lower ductility
    (d) higher ductility
16. The majority of Grey Iron castings will have a total % of Carbon varying from
    (a) 6.0 – 8.0% (b) 2.5 – 4.5%
    (c) 0.5 – 1.0% (d) 0.02 – 0.04%
17. Gibb's phase rule is given by the expression F is equal to
    (a) $C + P$ (b) $C - P$
    (c) $C - P - 2$ (d) $C + P - 2$
    (e) $C - P + 2$
    where F = no. of degrees of Freedom
    C = no. of Components, and
    P = no. of Phases
18. An important product manufactured by the rolling process is
    (a) Rollers (b) Discs
    (c) Metallic cans (d) Metal rolls
    (e) I-sections (Joists)
19. Seamless tubes in mass production are manufactured by the ______ process.
    (a) Metal rolling (b) Extrusion
    (c) Wire drawing (d) Spinning
    (e) Longitudinal welding

20. In malleable iron, the Carbon is present in the form of
    (a) nodular aggregates of graphite (b) flakes
    (c) spikes (d) cementite
21. Steel is made from cast iron by removing all excess
    (a) Ferrous carbide (b) Carbon (c) Silicon
    (d) Sulphur (e) Oxygen
22. The most important element which controls the physical properties of steel is
    (a) Silicon (b) Manganese
    (c) Tungsten (d) Carbon
    (e) Chromium.
23. Structural steel or mild steel are the terms commonly used to describe the steel containing
    (a) more than 0.40% Carbon
    (b) less than 0.30% Carbon
    (c) more than 0.50% Carbon
    (d) more than 1.00% Carbon
24. Large amounts of Silicon when added to steel will increase the ________ properties of the steel.
    (a) Mechanical (b) Refractory
    (c) Corrosive (d) Magnetic
25. The steel used in cutting tools of high quality have a _______ grain structure.
    (a) fine (b) medium
    (c) coarse (d) brittle
26. Steels containing low per centages of Nickel, Tungsten, or Chromium are classified as
    (a) plain carbon steels (b) alloy steels
    (c) tool steels (d) stainless steels
27. Steels containing high percentages of elements other than Carbon are classified as
    (a) alloy steels (b) stainless steels
    (c) structural steels (d) high carbon steels

28. High Carbon tool steels are alloyed with _____ to increase their resistance to shock.
    (a) Carbon (b) Tungsten
    (c) Nickel (d) Vanadium
    (e) Chromium
29. A test commonly applied to steel of unknown quality for identification purposes is the
    (a) Acid-etch test
    (b) spark test (on a grinding wheel)
    (c) fracture test
    (d) dye-penetrant test
30. Two alloying elements of steel commonly used as purifiers or as cleaning agents are
    (a) Manganese and Silicon
    (b) Vanadium and Chromium
    (c) Molybdenum and Nickel
    (d) Tungsten and Chromium
31. An alloying element commonly used in steel to control its properties is
    (a) Tungsten (b) Silicon
    (c) Nickel (d) Magnesium
    (e) Vanadium
32. Hot working steels are alloy steels having a relatively
    (a) Low Carbon content (b) High Carbon content
    (c) Low Sulphur content (d) High Oxygen content
33. The element most prominent in H.S.S. (High Speed Steel) is
    (a) Carbon (b) Tungsten
    (c) Chromium (d) Vanadium
34. High Speed Steel (H.S.S.) which is commonly used in the manufacturing of cutting tools belongs to the category of
    (a) Low-carbon steel (b) Medium-carbon steel
    (c) High-carbon steel (d) Alloy steel
    (e) Stainless steel

35. Which one of the following metals would you use to make a pair of calipers?
    (a) H.S.S. (b) CI
    (c) Mild steel (d) Aluminium
    (e) Brass
36. Which one of the following metals is used to make cold chisel?
    (a) Nickel silver (b) Mild steel
    (c) Zinc (d) High-carbon steel
    (e) CI
37. Good quality twist drills are made of
    (a) CI (b) Mild steel
    (c) Stainless steel (d) H.S.S. (High Speed Steel)
    (e) Zinc
38. Which one of the following metals could not be forged?
    (a) Wrought Iron
    (b) Mild steel
    (c) High Speed Steel (H.S.S.)
    (d) High-carbon steel
    (e) CI
39. Stainless steel contains
    (a) Chromium, Iron and Nickel
    (b) Chromium and Nickel
    (c) Iron and Carbon
    (d) Chromium, Nickel, Iron and Carbon
40. What is the material of the cutting tool generally used on a lathe?
    (a) Cast Iron
    (b) Mild steel
    (c) Low-Carbon steel
    (d) High Speed Steel (H.S.S.)
    (e) Brass

41. Hammers should be made from
   (a) Cast Iron
   (b) High-Carbon steel
   (c) Mild steel
   (d) Case hardened mild steel
   (e) H.S.S. (High Speed Steel)
42. When observed unetched the Carbon in gray cast iron will be seen in the form of
   (a) Austenite (b) Cementite
   (c) Ferrite (d) Martensite
   (e) Graphite
43. Reinforcing steel used in R.C.C. (Reinforced-Cement-Concrete) work is made of
   (a) medium carbon steel (b) alloy steel
   (c) wrought iron (d) tool steel
44. Lead is poured into the joint between two pipes. The material of each pipe is most likely to be
   (a) Cast iron (b) Vitrified clay
   (c) Asbestos cement (d) Concrete
45. Some of the elements found in ordinary structural steel are
   (a) Silicon, Boron, and Nickel
   (b) Carbon, Manganese and Sodium
   (c) Carbon, Fluorine and Tungsten
   (d) Carbon, Silicon and Manganese
46. The frame and cover of a sewer manhole is usually made of
   (a) Stainless steel (b) Cast iron
   (c) Monel metal (d) Structural steel
   (e) Mild steel
47. The hardness of carbon tool steel will be increased when alloyed with ______ and Vanadium.
   (a) Tungsten (b) Chromium
   (c) Silicon (d) Manganese
   (e) Silver

48. The element when added to steel that will give the steel high strength and toughness is
    (a) Magnesium (b) Manganese
    (c) Phosphorous (d) Sulphur
49. Among the following materials, the most suitable material for withstanding shock and vibrations without danger of cracking is
    (a) Chilled cast iron (b) Gray cast iron
    (c) Malleable iron (d) White cast iron
50. The maximum number of cubic structures which crystalline solids may assume is
    (a) 10 (b) 2 (c) 8 (d) 3
    (e) 5
51. Ferrites are most likely to be found in which of the following?
    (a) Analogue computers
    (b) Computer programmes
    (c) Computer output printers
    (d) Key-punch machines
    (e) Computer memories
52. The property of material by virtue of which it can be drawn into fine wire is known as
    (a) Elasticity (b) Malleability
    (c) Ductility (d) Rigidity
53. Which one of the following metals will readily fracture if hit with a hammer?
    (a) Mild steel (b) Nickel silver
    (c) Brass (d) Cast iron
    (e) Lead
54. From which one of the following materials is a foundry crucible made?
    (a) Wood (b) Polystyrene
    (c) Graphite (d) Aluminium
    (e) Lead

55. Which one of the following is a structural steel shape?
    (a) H (b) T (c) V (d) I
    (e) All of the above
56. A special steel which has a temperature coefficient of almost zero and which is widely used for making measuring devices such as tapes is called
    (a) Stainless steel (b) Platinum steel
    (c) Cobalt steel (d) Invar steel
    (e) Nickel-Chrome steel
57. The unique property of cast iron is its
    (a) high tensile strength
    (b) high ductility
    (c) high density
    (d) good damping characteristics
    (e) good surface finish
58. Pearlite is a combination of
    (a) Ferrite and Austenite
    (b) Ferrite and Graphite
    (c) Graphite and Austenite
    (d) Ferrite and Cementite
    (e) Cementite and Austenite
59. In nodular iron, graphite is in the form of
    (a) Flakes (b) Spikes
    (c) Stars (d) Spheroids
60. Austenite consists of
    (a) Cementite and Alpha-Iron
    (b) Cementite and Beta-Iron
    (c) Cementite and Gamma-Iron
    (d) Cementite and Pearlite
    (e) Graphite and Ferrite
61. Brass is an alloy of two metals. They are
    (a) Chromium and Carbon
    (b) Tin and Aluminium

(c) Copper and Tin
(d) Copper and Zinc
(e) Zinc and Tin

62. The crystal structure of brass is
(a) F.C.C.
(b) B.C.C.
(c) H.C.P.
(d) Orthorhombic crystalline structure
(e) None of the above

63. Which one of the following metals would work-harden more quickly than the others?
(a) Copper (b) Brass
(c) Lead (d) Silver
(e) Aluminium

64. Bronze is an alloy of
(a) Brass and Tin (b) Zinc and Tin
(c) Sulphur and Tin (d) Copper and Zinc
(e) Copper and Tin

65. Babbitt metal is used primarily in
(a) Ball bearings (b) Roller bearings
(c) Sleeve bearings (d) Thermocouples
(e) Bridge construction

66. A specimen of Aluminium metal when observed under microscope shows:
(a) B.C.C. crystalline structure
(b) F.C.C. crystal structure
(c) H.C.P. structure
(d) A complex cubic structure
(e) Orthorhombic crystalline structure

67. Which one of the following metals is sufficiently ductile to enable it to be drawn into wire?
(a) Tin (b) Copper
(c) Lead (d) Zinc

68. Zinc has got the following crystal structure
    (a) Face Centred Cubic structure
    (b) Body Centred Cubic structure
    (c) Hexagonal Close Packed structure
    (d) Orthorhombic crystalline structure
69. Powder Metallurgy (PM) techniques are used in the production of
    (a) High Carbon Steel
    (b) HSS tools
    (c) Tungsten carbide tool bits
    (d) Twist drills
70. Asbestos is a
    (a) mineral fibre
    (b) fibre reinforced plastic
    (c) mixture of fibre and glass wool
    (d) mixture of mica and fibre
71. Mica is
    (a) a conductor of current
    (b) an insulator
    (c) a mixture of plastic and rubber
    (d) a petroleum by-product
72. Vulcanizing is usually applied to
    (a) plastic products
    (b) rubber products
    (c) mixture of plastic and rubber
73. Poly-Vinyl-Chloride (PVC) is
    (a) a compact form of mica
    (b) a mixture of wool and cork
    (c) a thermo plastic material
    (d) a thermo setting plastic material
    (e) a rubber product
74. With reference to the plastics listed, which one of the following statements is true?

I. Polystyrene
II. Poly-Vinyl-Chloride (PVC)
III. Phenol formaldehyde
IV. Tetra fluoro-ethylene
V. Polyethylene

(a) All are thermoplastic materials
(b) All are thermosetting materials
(c) I and II are both thermoplastic materials
(d) Only III is thermosetting material
(e) I and III are both thermosetting materials

75. The plastic material 'teflon' is known primarily for its

(a) great mechanical strength
(b) extreme low coefficient of friction
(c) great heat resistance
(d) great hardness
(e) good electrical conductivity

76. Which statement best describes thermoplastic materials?

(a) they soften when cooled and harden when heated
(b) they become permanently hard when heated
(c) they relate to one particular trade name
(d) they soften when heated and harden when cooled
(e) they are phenol formaldehydes

77. In specifications containing the statement "Scrubbers are constructed of P.V.C.", the term P.V.C. refers to

(a) Plain Vitrified Clay
(b) Poly Vinyl Chloride
(c) Patented Vanadium Copper
(d) Pliable Varnished Canvas

78. The soft material placed between two flanges of a CI water pipe to prevent leakage is called

(a) Gasket (b) Shim
(c) Plinth (d) Spelter

79. A process commonly used to preserve wood against decay is
    (a) Anodizing (b) Creosoting
    (c) Grouting (d) Guniting
80. Joints in cast iron water lines are generally sealed with
    (a) mortar
    (b) lead and jute fibres
    (c) couplings
    (d) quick setting cement mixture
81. The two important Copper base alloys are
    (a) Monel and Brass
    (b) Brass and Bronze
    (c) Monel and Bronze
    (d) Cupronickel and Bell metal
    (e) Tin and Phosphorus Bronze
82. The Babbitt metals are used
    (a) for dies and cutting tools
    (b) for high speed shafts
    (c) in the manufacture of gears
    (d) as bearing metals
83. Aluminium is extracted from
    (a) Anthracite ores (b) Bauxite ores
    (c) Hematite ores (d) Magnesite ores
84. Monel metal is an alloy of
    (a) Molybdenum and Aluminium
    (b) Chromium and Silicon
    (c) Nickel and Chromium
    (d) Nickel and Copper
85. Gun metal is an alloy of
    (a) Copper and Zinc
    (b) Copper and Tin
    (c) Copper, Tin and Phosphorus

(d) Copper, Tin and Silicon
(e) Copper, Zinc and Tin

86. Phosphorus Bronze is an alloy of
(a) Phosphorus and Copper
(b) Phosphorus, Copper and Tin
(c) Phosphorus, Copper and Zinc
(d) Phosphorus, Tin and Zinc

87. A process of applying outside coating on aluminium is known as
(a) Oxidizing (b) Anodizing
(c) Galvanizing

88. A process of coating zinc by hot dipping is called
(a) Anodizing (b) Galvanizing
(c) Brazing

89. The main purpose of heat treatment is to change the
(a) chemical composition of the metal
(b) mechanical properties of the metal
(c) corrosion properties of the metal
(d) surface finish on the material

90. An important process used in case-hardening steel is
(a) Tempering (b) Cyaniding
(c) Annealing (d) Spherodizing

91. All parts which have been hardened must be—to relieve the hardening strains and to increase the toughness of the part.
(a) Annealed (b) Carburized
(c) Tempered (d) Anodized
(e) Galvanized

92. To anneal hard steel so that it can be machined, the steel must be heated above its critical temperature and then
(a) cooled slowly
(b) cooled rapidly

(c) left in the furnace itself at the same temperature
(d) quenched in oil

93. It is necessary to temper all parts after hardening
(a) to reduce the hardening strains
(b) to increase the hardness
(c) to reduce the hardness
(d) to increase the tensile strength
(e) to reduce the scale formation

94 ______ can be hardened only by carburizing or cyaniding.
(a) Low carbon steel (b) Medium carbon steel
(c) High carbon steel (d) Alloy steel
(e) Tool steel

95. Carbon steels are hardened by heating the steel to its critical temperature and then
(a) cooling it slowly
(b) quenching it in water
(c) leaving it in the furnace itself at the same temperature
(d) leaving it outside the furnace

96. The element in steel which directly affects the critical temperature of the steel to be heat-treated is
(a) Sulphur (b) Phosphorus
(c) Carbon (d) Chromium

97. The highest rate of heat transfer, or the ability to heat metal most rapidly, is obtained in
(a) Molten baths (b) A fuel-fired furnace
(c) Electric furnaces (d) Pit furnace

98. High alloy steels must be heated slowly and uniformly for hardening, to avoid
(a) Scaling (b) Shrinkage
(c) Warpage (d) Segregation

99. Overheating high alloy steel when pack hardening must be avoided to prevent

(a) Low hardness and shrinkage
(b) Extreme hardness and shrinkage
(c) Distortion
(d) Scale formation

100. Localized hardening of jobs which require that only a small selected portion of the job be hardened, can be accomplished only by
(a) Flame and induction hardening
(b) Pack hardening
(c) Cyaniding
(d) Nitriding

101. Case hardening is the only method suitable for hardening
(a) High alloy steel (b) High-carbon steel
(c) Low-carbon steel (d) High speed steel

102. When hardening steel by the carburizing process, after the steel has been heated to the correct temperature and for the correct amount of time, the furnace is shut off and the steel is
(a) removed and quenched in water
(b) retained in the furnace itself to cool
(c) removed and cooled in air
(d) removed and quenched in oil

103. Hardening strains created in steel after it has been heated and quenched must be removed by
(a) Tempering (b) Annealing
(c) Normalizing (d) Spherodizing

104. The heat-treatment process used to soften hard alloy and tool steels so that they can be more easily machined is called
(a) Carburizing (b) Annealing
(c) Normalizing (d) Tempering

105. Carbon steels are generally classified as
(a) Air-hardening steels
(b) Oil-hardening steels

(c) Water-hardening steels
(d) Hydrogen-hardening steels

106. Proper control of the ______ in a heat-treating furnace is necessary to prevent excessive scaling of parts being hardened.
(a) Atmosphere (b) Temperature
(c) Fuel (d) Air-fuel ratio
(e) Draft

107. The temperature of furnace used for heat-treating steel is measured and controlled by an instrument called
(a) Hydrometer (b) Thermometer
(c) Dilatometer (d) Pyrometer
(e) Orsat-Apparatus

108. The product which results from annealing white cast iron is called
(a) Malleable iron (b) Nodular iron
(c) Spheroidal iron (d) Grey cast iron

109. A more scientific and reliable method of measuring furnace temperatures is by an instrument called
(a) Thermometer (b) Voltmeter
(c) Pyrometer (d) Planimeter

110. The element in steel which determines whether or not steel will harden when heated to its critical temperature and then quenched in oil or water is
(a) Iron (b) Carbon
(c) Vanadium (d) Tungsten
(e) Oxygen

111. The surface hardness in a low-carbon steel can be generally improved by the heat treatment process known as
(a) Hardening (b) Nitriding
(c) Carburizing (d) Tempering
(e) Normalizing

112. ______ is a process in which an iron based alloy is heated to a temperature above the upper critical temperature range and subsequently cooled in air.
(a) Hardening (b) Tempering
(c) Annealing (d) Quenching
(e) Normalizing

113. A process involving heating and cooling, usually applied to induce softening in steel is called
(a) Hardening (b) Tempering
(c) Austempering (d) Quenching
(e) Anealing

114. In nitriding steel components, the following atmosphere is generally used in the furnace
(a) Inert (b) Nascent Nitrogen
(c) Liquid Nitrogen (d) Carbon
(e) Ammonia

115. Galvanizing is the name of a process by which
(a) steel is coated with molten zinc
(b) steel is coated with molten tin
(c) copper is electroplated with silver
(d) mild steel is coated with enamel paint
(e) soldering irons are coated with solder

116. After anealing a non-ferrous metal, surface oxides formed on the metal are
(a) removed with coarse emery cloth
(b) left on the metal to protect the surface
(c) pickled in acid and then removed
(d) hammered into the surface
(e) polished to give a good colour

117. Case hardening of mild steel is carried out in order to
(a) colour the surface (b) enable it to resist wear
(c) render it stainless (d) use it for cutting tools
(e) make it malleable

118. Tempering is carried out on a part of high-carbon steel to
   (a) colour the metal
   (b) make the metal less brittle
   (c) make the metal harder
   (d) convert the metal to mild steel
   (e) allow the metal to be forged
119. When a part of high-carbon steel is made red hot, then plunged into cold water it will be
   (a) Tempered (b) Hardened
   (c) Unaffected (d) Normalized
   (e) Case-hardened
120. To eliminate the brittleness which occurs due to welding of saw blades, the weld must be
   (a) toughened (b) annealed
   (c) work hardened (d) forged
121. Slow-cooling of high carbon steels prevents the formation of ______ and the phase composition at room temperature is almost entirely pearlite.
   (a) Martensite (b) Cementite
   (c) Ferrite (d) Austenite
   (e) Carbide
122. If a high carbon steel (0.85%C) is cooled very slowly from 1000°C to room temperature, it will consist almost entirely of which one of the following?
   (a) Pearlite (b) Austenite
   (c) Ferrite (d) Cementite
   (e) Martensite
123. The softening of a metal component and the removal of internal stresses by heating the metal to its critical temperature and allowing it to cool very slowly is called
   (a) Anealing (b) Burnishing
   (c) Planishing (d) Spheroidizing

124. After heating, carbon steels are generally quenched in the medium of
    (a) Water (b) Oil
    (c) Air (d) Hydrogen

125. The temperature to which steel must be heated so that it will harden when quenched is called the _______ temperature.
    (a) Eutectic (b) Eutectoid
    (c) Critical (d) Melting

126. If steel is galvanized, it is coated with
    (a) Copper (b) Zinc (c) Tin (d) Lead

127. Galvanized iron is iron coated with
    (a) Copper (b) Aluminium
    (c) Zinc (d) Tin

128. Which test is used to check the evenness of the coating on galvanized wire?
    (a) Copper Sulphate dip
    (b) Hydrochloric Acid dip
    (c) Reverse bend test
    (d) Wrapping test

129. The following list comprises individual steps in a Spheroidizing cycle for a carbon steel in a batch furnace:
    1. Load the furnace and install inner cover
    2. Heat to desired soak temperature
    3. Cool to desired unloading temperature and unload
    4. Fast cool followed by slow cool
    5. Purge with atmosphere gas
    6. Soak for desired time

    The correct order of operation is
    (a) 152643 (b) 124643
    (c) 126453 (d) 152463

130. Acids used in pickling tanks after their usage must be _______ before they are emptied into an open tank to prevent environmental pollution.
   (a) agitated with hot steam
   (b) cooled to room temperature
   (c) neutralized with iron sulphate
   (d) neutralized with lime

ANSWER KEY

1. (d)	2. (c)	3. (a)	4. (a)	5. (b)
6. (e)	7. (b)	8. (b)	9. (a)	10. (b)
11. (e)	12. (b)	13. (b)	14. (c)	15. (d)
16. (b)	17. (e)	18. (e)	19. (b)	20. (a)
21. (b)	22. (d)	23. (b)	24. (d)	25. (a)
26. (a)	27. (a)	28. (d)	29. (b)	30. (a)
31. (b)	32. (b)	33. (b)	34. (d)	35. (c)
36. (d)	37. (d)	38. (e)	39. (d)	40. (d)
41. (b)	42. (e)	43. (a)	44. (a)	45. (d)
46. (b)	47. (a)	48. (b)	49. (c)	50. (d)
51. (e)	52. (c)	53. (d)	54. (c)	55. (e)
56. (d)	57. (d)	58. (d)	59. (d)	60. (c)
61. (d)	62. (a)	63. (b)	64. (e)	65. (c)
66. (b)	67. (b)	68. (c)	69. (c)	70. (a)
71. (b)	72. (b)	73. (c)	74. (d)	75. (b)
76. (d)	77. (b)	78. (a)	79. (b)	80. (b)
81. (b)	82. (d)	83. (b)	84. (d)	85. (e)
86. (b)	87. (b)	88. (b)	89. (b)	90. (b)
91. (c)	92. (a)	93. (a)	94. (a)	95. (b)

96. (c)	97. (c)	98. (c)	99. (c)	100. (a)
101.(c)	102. (b)	103. (a)	104. (b)	105. (c)
106.(b)	107. (d)	108. (a)	109. (c)	110. (b)
111.(c)	112. (e)	113. (e)	114. (e)	115. (a)
116.(c)	117. (b)	118. (b)	119. (b)	120. (b)
121.(a)	122. (a)	123. (a)	124. (a)	125. (c)
126.(b)	127. (b)	128. (a)	129. (a)	130. (d)

3

Shaping and Planing

Shaping and planing are two primary machining processes employed for machining of flat surfaces. Shaping is generally for machining of small flat surfaces and planing is for machining of large flat surfaces. Flat surfaces are machined to required dimensions by removing the excess material in the form of chips. Various types of machining operations are shown in Fig. 1.

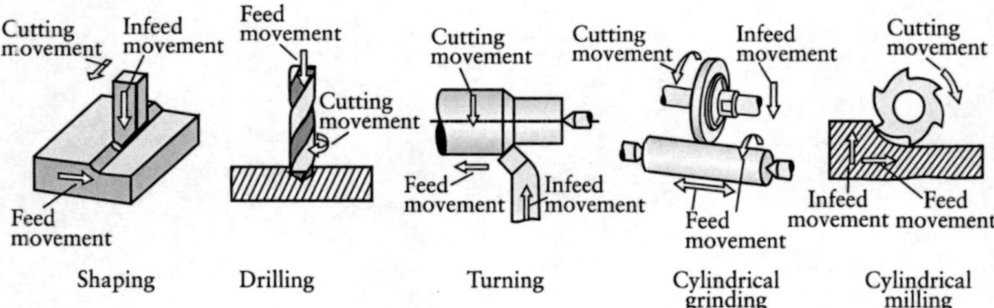

Fig. 1: Types of machining operations

Machine Tools and Machining Processes

Flat, cylindrical, tapered and many curved surfaces can be machined on workpieces or jobs using several types of machine tools. In all metal cutting processes, a tool harder than the workpiece is used and tool geometry is such that it penetrates into the workpiece and metal is removed in the form of chips. There is always a relative movement between the tool and the workpiece. The relative speed between the tool and the workpiece depends on the type of machining process employed, the tool and the work material hardness and the desired rate of metal removal and the surface quality.

There are three types of movements involved in any machining process:

(i) Main cutting movement or cutting speed or cutting velocity,

(ii) Feed movement, and

(iii) Depth of cut movement (sometimes called in feed movement or tool advance in certain types of machining processes).

(i) Cutting Velocity

The main cutting movement is performed either by the workpiece or by the tool. The speed at which the main movement takes place is referred to as the cutting speed or cutting velocity. In turning process the workpiece rotation gives the desired cutting speed or velocity. In drilling operations the drill rotation provides the cutting velocity. In milling operations the rotation of the milling cutter gives the cutting speed. In grinding operations, the rotation of the grinding wheel provides the necessary cutting velocity.

The main cutting movement may be circular or rotary as in turning, drilling, milling and grinding operations. It can be rectilinear or reciprocating as in shaping, slotting, planing or broaching.

Cutting speed is the speed at which metal removal takes place. Cutting speed is normally represented as v (cutting velocity) in metres/min. If d is the diameter of the workpiece (in mm) in a turning operation and N is the r.p.m. (revolutions per minute of the workpiece), then cutting velocity.

$$v(\text{m/min}) = \frac{\pi dN}{1000}$$

Cutting velocity is the tangential velocity at which the workpiece surface passes or moves past the stationary cutting tool tip.

The magnitude of the cutting velocity depends on the type of machining process, material from which the workpiece is to be machined, the tool material, the production rate required (i.e. metal removal rate), the desired surface quality and type of cooling of the tool tip/cutting edge.

The most suitable cutting speeds for various work materials and tool materials have been determined based on experiments and metal cutting Research and Development.

(i) *Feed movement (Feed rate)*

The feed movement moves the tool or the workpiece in successive increments (steps) in the case of rectilinear cutting movement or continuously in the case of rotary cutting movement.

Feed movement can be given manually or mechanically actuated by the machine (i.e. automatic feed). The magnitude of the feed movement depends on the surface quality desired. Higher feed rates results in rough surface, but machining takes place quickly; Finer feed rates result in good and smooth surface finish but machining takes place slowly.

Feed rate is expressed in mm/revolution for turning and drilling operations; whereas it is expressed in mm/min for milling and grinding operations and in mm/stroke for shaping and planing operations. Optimum feed rates for each machining situation can be obtained from Machining data Handbook and Tables.

(ii) *Depth of cut*

Depth of cut or tool advancement movement or in feed is the relative movement between the workpiece and tool which determines the thickness of chip to be removed. Depth of cut depends on the machining process, rigidity of construction of machine tool, the material being machined, the tool material and the required surface finish.

In shaping and planing, the main metal cutting movement involved is the rectilinear movement, to and fro. The forward movement is the cutting movement which is called the working stroke and is followed by an idle stroke (return movement or backward movement).

Shaping Operation

In shaping, the tool is having a single cutting face. It is attached to the tool side with the aid of the tool holder (tool post). During the idle stroke or return stroke the shaping tool is lifted off the workpiece automatically by a suitable lifting device. This prevents the tool cutting edge from scratching/rubbing the workpiece already machined and prolongs the tool life. A quick return mechanism provided in the shaping machine reduces the idle time and return movement can be speeded up.

Shaping tool should be mounted such that it is projecting as little as possible to avoid vibration and chatter marks on the workpiece.

In shaping operation, workpiece is held stationary, in the vice jaws and tool reciprocates, to and fro. The operation is suitable for small and medium sized parts. The feed movement is given to the workpiece at the end of the return stroke. In shaping operation, the main cutting movement is performed by the shaping tool.

Planing Operation

In planing operation, workpiece is clamped on the planing table having T-slots and reciprocates to and fro and tool is held stationery in the cross arm of the planing machine. Planing operation is suitable for heavy jobs like casting (machine tool beds). The feed and in feed movement is provided by the planning tool. Long workpiece (up to 10 m) can be machined by planing.

Depending on the direction of the main movement, both types of machine, viz. vertical shaping machines and horizontal shaping machines are available. The stroke of shaping machines are usually small since, they are meant for small jobs. Horizontal shaping machine is shown in Fig. 2.

Workpiece feed movement (i.e. work table movement) takes place with the aid of a pawl and ratchet mechanism located on the cross spindle of the cross slide (Fig. 3). Depending on the setting of the crankpin at the stroke adjusting disc, the pawl is moved over one tooth or more than one tooth, and actuates the cross spindle of the work table accordingly, to provide the desired cross feed movement. The pawl is turned through 180° to reverse the direction of feed movement.

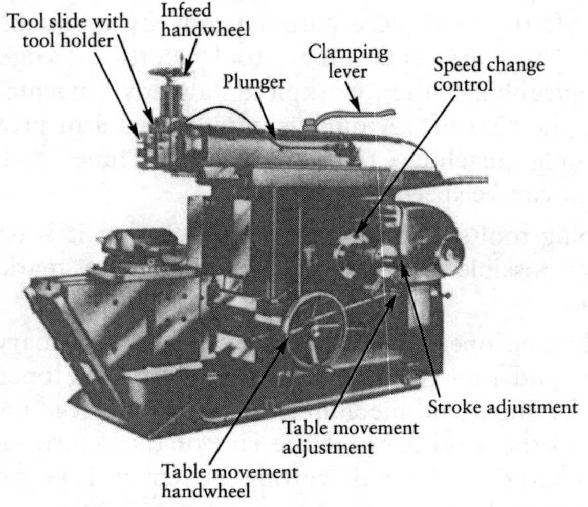

Fig. 2: Horizontal shaping machine

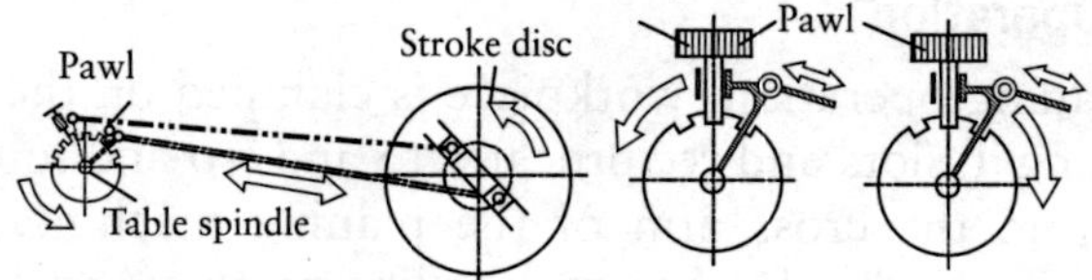

Fig. 3: Cross feed adjustment in a shaping machine

Slotting Machines

Vertical shaping machines are sometimes called slotting machines. The tool performs the main movement in the vertical direction. Slotting machines are used to cut internal and external slots (key way slots) or faces which are not easily accessible.

Gear Shaping Machines

Gear shaping machines have a cutting wheel similar to a gear wheel instead of the shaping tool. This cutting wheel is used to shape the spaces between the teeth of external and internal gears. The cutting wheel runs along the surface of the gear tooth being shaped, i.e. the tool and the workpiece mesh together like a pair of gears. For the production of external teeth a comb-type shaper can be used instead of the cutting wheel.

REVISION QUESTIONS

1. Describe the working principle of a shaping machine.
2. Sketch a planing machine and label the parts.
3. For which type of operations is the shaping machine specially suitable?
4. How can the cross feed be varied on a shaping machine?
5. Why is the planing or shaping tool lifted off during the return stroke?
6. Distinguish between a shaping machine and a planing machine.
7. For what types of jobs slotting machines are used?
8. Sketch and explain the working of a gear shaper?

OBJECTIVE TYPE QUESTIONS

Choose the correct alternative

1. The cutting action of the shaper occurs only on the ______ stroke of the ram.
   (a) feed movement (b) forward
   (c) return or backward (d) middle
   (e) end
2. The primary function of the shaper is to
   (a) produce slots
   (b) cut angles
   (c) cut key ways
   (d) produce a flat machined surface
   (e) cut dovetail
3. Shaper size is determined by the
   (a) Maximum weight of the workpiece that can be mounted
   (b) H.P. of the motor

(c) Diameter of the bull wheel
(d) Maximum stroke length of the ram

4. For proper seating of the work in a shaper vise for machining, the work should be supported on
   (a) jaws (b) clamps (c) shims (d) parallels
   (e) flats
5. Vee blocks are commonly used as work holding devices for ______ shafts which require shaping
   (a) square (b) splined (c) elliptical (d) cylindrical
6. To obtain an equal spacing of splines around the circumference of a shaft by shaping, the shaft is mounted between ______.
   (a) live centers (b) dead centers
   (c) differentials (d) indexing centers
7. Shapers equipped with a table that can be swivelled for angle cutting are referred to as ______.
   (a) Vertical shapers (b) Universal shapers
   (c) Gear shapers (d) Hydraulic shapers
   (e) Rotary shapers
8. Rotary tables for circular cuts are standard equipment on a ______.
   (a) Vertical shaper (b) Universal shaper
   (c) Rotary shaper (d) Hydraulic shaper
9. Shapers equipped with universal table are preferred when performing operations such as
   (a) vertical shaping (b) angle cutting
   (c) slotting (d) form cutting of gear teeth
10. The hinged part on the front face of a shaper tool head on which the tool post is mounted is called a
    (a) Box tool (b) Clapper box
    (c) Parallel (d) Vise
11. To prolong the life of shaper tools after they are ground, they should be ______.

(a) Lapped (b) Sanded
(c) Stoned (d) Hardened

12. The smoothly finished square or rectangular bars used as seat supports for work when shaping work in a vise are called
    (a) Clamps (b) Height adjusters
    (c) Jaws (d) Parallels
13. What work holding device would be recommended for use when shaping a key-way in a cylindrical shaft?
    (a) a V-block (b) an angle plate
    (c) a dividing head (d) a shaper vise
14. To shape splines in a shaft which must be accurately spaced, the work is mounted
    (a) in a shaper vise (b) between indexing centers
    (c) in V-blocks (d) in a special fixture
    (e) in an independent chuck
15. The factor to be considered in determining shaper size is
    (a) the overall height of the machine
    (b) the table size with which the machine is equipped
    (c) the maximum table feed movement
    (d) the maximum stroke length of the ram
16. Whitworth mechanism is used in
    (a) a shaping machine (Mechanical type)
    (b) an hydraulic shaper
    (c) a milling machine
    (d) a grinding machine
    (e) a broaching machine
17. In the case of a shaper equipped with Whitworth mechanism
    (a) the cutting stroke is faster than the return stroke
    (b) the return stroke is faster than the cutting stroke

(c) both the cutting stroke and the return stroke take the same time
(d) the return stroke is slower than the cutting stroke

18. The cutting speed of the tool in mechanical shaper is
(a) maximum at the beginning of the cutting stroke
(b) maximum at the end of the cutting stroke
(c) maximum at the middle of the cutting stroke
(d) minimum at the middle of the cutting stroke

19. The velocity of the cutting tool (or cutting speed) is uniform throughout the cutting stroke in the case of
(a) a mechanical shaper equipped with Whitworth mechanism
(b) an hydraulic shaper

20. The feeding of the job in a shaper is done by
(a) movement of the clapper box
(b) table movement
(c) V-block
(d) ram movement
(e) tool movement

21. In the case of a shaping machine, feeding of the job is done
(a) at the beginning of the cutting stroke
(b) at the middle of the cutting stroke
(c) at the end of the cutting stroke
(d) at the end of the return stroke

22. The feed rate in a shaping operation is usually expressed as
(a) mm/stroke (b) mm/double stroke
(c) mm/triple stroke (d) mm
(e) m/minute

23. In a mechanical shaper, the lifting of the tool during the idle stroke is ensured by
(a) tool head
(b) ram adjustment

(c) bull wheel movement
(d) ratchet and pawl mechanism
(e) clapper box mechanism

24. More than one tool head can be put into operation in the case of
    (a) a planing machine
    (b) a shaping machine
    (c) a slotting machine

ANSWER KEY

1. (b)	2. (d)	3. (d)	4. (d)	5. (d)
6. (d)	7. (b)	8. (a)	9. (b)	10. (b)
11. (a)	12. (d)	13. (a)	14. (b)	15. (d)
16. (a)	17. (b)	18. (c)	19. (b)	20. (b)
21. (d)	22. (b)	23. (e)	24. (a)	

4

Turning (Lathe Operations)

Turning work is carried out on a machine tool called Lathe. The main movement (rotary movement) is performed by the workpiece, the feed and the in feed movement (for setting the required depth of cut) is given by the tool.

If feed movement of the tool is performed along the axis of rotation, cylindrical parts are produced (longitudinal turning). If the feed movement is across the axis of rotations flat workpieces are produced (facing, or parting-off).

In addition, tapers can be produced by taper-turning. Threads can also be machined and contoured parts can be machined (contour turning).

If the outside faces of the workpieces are machined, it is called external turning; if the workpieces are machined internally it is called internal turning.

Constructional details of the Lathe

Lathe and its various parts are shown in Fig. 1. A lathe is provided with a lead screw and a feed rod.

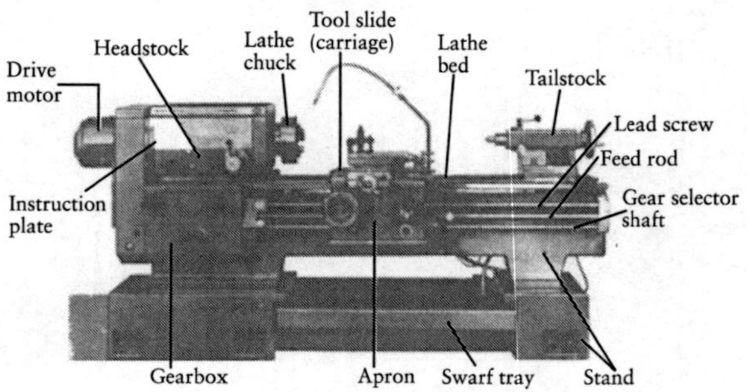

Fig. 1: Lathe and its various parts

Lathe Bed

The base supports the lathe bed and carries the fixed and moving elements. The bed absorbs all the drive forces and the cutting forces. Box type of construction is used for lathe bed. Box type construction provides good rigidity to the machine tool bed. The bed material is either CI or bed is fabricated by welding of steel plates.

Lathe bed consists of two vertical plates for mounting Vee guides and carries tool carriage (top slide and cross slide) and tail stock.

Head Stock

The head stock is mounted on the left side of the lathe bed. Head stock houses the gear box (for changing spindle speeds and for changing the feed rates). The head stock also houses the main spindle supported by bearings. The work spindle is bored for its entire length and has an internal taper at its head end to insert the centre and has a right hand thread on the projecting part of carry the 3-jaw chuck, and face plate. Work spindle is rigidly supported on the bearings in the head stock.

Tool Slide

Tool slide is also called as carriage. Carriage consists of the

- movable bed slide (saddle),
- cross slide, and
- the top slide (Tool post is accommodated on the top slide) (Fig. 2).

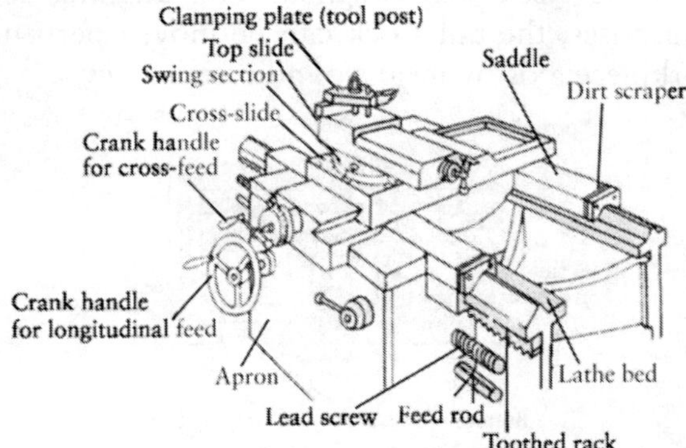

Fig. 2: Top slide and cross-slide of lathe

The top slide can be swivelled, and is used to mount the turning tools. If square tool post is mounted, four tools can be simultaneously accommodated in the tool post. The saddle and the cross slide can be moved manually by rotating the respective crank handles for the longitudinal feed and lateral (cross feed) or automatically by means of feed rod or lead screw (for thread cutting). These feed movements are obtained from the gear box installed in the apron. The feed rod and lead screw are driven by the main spindle at a speed determined by the proper selection of the gear box ratio.

Lead Screw

Lead Screw is used only for cutting or forming of threads on the workpiece, never for the longitudinal movement in longitudinal turning. Lead screw is driven by the lathe spindle at a speed depending on the pitch of the screw to be cut.

Tail Stock

Tail stock (Fig. 3) is mounted on the right side of the lathe. It acts as a support between the centres and also for holding the tools depending on the requirement. The tail stock has a V-guide and can be moved longitudinally on the lathe bed, and can be fixed in any desired position by the clamping lever. The upper part of the tail stock contains the centre sleeve which can be moved by the screw and the hand wheel. The centre or any tool with the Morse taper can be inserted into the centre sleeve. When turning cylindrical workpieces, the tail stock and head stock must be accurately aligned. For aligning and taper turning purposes, the tail stock can be moved perpendicularly to the workpiece axis by means of adjusting screws.

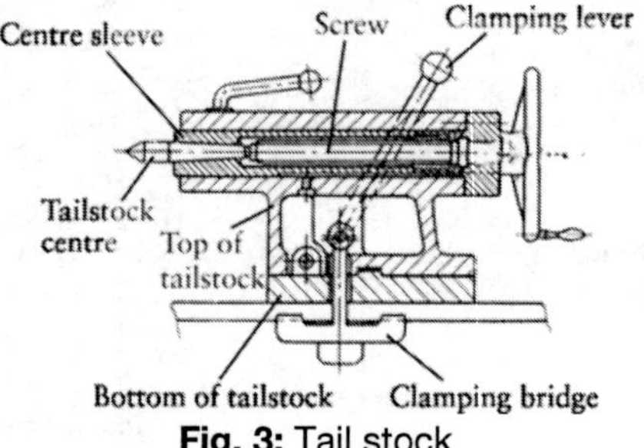

Fig. 3: Tail stock

Three Jaw Chuck

Three Jaw Chuck (Fig. 4) is screwed into the work spindle by its internal thread. Short and round shaped workpieces can be clamped into the three jaw chuck. The chuck transmits the rotary movement from the spindle to the workpiece. The internal thread of the chuck should be cleaned before screwing the chuck on to the spindle. The chuck is screwed manually.

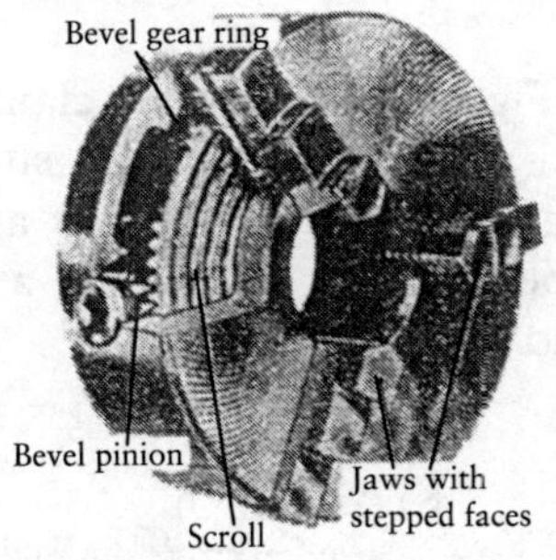

Fig. 4: Three jaw chuck

A chuck key is used to move all the 3-jaws uniformly inside or outside so that the job is clamped centrally. For large diameter workpieces the jaws are reverse clamped. The jaws should be inserted according to their numbering. Use the chuck belonging to the particular lathe. Chucks of other machines will not run smoothly.

Four Jaw Chuck

This chuck is used to clamp irregularly shaped external workpieces. The jaws are moved independently suiting the external surface for the workpeice.

Faceplate

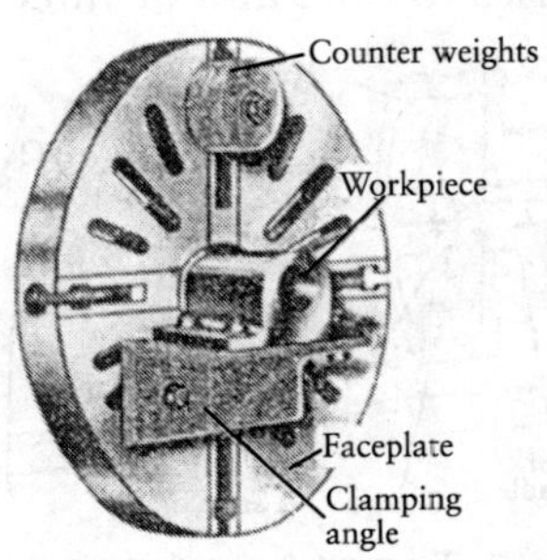

Fig. 5: Faceplate

Faceplate (Fig. 5) is used for clamping larger and irregularly shaped workpieces. Faceplate is screwed on to the work spindle. The faceplate has four clamping jaws which can be adjusted independently. Workpieces which cannot be held with the jaws are clamped into position with the aid of clamping dogs and screws. Projecting workpieces, bearing blocks, and levers can be clamped using an angle plate.

Collet Chucks

Collet chucks (Fig. 6) are used to clamp short and small diameter cylindrical workpieces. Collet surface is split upto certain length along the longitudinal axis and the tongues are pressed into the tapered work spindle by a collar nut, so that the workpiece is held firmly.

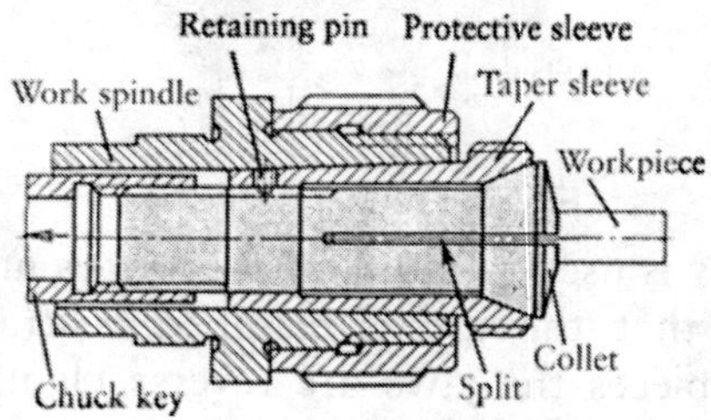

Fig. 6: Collet chuck

Turning between Centres

Long workpieces are clamped between the work spindle centre and the tail stock centre (Fig. 7). The rotary movement is transmitted to the workpiece through the carrier plate which is screwed on to the work spindle and the lathe dog. The carrier plate, is sometimes provided with a protective ring to avoid accidents/injuries to the hand or fingers.

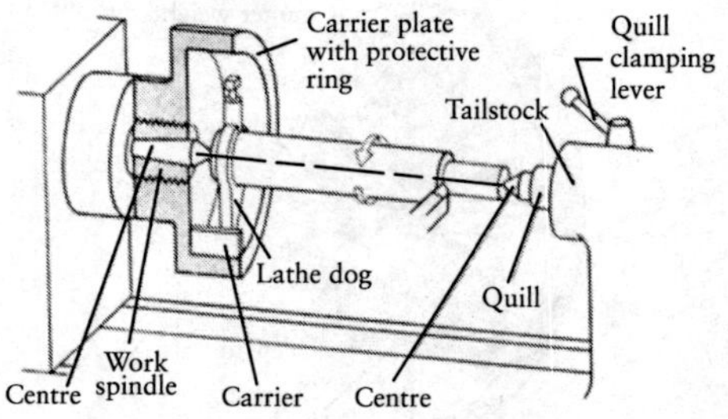

Fig. 7: Turning between centres

For turning between centres the workpiece must have a centre hole in each of the end faces. The centre hole is produced by drilling (with a twist drill) and counter-sinking (with a 60° countersink bit). A centre drill (Fig. 8) also can be used which makes the centre hole in a single operation. The centre hole must be drilled at low feed rate and at high spindle speed.

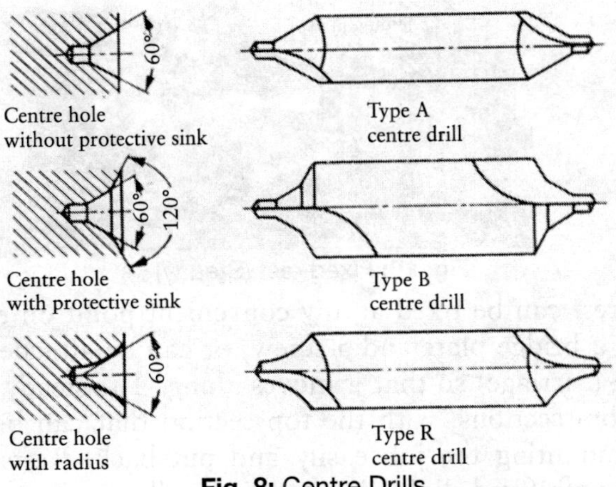

Fig. 8: Centre Drills

Centre hole must be drilled correctly without any errors (Fig. 9). The centre must run accurately and true in the work spindle. The spindle inside the taper and outside taper surfaces must be cleaned before inserting the centres into the work spindle and tail stock quill.

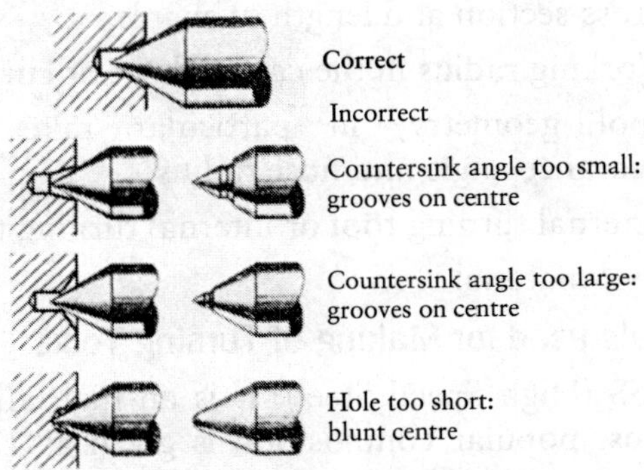

Fig. 9: Correct and Incorrect methods of drilling centre holes

Fixed Rest (steady)

Fixed rest (steady) (Fig. 10) is mounted on the bed when turning long and slender cylindrical workpieces to prevent the workpieces from bending due to the action of cutting forces.

Fig. 10: Fixed rest (Steady)

The rest can be fixed at any convenient point on the lathe bed with a bridge plate and a screw, or can be attached to the tool slide (carriage) so that it moves along. The steady has top and bottom sections, with the top section that can be swung out for mounting the job easily and put back. Rests can be equipped with fixed sliding jaws or with rollers.

Types of Turning Tools

Turning tools are specified by giving the following details:

- Shape of tool (straight, off-set or bent tool; right hand turning tool or left hand turning tool)
- Cross-section and length of shank
- Working radius in the case of bent or curved tools
- Tool geometry, in particular rake angle and clearance angle and nose radius
- External turning tool or internal turning tool
- Tool material

Tool Materials used for Making of Turning Tools

1. HSS (High Speed Steel). It is an high alloy steel; its most popular composition is given as 18-4-1 (18% Tungsten, 4% Chromium, 1% Vanadium, 1% Carbon and rest being Iron).

2. For machining very hard materials such as, stainless steel, hardened steels like die steels carbide tipped tools (carbide inserts) are used. The carbide tips are brazed or screwed on to shank material.
3. Ceramic tools made of Aluminium Oxide in the form of inserts clamped to shank materials.
4. Diamond tools in the form of tips.

Tool Grinding Gauge

A plate type tool grinding gauge (Fig. 11) is used to check whether clearance angles and wedge angles are ground properly after their regrinding.

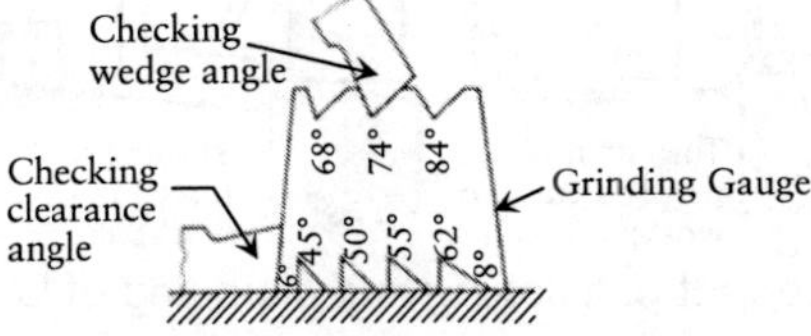

Fig. 11: Grinding Gauge for Lathe tools

Regrinding of lathe tools are carried out using a tool and cutter grinder:

Clamping and adjusting of turning tools

The turning tools must be clamped firmly to the topslide. In the normal position, the turning tool is aligned with the centerline/axis of workpiece. The height can be adjusted, if required, using flat metal shim plates (thin plates) below the tool. The centre of the tailstock can be used to check that the height is correct (Fig. 12).

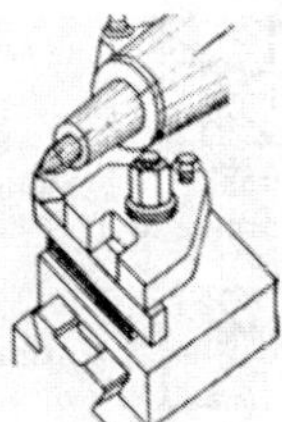

Fig. 12: Vertical Positioning of Lathe tool using tailstock centre

The turning tool must be inserted and clamped such that it does not project more than what is required (Fig. 13). If the overhang is more, the tool may deflect due to the cutting force and may result in wavy surface on the job. In addition, the turning tool must be set perpendicular to the workpiece axis.

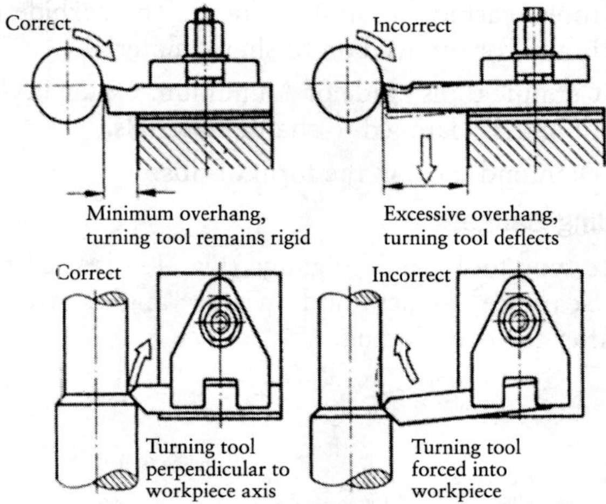

Fig. 13: Correct positioning and clamping of Lathe tools

Right hand and Left hand turning tool

Right hand tool works from right to left on the lathe. During machining (or cutting action), this tool moves from right to left.

Left hand tool works from left to right on the lathe. During machining (or cutting action), this tool moves from left to right.

Right-hand and left-hand tools are shown in Fig. 14. Various types of lathe tools are shown in Fig. 15.

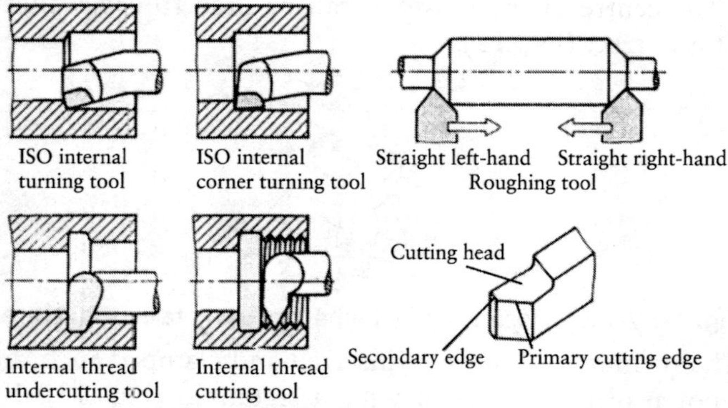

Fig. 14: Right hand and Left hand tools

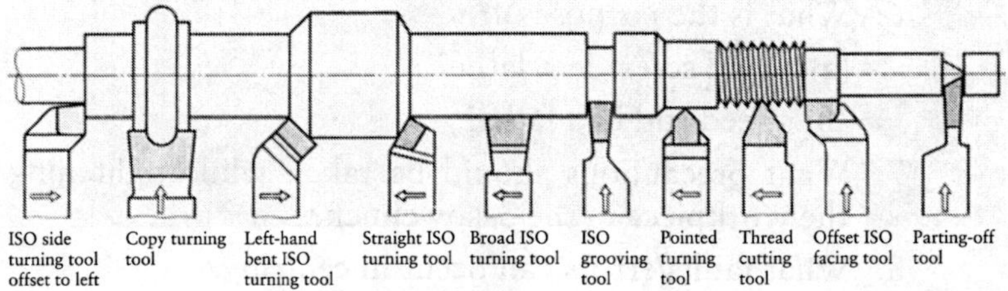

Fig. 15: Various types of Lathe Tools

Precautions while working with lathes

1. Change the gears (for speed variation and for setting of different feed rates) only after stopping the machine.
2. Clean the centres and the tapered surfaces before inserting and positioning the centres.
3. Oil the parts of the lathe only after switching off the power supply.
4. Use the 3-jaw chuck belonging to that particular machine only.
5. Reversible jaws must be reclamped in the correct numbered order only.
6. Check the tool angles ground using tool grinding gauge.
7. Clean and put oil in the centre holes before clamping the workpiece between centres.
8. Clamp the tools firmly with minimum projection from the tool holder.
9. Do not loosen or clamp the tool when the machine is running.

REVISION QUESTIONS

1. What are the main parts of the Centre lathe?
2. How is a Centre lathe specified?
3. What is the material of the lathe bed?
4. What is the function of the lead screw?
5. What is the function of the feed rod?

6. What is the purpose of
   (a) a lead screw in a lathe?
   (b) a feed rod in a lathe?
7. What precautions should be taken while tightening the workpiece in the 3-jaw chuck?
8. What faults/errors can occur in centring?
9. What care/precaution must be taken while inserting the centres in the bore?
10. What precautions must be taken when clamping workpieces between centres?
11. What are the various type of tools that can be used in lathe operation?
12. How are RH (Right-Handed) and LH (Left-Handed) lathe tools are distinguished?
13. How will you check whether the lathe tool angles are ground properly?
14. Why should the lathe tool be fixed with as small projection as possible?

OBJECTIVE TYPE QUESTIONS

Choose the correct alternative

1. Turning of metals in a machine shop is usually performed on a:
   (a) radial drill press (b) lathe
   (c) milling machine (d) shaper
   (e) cylindrical grinding machine
2. A lathe bed is usually made of
   (a) Mild steel (b) High-carbon steel
   (c) Cast iron (d) Aluminium
3. With HSS tools, highest cutting speed can be employed while machining ________.
   (a) Cast iron (b) Mild steel
   (c) Brass (d) Bronze
   (e) Aluminium

4. The cutting speed in any machining operation is usually expressed in
   (a) revolutions per minute of workpiece
   (b) mm/min.
   (c) surface metres/min.
   (d) mm/revolution of job
5. The surface speed in metres/minute of a workpiece of diameter D mm rotating at N r.p.m. is given by _______.
   (a) π DN (b) π DN/1000
   (c) 1000/ π DN (d) DN/1000
   (e) 1000/DN
6. The form of chip produced when machining a hard and brittle metal like cast iron is the
   (a) Continuous chip
   (b) Discontinuous chip
   (c) Continuous chip with built-up-edge (B.U.E.)
   (d) No chips are produced.
7. Powdery chips are produced while machining or drilling
   (a) Mild Steel (b) Hardened Tool Steel
   (c) Cast Iron (d) Stainless Steel
   (e) High Speed Steel (HSS)
8. The "feed rate" in lathe operations is usually expressed in
   (a) r.p.m. of workpiece (b) mm/minute
   (c) surface metres/minute (d) mm/revolution of job
9. Turret lathes which are constructed such that long turning cuts are made by moving the saddle along the bedways of the machine are classified as
   (a) drum type turret lathes
   (b) ram type turret lathes
   (c) saddle type turret lathes
   (d) automatic screw machines

10. External and internal tapers are machined on turret lathes
    (a) by using formed tools
    (b) by using taper turning attachment
    (c) by off-setting the tailstock center
    (d) by setting the compound rest at an angle

11. On automatic screw machines having vertical and cross slides, the vertical slide is normally used for _______ operations.
    (a) thread cutting (b) turning
    (c) drilling (d) parting-off

12. Turret lathes equipped with spindles which can be fitted with a universal 3-jaw chuck are referred to as _______ machines.
    (a) drum type (b) saddle type
    (c) chucking (d) universal lathe

13. The important difference between ordinary center lathes and turret lathes is that
    (a) center lathes are cheaper and simpler
    (b) turret lathes are costly and very complicated
    (c) tungsten carbide tools are used in turret lathes
    (d) turret lathes are constructed and tooled so that several different operations can be performed at the same time

14. On bar-type turret lathes, work to be machined is held or gripped in _______ chucks.
    (a) three-jaw (b) four-jaw
    (c) pneumatic (d) collet

15. Automatic screw machines capable of machining more than one workpiece at a time are classified as
    (a) single spindle automatic machines
    (b) multi-spindle automatic machines
    (c) drum type machines

(d) ram-type machines
(e) bar turners

16. Angles which are ground on the top surface or face of a lathe tool bit and which slant downward from the side cutting edge of the tool are called
(a) Positive side rake angles
(b) Negative side rake angles
(c) Side clearance angles
(d) End cutting edge angles

17. Which one of the following arrangements places the metals in correct order of turning speeds, starting with the slowest?
(a) Mild steel, aluminium, brass
(b) Mild steel, brass, aluminium
(c) Brass, mild steel, aluminium
(d) Aluminium, brass, mild steel

18. When taking a finish cut, the lathe should be operated at
(a) a lower cutting speed
(b) a medium cutting speed
(c) a higher cutting speed
(d) any cutting speed

19. The unit of a lathe which houses the lathe spindle and control levers for speed selection is called a
(a) head stock (b) tail stock
(c) feed box (d) carriage
(e) tool post

20. During drilling work on a lathe, mount the drill in the spindle of the
(a) Head stock (b) Tool post
(c) Tail stock (d) Carriage
(e) Gear box

21. Work which cannot be held in a lathe chuck can be clamped to a ______ mounted on the head stock spindle.

(a) Camplate (b) Faceplate
(c) Carriage (d) Tailstock
(e) Surface plate

22. To take up wear on the cross slide or compound rest, the adjustable _______ must be tightened.
(a) jigs (b) plates (c) gibs (d) pins
(e) screws

23. The angle on the face or top surface of a lathe tool which slants downward from the side cutting edge is called
(a) Side rake angle (b) Back rake angle
(c) Front clearance angle (d) Side clearance angle
(e) Side cutting edge angle

24. The angle on the face of a tool which slants backward from the nose of the tool is called
(a) Side rake angle (b) Back rake angle
(c) Side clearance angle (d) Front clearance angle
(e) End Cutting Edge Angle (ECEA)

25. The angles which are ground on the sides of a lathe tool bit are called
(a) Rake angles
(b) Clearance angles or relief angles
(c) Cutting edge angles
(d) Point angles

26. During facing work in a lathe, it is important that the tip of the tool be set
(a) on the exact centre of the workpiece
(b) below the centre of the workpiece
(c) above the centre of the workpiece
(d) any where on the workpiece

27. A round nose tool ground without _______ can be used as a right or left hand tool.
(a) Back rake angle (b) Side rake angle
(c) Front clearance angle (d) Side clearance angle
(e) End cutting edge angle

28. Work which cannot be held on a chuck because of its shape can be mounted on a ______ for facing.
(a) collet (b) vise (c) V-block (d) faceplate

29. For a turning operation, the tool bit should be set
(a) slightly below the centre of the workpiece
(b) slightly above the centre of the workpiece
(c) exactly at the centre of the workpiece
(d) at an angle

30. Which one of the lathe parts mentioned below is provided with a power feed?
(a) Carriage (b) Compound rest
(c) Cross slide (d) Feed screw

31. Parts having a square or irregular shape to be machined on a lathe can only be mounted in
(a) universal chucks (b) independent chucks
(c) collet chucks (d) cam lock chucks

32. The preferred chuck for mounting small cylindrical work to be turned is ______.
(a) the Cam lock chuck (b) the Universal chuck
(c) the 4-jaw chuck (d) the collet chuck

33. Tail stock centers which do not revolve with the workpiece are referred to as ______.
(a) Revolving centers (b) Dead centers
(c) Live centers (d) Independent centers
(e) Magnetic centers

34. To refinish the points of lathe centers which have become worn, grind them to an included angle of
(a) 30° (b) 60° (c) 90° (d) 120°
(e) 118°

35. The dovetail surfaces of the compound rest, the carriage, and the cross slide are provided with adjustable ______ to take up wear.
(a) screws (b) plates (c) gibs (d) jigs
(e) pins

36. Work to be faced in a lathe chuck requires that
    (a) the compound rest be set at a certain angle
    (b) the tip of the tool is set exactly on the work center line
    (c) the tip of the tool be set below the work centerline
    (d) the tool be set slightly above the work center
37. During plain turning of a shaft mounted between lathe centers:
    (a) Set the compound rest parallel to the work axis
    (b) Use a lubricant
    (c) Set the tool bit slightly above the work center line
    (d) Set the tool bit below the work centerline
38. The included thread angle of a standard form of thread is
    (a) 30 degrees (b) 45 degrees
    (c) 60 degrees (d) 90 degrees
    (e) 120 degrees
39. The width of the tool bit point for cutting a _______ thread should be ground to half-the-pitch of the thread.
    (a) ACME- (b) SQUARE-
    (c) V- (d) STANDARD-
40. Taper turning on a lathe can be done by
    (a) One method only
    (b) Two different methods
    (c) Three different methods
    (d) Four different methods
41. Taper turning can be performed on a lathe when work is mounted between centers by off-setting the _______ center.
    (a) Head stock (b) Tail stock
    (c) Live (d) Drill

42. To machine a 60 degree included angle on a lathe center with the compound rest, the compound rest must be set at ______ to the work centerline.
 (a) 15 degrees (b) 30 degrees
 (c) 60 degrees (d) 120 degrees
43. The operation of producing an accurately-sized hole in a lathe using a tool bit is called
 (a) Drilling (b) Reaming
 (c) Boring (d) Burnishing
44. The use of a ______ tap when large size holes are threaded, eliminates the need to reverse the rotation of the lathe spindle when withdrawing the tap from the hole.
 (a) Square (b) Solid
 (c) Hand (d) Collapsible
45. "The lead of any thread is always the same as the pitch." This statement is true
 (a) in the case of single-thread screw only
 (b) in the case of double-thread screw only
 (c) in the case of multiple-thread screw only
 (d) under all conditions
46. The only automatic screw machine having sliding ______ which can be actuated longitudinally is the Swiss-type.
 (a) Carriage (b) Tailstock
 (c) Compound rest (d) Spindle
47. The setting of the lathe gear box levers in thread-cutting operations is determined by the ______.
 (a) Pitch of threads to be cut
 (b) Material being threaded
 (c) Length of the thread to be cut
 (d) R.P.M. of the lathe spindle
48. While cutting a standard V-thread the compound rest of the lathe must be set at an angle of ______.

(a) 14.5 degrees (b) 29 or 30 degrees
(c) 45 degrees (d) 60 degrees

49. A good lubricant to use in thread-cutting operations is
(a) Graphite (b) White lead
(c) Mineral Lard oil (d) Water soluble oil

50. A lubricant is not needed when cutting threads in
(a) Tungsten carbide (b) Mild steel
(c) Titanium (d) Brass or cast iron
(e) High Speed Steel (H.S.S.)

51. What would be the finish depth of a square thread having a pitch of 4 mm?
(a) 16 mm (b) 4 mm (c) 2 mm (d) 1 mm
(e) 8 mm

52. Short or sharp angle tapers are machined using
(a) a taper attachment
(b) the compound rest
(c) the tailstock set over method
(d) a form tool ground to the taper angle

53. On single-spindle automatics, stock cut off operations are normally performed from the ______.
(a) Vertical tool slide (b) Rear tool post
(c) Square Turret (d) Hexagonal turret
(e) Cross slide

54. The movement of the various slides as well as the feeding of the stock is entirely automatic on screw machines and is obtained by _______ action.
(a) gear (b) collet
(c) roller (d) spring
(e) cam

55. Long turning cuts are made on _______ type turret lathes by longitudinal movement of the saddle along the bedways of the machine.

(a) Drum (b) Saddle
(c) Universal (d) Chucking

56. A tool which is mounted in the hexagonal turret when taking heavy turning cuts at high speeds is called a _______.
(a) box type of tool
(b) bar turner
(c) chaser
(d) heavy turning attachment

57. Square or irregular shaped workpiece for turning is usually mounted in
(a) Three jaw chuck (b) Independent chuck
(c) Collet chuck (d) Bar chuck
(e) Mandrel

58. A device which is fastened to the headstock end of work to be turned between centres is called a
(a) faceplate (b) lathe dog
(c) vise (d) work-steady
(e) independent chuck

59. Previously drilled and reamed parts can be turned between centres by mounting such parts on a
(a) Three jaw chuck (b) Four jaw chuck
(c) Steady (d) Mandrel
(e) Faceplate

60. The angles ground into the face or top surface of lathe tool bits are referred to as
(a) Clearance angles
(b) Relieving angles
(c) Rake angles
(d) Side/Face Cutting Edge angles

61. The angles which are ground on the sides of a lathe tool bit are called
(a) Rake angles
(b) Clearance angles

(c) Side Cutting Edge Angles (SCEA)
(d) Point angles

62. Which of the following lathe operations requires that the cutting edge of a tool bit be placed exactly on the work centerline?
(a) Boring (b) Drilling
(c) Facing (d) Turning
(e) Chamfering

63. All moving elements used in a machining operation on a screw machine are positioned for cuts automatically by ______ action.
(a) gear (b) collet
(c) spring (d) roller
(e) cam

64. On screw machines having cross and vertical slides, cutting off operations are usually performed from the ______.
(a) Horizontal slide
(b) Vertical slide
(c) Cross-slide
(d) Cross-drilling attachment

65. Drill chucks when used for drilling operations in a lathe are mounted in the ______ spindle.
(a) Tool post (b) Carriage
(c) Head stock (d) Tail stock

66. The automatic performance of all operations on screw machines is obtained by
(a) Dowels (b) Cams
(c) Bushes (d) Fillets

67. When machining internal or external tapers on a turret lathe ______ is used.
(a) Face turning attachment
(b) Taper turning attachment
(c) Sliding attachment
(d) Morse Taper attachment

68. In lathe work, the cutting speed is usually expressed in

(a) r.p.m. (b) mm/min.
(c) m/min. (d) cm/min.
(e) mm/rev.

69. In lathe work, the cutting tool feed is expressed in

(a) r.p.m. (b) m/min.
(c) m/rev. (d) inches/min.
(e) mm/rev.

70. Discontinuous chips are normally obtained when turning metals such as

(a) Mild Steel (b) Aluminium
(c) Cast Iron (d) High Speed Steel (H.S.S.)

71. Continuous chips are normally produced when turning

(a) Ductile metals (b) Brittle metals
(c) Brass (d) Diamond

72. In a taper turning exercise on a lathe, the following details are given on the component drawing:

(i) The amount of taper is 1/20.
(ii) The difference in the diameters of big end and small end of the tapered portion is 8 mm.

What is the length of the tapered portion?

(a) 160 mm (b) 320 mm
(c) 80 mm (d) 2.5 mm
(e) 5 mm

73. A lathe tool is considered as a

(a) single-point cutting tool
(b) multi-point cutting tool
(c) single edged cutting tool
(d) multi-edged cutting tool

74. A shaft 1000 mm long has a taper of 1:100 for a length of 300 mm. The maximum diameter of the

tapered portion is 75 mm. What is the minimum diameter of the tapered portion?

(a) 81 mm (b) 78 mm
(c) 72 mm (d) 69 mm
(e) 66 mm

ANSWER KEY

1. (b)	2. (c)	3. (e)	4. (c)	5. (b)
6. (b)	7. (c)	8. (d)	9. (c)	10. (b)
11. (d)	12. (c)	13. (d)	14. (d)	15. (b)
16. (a)	17. (b)	18. (c)	19. (a)	20. (c)
21. (b)	22. (c)	23. (a)	24. (b)	25. (b)
26. (a)	27. (b)	28. (d)	29. (b)	30. (b)
31. (b)	32. (d)	33. (b)	34. (b)	35. (c)
36. (b)	37. (a)	38. (c)	39. (b)	40. (c)
41. (b)	42. (b)	43. (c)	44. (d)	45. (a)
46. (d)	47. (a)	48. (a)	49. (c)	50. (d)
51. (c)	52. (b)	53. (a)	54. (e)	55. (b)
56. (b)	57. (b)	58. (b)	59. (d)	60. (c)
61. (b)	62. (c)	63. (e)	64. (b)	65. (d)
66. (b)	67. (b)	68. (c)	69. (e)	70. (c)
71. (a)	72. (a)	73. (a)	74. (c)	

5

Drilling, Reaming and Tapping

Drilling is a machining processes intended for the production of cylindrical holes in metals by cutting. The main movement is always rotary with the drilling tool (called a drill) rotating and simultaneously the drill performs an axial feed movement towards and into the stationery workpiece.

The basic cutting action is by a wedge penetrating into the tool and scooping of the metal (chips) upwards. Drilling tools are called as twist drills and centre drills.

Twist drills are made from HSS (High Speed Steel).

Advantages of twist drills are

- Good cutting point angles,
- Diameter do not change when drills are reground,
- Good guidance in the workpiece,
- Automatic chip removal from the hole, and
- Ease of clamping.

Some drills are of carbide tips mounted on carbon steel shank. Carbide tipped drills are used for drilling very hard work materials like chilled CI, Manganese alloy steel, and other hard materials like concrete, rock, marble and stone. For drilling in glass, carbide tipped drills or diamond tipped drills are used.

Parts of a Twist Drill

Twist drills are standardized. The nomenclature of twist drill is given in Fig. 1. Elements of twist drills are explained below: The shank helps for fitting in the machine spindle or for holding in the drill chuck. The shank can be parallel or tapered. The machining part, at the tip of the drill, gets its basic form/shape by two helical flutes.

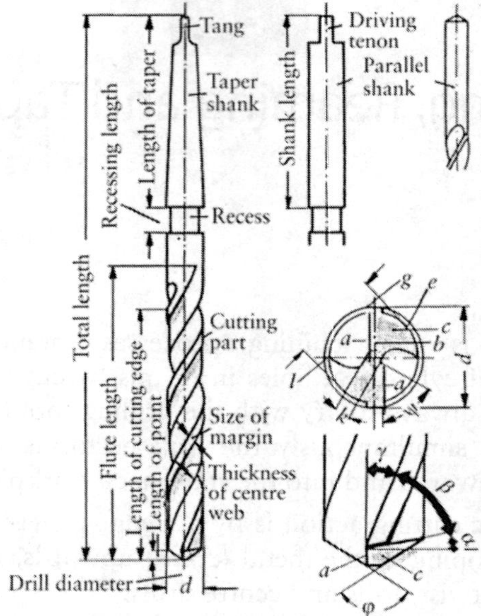

Fig. 1: Nomenclature of a Twist Drill

Web: The material portion remaining between the flutes is called centre web.

In Fig. 1, *k* represents thickness of centre web. The web edge doesn't cut, it only scoops or extrudes the metal. It pushes the material out of the centre of the hole infront of the main cutting edges and in this action it consumes nearly 40% of the feeding power.

Lips: The two main cutting edges are called lips and are formed by grinding of the point.

Point angle: The point angle is formed by the two main cutting edges.

Margins: The margins guide the drill by providing a support at the surface of the hole and prevents rubbing of the heel in the drill hole. To avoid damaging of the margins when drilling deep holes, the diameter of the drill is reduced towards the shank by about 0.05 mm in 100 mm length.

As in all cutting tools twist drills also have clearance angle, rake angle and wedge angle. In addition, drills have point angles; clearance angle, rake angle and wedge angles are on each of the two main cutting edges.

Clearance angle: The clearance angle, measured at the face edge is about 5° to 8°. The clearance faces, slope backwards in a curve.

Rake angle: Rake angle is provided by the helix angle of the flutes. Rake angle at the face is equal to the helix angle. Rake angle decreases towards the centre of the drill and almost zero degree at the centre of the drill. That is why the metal removal action at the centre is entirely by scooping or by extrusion action.

Wedge angle: This is equal to 90°—(clearance angle + rake angle).

Point angle: The included angle between the two main cutting edges.

Chamfer: The chamfer (polished portion at the 2 edges of the drill) provides guidance of the drill in the workpiece. The width of chamfer: 6 mm (Higher the diameter of the drill, wider chamfer is provided). Chamfer is formed by providing relief along the flutes of the drill while flutes are milled by formed cutters. This relief (obtained by relief milling) reduces the friction area of the drill and therefore prevents the drill edges from sticking to the hole walls during drilling.

Helix angle: Angle between the flutes (i.e. helix angle of the rake faces) and the drill axis is known as helix angle. High helix angle drills are used for drilling softer materials like aluminium which will give long continuous chips going around the flutes. Low helix angle drills are used for drilling hard materials like CI and plastics which will produce powdery chips which will rise up along the flutes.

Nomenclature of Twist drill includes important angles and dimensions of the drill.

£ : Clearance angle at the face edges
β : wedge angle
r : rake angle at the face edges
φ : Point angle (Phi)
Ψ : Web angle (Psi)
a : Main cutting edge
b : Centre line between the two cutting edges
c : Clearance face
d : diameter of drill
e : body clearance
g : face edge
k : thickness of centre web

Helix angle: It is equal to rake angle of the cutting edge at the periphery. As you move towards the centre of the drill, rake angle reduces gradually.

A large angle helix, results in a low helix pitch and a larger rake angle. A smaller helix angle results in a steep helix pitch. A steeper helix pitch means a smaller rake angle.

Rake angle: Rake angle is same as helix angle at the periphery (i.e. outside diameter) of the cutting edge. At other points along the cutting edge the rake angle varies, minimum at the centre, as you move outwards rake angle gradually increases. Rake angle of a twist drill depends on the material being machined.

For steel, CI, malleable iron, helix angle is 16° – 30° (These drills are called medium helix angle drills).

For Copper-Zinc alloys, marble and concrete, helix angle is 10° – 13° (These drills are called low helix angle drills).

For aluminium and copper helix angle is 35°– 40° (These drills are called high helix angle drills).

Point angle: In plan view, the two main cutting edges of the twist drill appear as two parallel lines if the point is ground to the correct angle. The included angle between the two cutting edges (as seen in the elevation) is known as point angle. Values of point angle depends on the materials to be machined.

	Work materials to be machined	Point angle (also known as tip angle)
1.	Hard rubber, marble, slate, concrete	80°
2.	Steel, Cast steel, (700 N/mm^2 tensile strength) CI, malleable iron, copper-zinc alloy	118°
3.	Zinc alloys, Lead-Tin alloys (white metal)	118°
4.	Steel, cast steel (above 700 N/mm^2 tensile strength)	130°
5.	Stainless steels, Aluminium alloys and Magnesium alloys	140°

The cross-cutting edge is the line on the centre point connecting the two main cutting edges. Normally, it forms 55° with the main cutting edges. The cross-cutting edge helps in scooping the metal out. The cross-cutting edge is obtained by relieving the clearance faces (by grinding).

Twist drill grinder is used for accurate grinding and for maintaining the correct geometry of various tool angles. Sometimes grinding fixtures are used to hold twist drills properly.

Errors in Holes due to Improper drill Grinding

If grinding of twist drills is not carried out properly several errors can occur (Fig. 2).

1. Cutting edges become unequal in length.
2. Cutting angles become unequal.

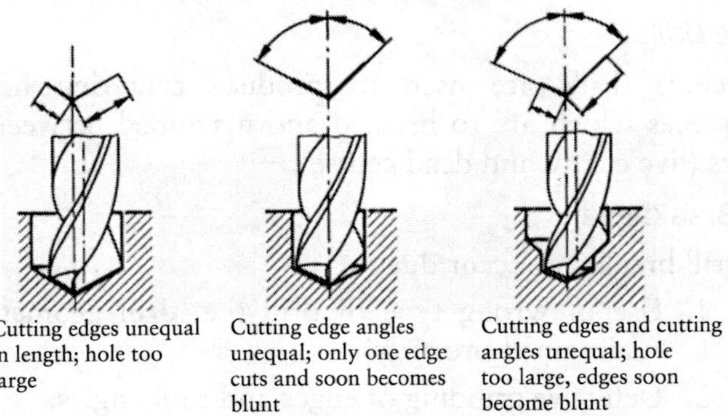

Fig. 2: Errors likely to occur due to improper grinding of twist drills

The consequences of these faults are that holes drilled become oversized and leads to shorter drill life.

Clearance angle should also be ground properly. If clearance angle is too much, it weakens the drill leading to drill breakage. After grinding, the angles should be checked with a grinding gauge.

Pilot Holes

For holes of more than 15 mm diameter, a pilot hole using a small drill is drilled first. If pilot hole is not made, large diameter drills will run out of line. Also, large cutting force (drill thrust) is required and sometimes drilling machine motor may be overloaded. Pilot hole reduces the axial thrust considerably; however the drilling torque is reduced marginally only.

Holding of Drills

Small diameter drills (up to 10 mm) have cylindrical shanks and can be held by means of self-centering drill chucks. Drill must be properly clamped in the drill-chuck.

Large drills have tapered shank (called Morse taper). A tang is provided at the top end of the drill. The purpose of the tang is that it provides positive transmission of torque to the drill from the machine tool spindle. Drilling machine spindle is provided with a slot (through hole) into which the tang fits. A suitable drift is used to tap the drill out of the spindle or sleeve (reducing or connecting adapter).

Centre Drills

Centre drills are used to produce centering holes in workpieces which are to be held and machined between lathe centers (live centre and dead centre).

Drill Breakages

Drill breakages occur due to

1. Use of wrong type of drill (i.e. drill geometry not maintained properly),
2. Defective grinding of edges and tool angles,
3. Feed rate is very high,

4. Cutting speed is very high,
5. Hole axis not properly aligned with drill centre,
6. Inadequate cooling, and
7. Proper coolant is not used.

Countersinking and Counterboring

Countersinks (Fig. 3) and counterbores (Fig. 4) are used for deburring holes and to produce the necessary recesses in the pre-drilled holes. Countersinking and counterboring of the pre-drilled holes are required for inserting rivets and screw heads.

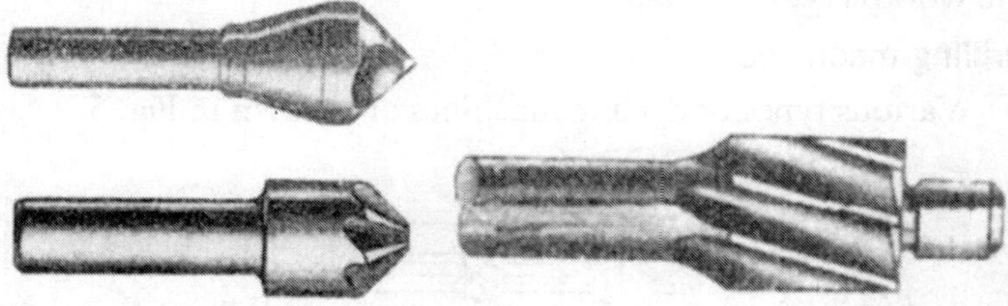

Fig. 3: Countersinks **Fig. 4:** Counterbore with central pilot

Countersinks for screw holes must be of the same diameter as the screw heads to be countersunk.

The point angles of countersinks are 60° for deburring, 75° for rivet heads, 90° – 120° for counter sunk-head, raised counter sunk-head screws and for sheet metal rivets.

Counter Bore

Counter bore has two or more cutting edges and is used for making counter bores. Counter bores are required to insert cylindrical screw heads and also to produce flat contact faces for screw heads and nuts in castings/work pieces.

The central pilot acts as a guide by locating in the pre-drilled holes.

Clamping the Workpiece while Drilling

Workpieces should be rigidly held using clamps, to avoid tendency of workpiece becoming loose and start moving.

Also, as the drill emerges out from the hole, sudden loosening of the workpiece might occur which should also be prevented.

Small workpieces can be held in the machine vices. If sheet metal is to be held it can be held in the hand vice. Large workpieces are held on the drill table with dogs and screws using T-slots provided on the table. While drilling through holes, drilling of the table itself must be avoided by clamping the job using parallel/rectangular steel bars providing sufficient gap below the workpiece.

If the hole is to be drilled on an inclined surface, a suitable cross-mark/punch mark should be made using a chisel for the tip of the drill to locate and enter initially without sliding on the workpiece.

Drilling machines

Various types of drilling machines are shown in Fig. 5.

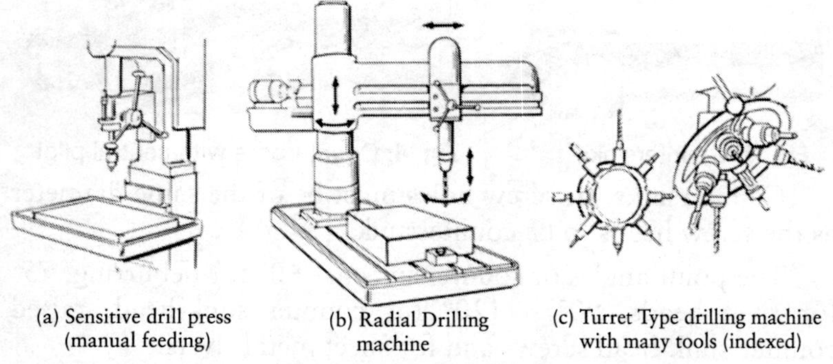

(a) Sensitive drill press (manual feeding) (b) Radial Drilling machine (c) Turret Type drilling machine with many tools (indexed)

Fig. 5: Various Types of Drilling Machines

Hand drilling machines can be powered by electric motor (or by compressed air). They are equipped with three jaw chucks; Drills from 0.5 mm – 12.0 mm diameter can be held in them. For drilling in rock or in concrete for construction work hand held drilling machines are equipped with percussion/vibratory (hammer drill) mechanism. Supply cables, their insulation and power plugs should be in good condition to avoid accidents and injuries while working on hand drilling machines.

Bench Drilling Machine

These are mounted on work benches or tables. Drills up to 20 mm can be handled using bench drilling machines.

Upright Drilling Machines

These are used for drilling up to 25 mm diameter and large sized holes. Sometimes, these are floor mounted and the drilling table can be adjusted vertically and to some extent can be tilted as well.

Column-Drilling Machines

The drilling spindle of round-column drilling machine is mounted on circular guides and can be moved axially (up and down) by a toothed rack. When drilling deep holes, the spindle has to extend a long way out of its bearings, leading to drill run-out. On box-column drilling machines, a slide moves up and down the column to provide automatic feed movement, keeping the spindle adequately guided even while drilling deep holes.

Precautions to be Observed in Drilling

1. While mounting tapers, sleeves (connecting adoptors) and drill shanks ensure that dirt and chips are removed from the surface by wiping.
2. Ensure that the drill is firmly seated, secured and runs true.
3. Do not clamp taper shank drills into normal drill chuck.
4. Use a suitable drift (wedge shaped steel piece) to tap the drill out of the spindle.
5. Make sure that the workpiece is held rigidly when drilling.
6. Make a pilot hole first with a smaller drill before a large size hole is drilled.

Reaming

Reaming produces high quality and accurate dimensions on the holes already drilled. Machining allowance for remaining should not exceed 0.3 mm.

Reamers can have straight teeth or helical teeth (Fig. 6). Helical teeth reamers are used for reaming holes which are interrupted by longitudinal slots. Helical teeth reamers have a left-hand helix so that they are pulled into the workpiece.

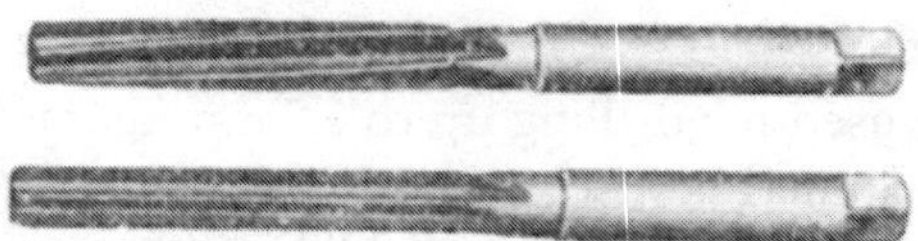

Fig. 6: Hand reamers with helical and straight flutes

Reamers with even no. of teeth and unequal pitch give good performance resulting in easy cutting and no vibrations/chatter (Fig. 7).

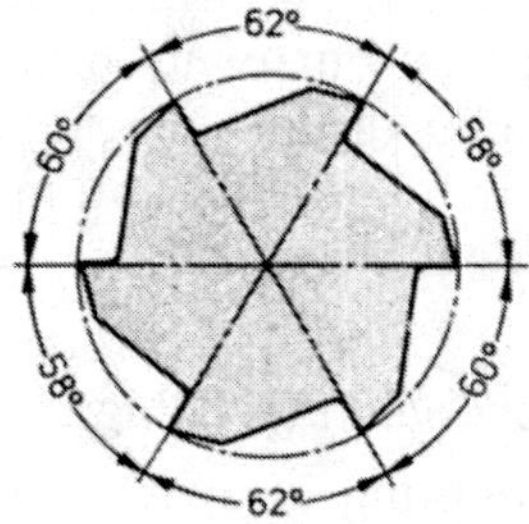

Fig. 7: Reamer with unequal pitch

Types of Reamers

Hand Reamers

Hand reamers have square-end shank which can be held with a wrench. To provide guidance in the workpiece, they have a long tapered end (Fig. 8).

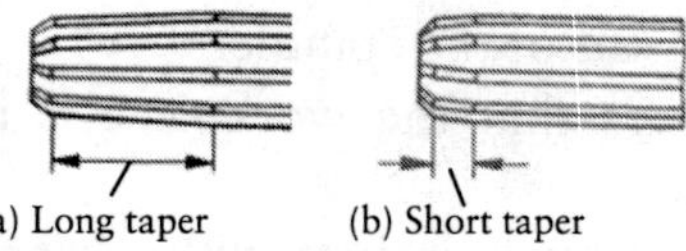

Fig. 8: Hand reamers with long and short end taper

Reamers should not be allowed to rotate in the reverse direction. Metal chips present in the hole, already removed will block them damaging the cutting edges.

Machine Reamers

Machine reamers (Fig. 9) have a shorter tapered end and are used in production work for reaming blind holes right up to the bottom. Machine reamers are available with cylindrical shanks and with tapered shanks. Cylindrical shank reamers are held in the drill chuck and tapered shank reamers can be inserted into the tapered holes of machine spindles.

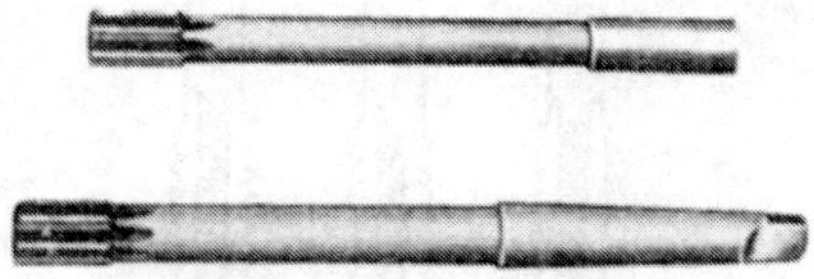

Fig. 9: Machine reamers

Slotted Reamers

Slotted reamers (Fig. 10) can be expanded to certain limits with the help of taperd pins. Reamers with inserted cutters can be expanded/adjusted to a larger extent (upto 3 mm). To adjust them, the cutters are moved by two threaded rings acting on tapered faces. Adjustable reamers as well as other reamers must be handled carefully; otherwise they will get damaged and the accuracy of holes reamed will be affected.

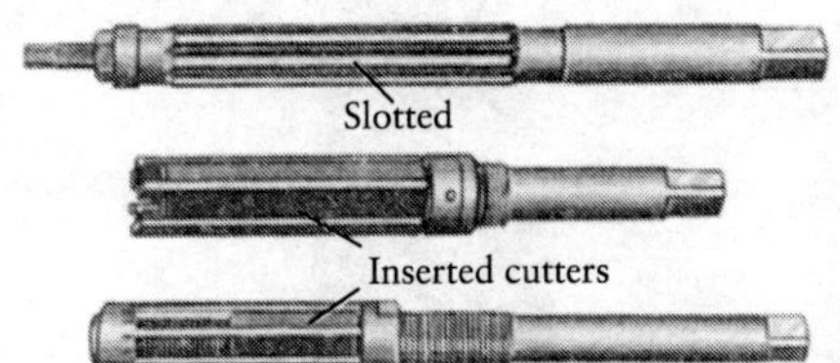

Fig. 10: Adjustable hand reamers (slotted reamers)

Tapping

Tapping is a machining process by which internal threads in the holes can be machined. Drilling, reaming and tapping is normally the sequence followed to make internal threading. Tapping can be done manually or by using machine taps.

Hand taps are available in sets of three (Fig. 11). These are known as (a) Rougher, (b) Intermediate, and (c) Finisher. The thread profile is made progressively by using (a), (b) and (c) in the sequence.

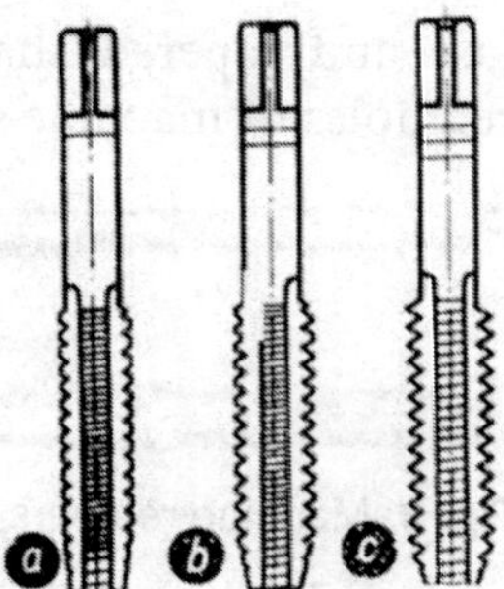

Fig. 11: Hand Taps available in sets of three: (a) Rougher, (b) Intermediate, (c) Finisher

Machine taps are mounted on the spindle/chuck of the drilling machine. When the taps are withdrawn from the hole, they are reversed so that thread profile is not damaged. A self-reversing tapping head is shown in Fig. 12.

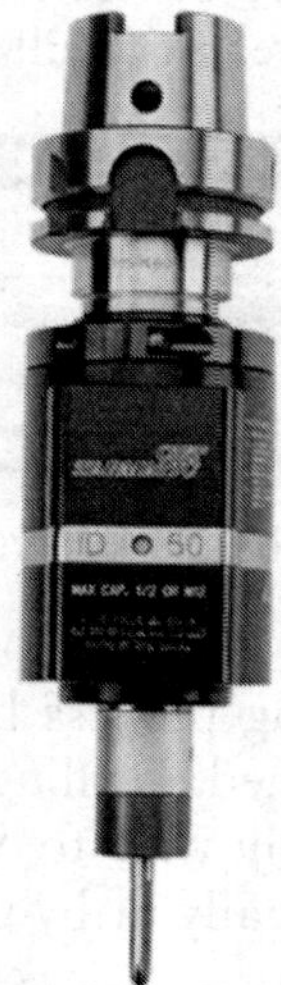

Fig. 12: Self-Reversing Tapping Head

Features:

(a) High Speed Tapping

(b) Quick Changing tap chuck

(c) Floating Tap Spindle fitted with ball and needle bearings

(d) Maximum durability and strength

(e) Infeed and Outfeed (Reversing) occurs at high speed.

REVISION QUESTIONS

1. List the various types of drills.
2. What are the advantages of the twist drill?
3. What is the important specification to designate a twist drill?
4. What is the point angle on the normal twist drill for drilling steel?
5. What is the point angle for drilling on aluminium?
6. What angle should the cross-cutting edge form with the main cutting edge?
7. In drilling, what are the causes for production of over-sized holes?
8. What precautions should be observed while clamping/holding drills in the drill chucks or in machine tapers?
9. What tools are used for
   (a) Countersinking of holes, and
   (b) Counter boring of holes
10. Why reamers have an even no. of flutes and unequal pitch?

OBJECTIVE TYPE QUESTIONS

Choose the correct alternative

1. The tool used to scribe a circle in laying out a hole is called
   (a) a scriber (b) a divider
   (c) a compass (d) a combination set
2. Center punches are used in layout work to locate and mark the centres of
   (a) shafts (b) holes
   (c) slides (d) bearings
3. The most common method of originating a hole in a metallic workpiece is by ______.

(a) shearing (b) punching
(c) countersinking (d) drilling
(e) boring

4. Cylindrical job should always be clamped on a ______ for drilling.
   (a) collet (b) socket
   (c) jaw (d) vise
   (e) V-block
5. The cutting edges of a standard twist drill are called ______.
   (a) flutes (b) lips
   (c) wedges (d) flanks
   (e) conical points
6. The helical grooves which extend to the full length of the drill body are called:
   (a) lips (b) cutting edges
   (c) margins (d) flutes
   (e) shanks
7. The number of helical grooves which are present in a standard twist drill is usually
   (a) one (b) two
   (c) three (d) four
8. The important purpose of flutes in a drill is
   (a) they enable the cutting fluid to enter the working zone.
   (b) they enable the chips to come out.
   (c) they increase the strength of the drill at the cutting point.
   (d) they reduce the weight of the drill used.
   (e) they enable the cutting edges to be formed.
9. The "depth-of-cut" in a drilling operation is equal to
   (a) length of hole drilled.
   (b) the feed rate used.
   (c) diameter of the drill used.

(d) 1/2 the diameter of the drill used.
(e) 1/4 the diameter of the drill used.

10. The usual point angle of the drill is
(a) 90° (b) 98°
(c) 118° (d) 135°
(e) same as the helix angle of the drill

11. The point angles and clearance angles of drills must be varied by grinding depending on the
(a) Drill Diameter
(b) Material to be drilled
(c) r.p.m. of the drill
(d) Depth of the hole to be drilled

12. To drill free machining stainless steel at a cutting speed of 15 m/minute with a 15 mm diameter drill, the drill must rotate at a speed of
(a) 320 r.p.m. (b) 640 r.p.m.
(c) 1000 r.p.m. (d) 3.1416 r.p.m.
(e) 32 r.p.m.

13. A hole which does not go entirely through a workpiece is referred to as
(a) an inclined hole (b) a blind hole
(c) a through hole (d) a countersink hole
(e) a pilot hole

14. All straight-shanked drills used in drill press work must be held ______.
(a) directly into the spindle of the drill press
(b) in a key type drill chuck
(c) in a taper sleeve
(d) in a vise

15. To provide a positive drive, all taper shanked drills are provided with a ______ to prevent them from slipping under a cutting load.
(a) sleeve (b) socket
(c) tang (d) neck

16. The purpose of the ______ on a taper shank drill is to prevent drill slipping and to ensure a positive drive.
    (a) sleeve (b) socket
    (c) neck (d) tang
    (e) flutes
17. To reduce the feeding pressure needed for drilling larger holes, it is a good practice first to ______.
    (a) anneal the workpiece (b) drill a countersink hole
    (c) drill a small pilot hole (d) drill a stepped hole
18. A drill having flat sides and two cutting edges for drilling large holes is called a ______.
    (a) Micro-drill (b) spade drill
    (c) boring tool (d) counter-boring tool
19. A cutting fluid need not be used when drilling or reaming ______.
    (a) Aluminium (b) Brass
    (c) Bronze (d) Mild Steel
    (e) Cast Iron
20. Kerosene is a good cutting fluid to use when drilling ______.
    (a) Cast Iron (b) Mild Steel
    (c) Aluminium (d) Brass
    (e) Bronze
21. The taper usually employed in drill sleeves in known as
    (a) Girlling taper
    (b) Morse taper
    (c) Brown and Sharp taper
    (d) 1:1 taper
22. A taper shank drill is removed from the drill spindle by
    (a) tapping the drill by a hammer
    (b) a screw driver

(c) a draft (d) a drift
(e) a tang

23. Oversized holes in drilling are caused by
(a) equal lengths of lips (b) unequal length of lips
(c) larger helix angle (d) smaller helix angle
(e) wornout drills

24. The operation of providing a smooth seat or bearing surface around a previously drilled hole for a washer or nut is called
(a) boring (b) counterboring
(c) counter sinking (d) spot-facing

25. The operation of making a recess at the top of a drilled hole for fixing a flat-head machine screw is called
(a) boring (b) counter sinking
(c) micro-boring (d) reaming

26. A cutting tool used to make a recess at the top of a drilled hole for fixing a flat head machine screw is called ______.
(a) a core drill (b) a spade drill
(c) a reamer (d) an end mill
(e) a countersink tool

27 ______ is the only drilling machine not equipped with power feeds.
(a) The Sensitive drill press
(b) The Multiple-spindle drilling machine
(c) The Radial drilling machine
(d) The Gang drilling machine

28. When a number of single spindle drilling machine columns are placed side by side on a common work table, the machine is known as
(a) Radial drilling machine
(b) Gang drilling machine
(c) Multiple-spindle drilling machine

(d) Universal drilling machine
(e) Pillar type drilling machine

29. The machine which is used to drill a number of holes simultaneously on a tractor engine housing is called
(a) Sensitive drilling machine
(b) Radial drilling machine
(c) Universal drilling machine
(d) Gang drilling machine
(e) Multi-spindle drilling machine

30. Reaming is done
(a) for making a hole initially.
(b) to enlarge the diameter of the hole.
(c) to improve the finish in the hole.
(d) to achieve correct diameter.

31. When reaming holes with solid reamers, mount the reamer in _____.
(a) a rigid holder (b) a floating holder
(c) a semi-rigid holder (d) a collet

32. A tool used to finish drilled holes accurately to size is called
(a) an end mill (b) a tap
(c) a die (d) a twist drill
(e) a reamer

33. _____ is the operation of sizing and finishing a hole accurately with a very little metal removal.
(a) Drilling (b) Tapping
(c) Piercing (d) Reaming
(e) Turning

34. To provide a close tolerance and a fine finish in a drilled hole, the hole must be _____.
(a) drilled again (b) tapped
(c) sheared (d) reamed

35. The cutting speed to be used in reaming a hole _____.

(a) should be same as the cutting speed used in drilling that hole
(b) should be more than the cutting speed used in drilling that hole
(c) should be less than the cutting speed used in drilling that hole
(d) could be any speed.
(e) should be as large as possible

36. The operation of threading a drilled hole is called
(a) Lapping (b) Reaming
(c) Broaching (d) Tapping

37. A fluted tool used to cut internal threads is called
(a) a tap (b) a drill
(c) a die (d) an end mill

38. Hand-tapping is an operation for generation of internal threads. The number of taps generally used for completing this operation is
(a) 1 (b) 2
(c) 3 (d) 4
(e) depends on the diameter of the hole

39. A tool used in threading a hole is called
(a) a tap (b) a die
(c) an end mill (d) a twist drill

40. Taps are used for
(a) generating internal threads
(b) generating external threads
(c) drilling holes
(d) enlarging the holes

41. A tool used in cutting an external thread is called a
(a) twist drill (b) tap
(c) die (d) nut

42. Dies are used for
(a) generating internal threads
(b) generating external threads

(c) drilling holes
(d) enlarging the holes

43. In a double-start thread the lead is equal to ______.
(a) ½ pitch (b) pitch
(c) twice the pitch (d) Thrice the pitch
(e) 4 times the pitch

44. 'Point thinning' a drill
(a) reduces the point angle of the drill
(b) makes the drill to produce an oversize hole
(c) reduces the effort needed to make the drill cut
(d) help the chips to come out easily

45. A twist drill has been ground so that the cutting edges are of unequal length. The effect of this in use will be to produce
(a) a hole larger in diameter than the drill
(b) an increase in the life of the drill
(c) breakage due to the over loading of the longer cutting edge
(d) softening due to overheating the drill

46. A drill is found to machine always an 'oversize' hole. Which one of the following is the cause of this fault?
(a) The drill is too large in diameter.
(b) The cutting edges are of unequal length.
(c) The drill has been allowed to get hot.
(d) Considerable amount of wear has occured on the drill.

47. For which one of the following purposes are screw-cutting taps used?
(a) to cut internal threads
(b) to clean out blind holes
(c) for light riveting
(d) to cut external threads
(e) to enlarge undersize holes

48. Screw threads may be cut on a 10 mm *dia*, steel rod by using

(a) a circular split die (b) a set of taps
(c) a drill and a reamer (d) knurling tool

49. The machining process which makes a cone-shaped recess at the top of a drilled hole for fixing a flat head machine screw is called
(a) counter-boring (b) die-sinking
(c) counter-sinking (d) taper drilling
(e) micro-boring

50. The groove in the body of the drill which allows the chips to come out is known as
(a) the chip breaker (b) the lip
(c) the flute (d) the margin
(e) chip follower

51. For a given work material, a large sized drill must rotate _____ than a small sized drill to maintain the same cutting speed (surface speed).
(a) faster (b) at a lower r.p.m.
(c) at an higher r.p.m.

52. A standard twist drill may be considered as a
(a) single point cutting tool
(b) multi-point cutting tool
(c) single edged cutting tool

53. The workpiece motion and tool motion required respectively, in a horizontal boring machine are
(a) rotational and translational
(b) stationary and rotational
(c) stationary and rotational with translation

54. A boring tool used for boring a hole over a large length should have
(a) one tool bit on a bar
(b) 2 tool bits fitted on diametrically opposite sides
(c) 4 tool bits

55. Boring is
    (a) enlargement of an existing hole
    (b) finishing of an existing hole
    (c) changing the shape of a hole
56. A Jig boring machine resembles in appearance to
    (a) vertical milling machine
    (b) vertical boring machine
    (c) horizontal boring

ANSWER KEY

1. (b)	2. (b)	3. (d)	4. (e)	5. (b)
6. (d)	7. (b)	8. (e)	9. (d)	10. (c)
11. (b)	12. (a)	13. (b)	14. (b)	15. (c)
16. (d)	17. (c)	18. (b)	19. (e)	20. (c)
21. (b)	22. (d)	23. (b)	24. (d)	25. (b)
26. (e)	27. (a)	28. (b)	29. (e)	30. (c)
31. (b)	32. (e)	33. (d)	34. (d)	35. (a)
36. (d)	37. (a)	38. (c)	39. (a)	40. (a)
41. (c)	42. (b)	43. (c)	44. (c)	45. (a)
46. (b)	47. (a)	48. (a)	49. (c)	50. (c)
51. (b)	52. (b)	53. (c)	54. (b)	55. (a)
56. (a)				

6

Milling and Milling Machines

In Milling Process

- The main rotary movement is always performed by the milling cutter.
- The feed and infeed movements are performed by the workpiece.

In cylindrical milling, the cutter axis is parallel with the milling face. The teeth are arranged on the periphery of the milling cutter (Fig. 1). Each tooth of the milling cutter removes a chip (of comma shaped cross-section) at its cutting edge. The chip thickness is unequal (i.e. not uniform) which means the cutting forces on the machine is continuously changing.

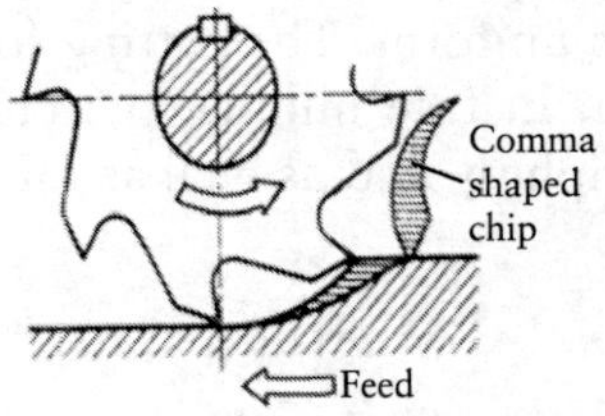

Fig. 1: Cylindrical milling

Cylindrical Milling

Cylindrical milling is divided into up-milling and down-milling.

Up-Milling

Feed movement is in the opposite direction to the cutter movement.

This results in

- More Toolwear
- Lower machining rate (Fig. 2).

Down-Milling

Feed movement and cutter movement are in the same direction. This results in

- Higher tool life of Cutter
- Higher machining rate
- Better Surface finish, (Fig. 2) (Down milling can only be performed on milling machines which have more rigidity).

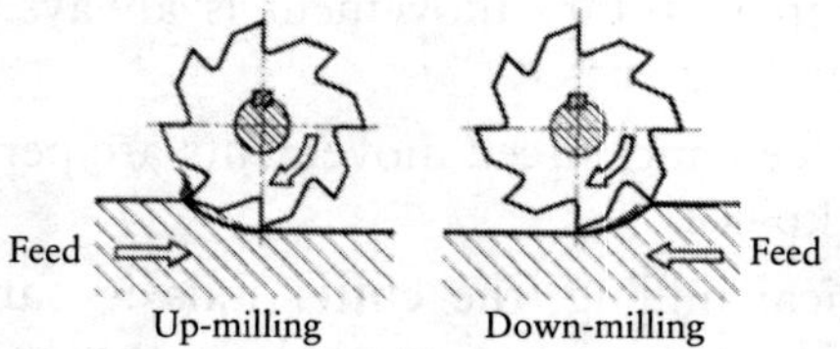

Fig. 2: Up-milling and Down-milling

Face Milling

In face milling (Fig. 3), the cutter axis is perpendicular to the milling face. Many teeth cut the job simultaneously. The chip cross-section is uniform. The cutting forces or load on the machine is uniform. In face milling, the cutter cuts both with the teeth on the periphery and as well as on the face.

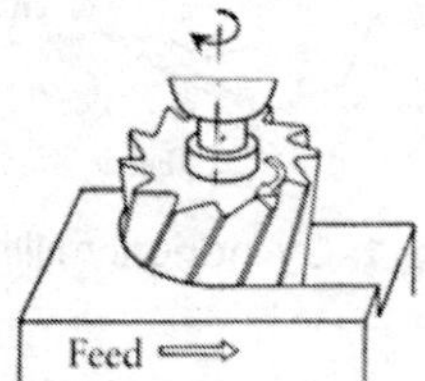

Fig. 3: Face milling

Milling Cutters

Milling cutters are basically multi-point cutting tools. The load on the cutting edges is lesser than in drilling or turning.

For hard and high strength work materials	Cutters with a large no. of teeth and with low rake angle are required
For lighter and less strength materials	Cutters with a small no. of teeth and with larger rake angles are required. Rounded gaps between teeth are provided to improve the chip disposal.

Types of Milling Cutters

There are two types of milling cutters classified on the basis of

- *Pointed*—tool milling cutters (manufactured by using form milling cutters) (used to produce flat surfaces).
- *Relief*—turned milling cutters (Relief surface of cutter tooth is made by turning process) (used to produce shapes, such as curves, threads and gears). These cutters are reground on the face only retaining the original rake angle.

There are different types of milling cutters depending on the shape or type of operation (Fig. 4):

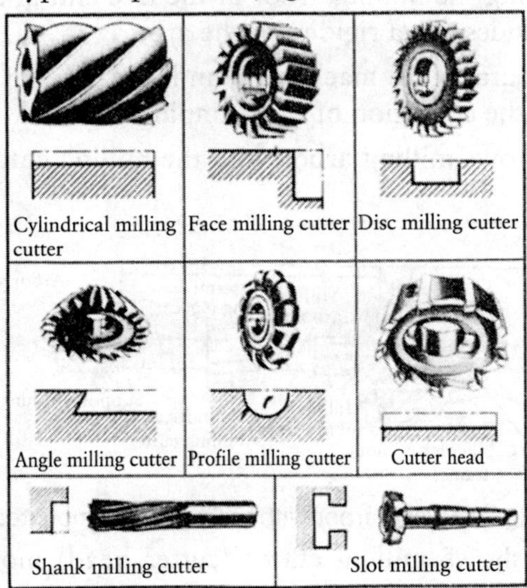

Fig. 4: Different types of milling cutters

1. Cylindrical milling cutter
2. Face milling cutter

3. Disc milling cutter
4. Angle milling cutter (for machining bevel edges)
5. Profile milling cutter (for concave shape)
6. Cutter head (for machining of large surface areas)
7. End milling cutter (end mill)/shank milling cutter
8. Slot milling cutter (for cutting T-slots)
9. Circular saw (for cutting off operations).

Milling cutters can be straight-toothed or helical-toothed depending on the direction of the tooth. Helical toothed milling cutters give better performance when compared with straight-toothed cutters. Many teeth cut simultaneously but gradually resulting in uniform cutting and the machine runs smoothly. Chatter (i.e. vibrations) are significantly reduced.

Milling arbors (milling machine tool spindles) are used to clamp milling cutters. Spacers are used to adjust the exact location of the cutter on the arbor.

Supporting the milling arbor at the free end in the machine bracket provides extra rigidity to the arbor.

The accuracy of machining and the life of the cutter depends on the condition of the clamping.

Fig. 5 shows milling arbor with the milling cutter mounted on it.

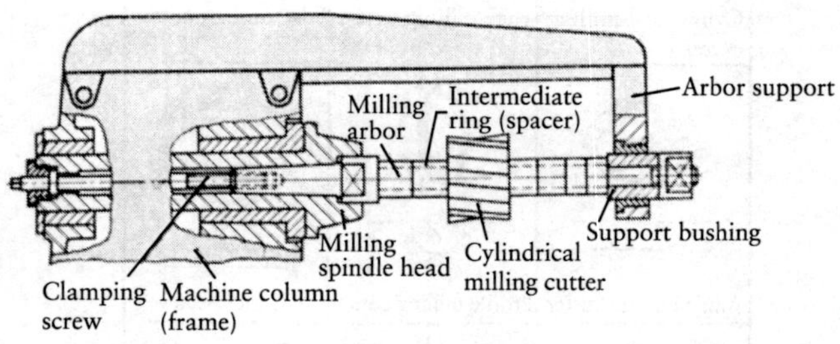

Fig. 5: Milling arbor with milling cutter mounted

Fig. 6 shows milling cutter (cutter head) mounted on a stub arbor.

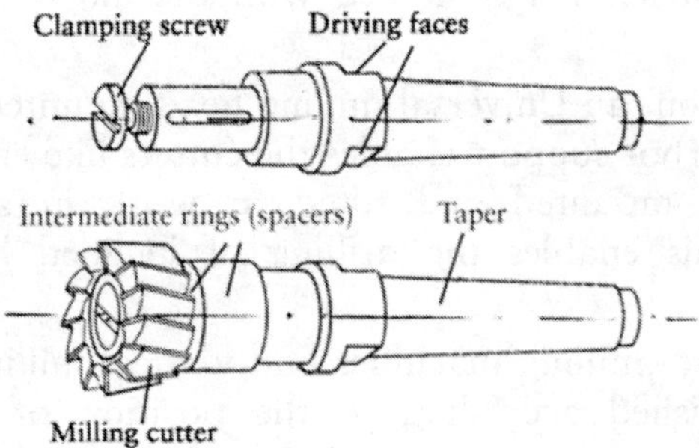

Fig. 6: Stub arbor

Milling Machines

The most important milling machine is known us Universal Milling Machine (Fig. 7). This type of machine is suitable for a wide range of operations.

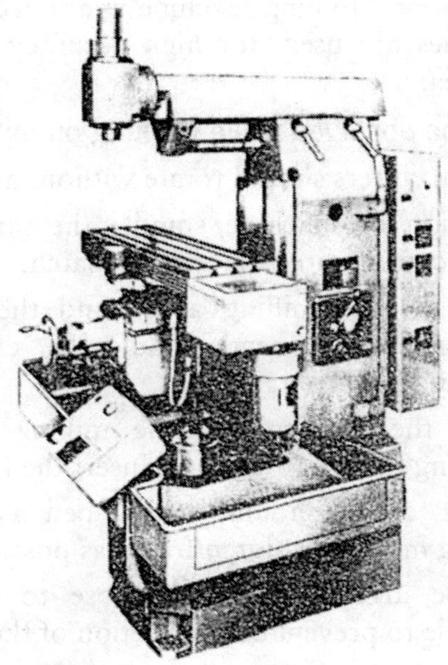

Fig. 7: Universal Milling Machine

Its main feature is a work table which can be swivelled or swing through 45° to the left or to the right. This enables spiral slots to be machined or milled with the aid of an indexing head.

In addition, an Universal milling head mounted on the top along with arbor support enables the cutters like end mills, and drills to be mounted and tilted to work in any inclined position. This enables the drilling of inclined holes in the workpiece.

Horizontal milling machines and vertical milling machines are distinguished according to the position of the milling spindle.

Limited production of gear wheels can be fabricated on horizontal milling machine, by mounting the gear blank in the index head and by indexing the blank from tooth to tooth. Form milling cutters are used in cutting gears.

However, for large scale production of gears, production of gears in horizontal milling machine is not economical; gear hobbing machines are used, for high accuracy and for mass production of gears.

Precautions to be observed while working on milling machine

1. Milling cutters should rotate without any run-out.
2. Direction of machine spindle (i.e. arbor) and the tool's cutting direction should match.
3. Taper on the milling arbor and the seats in the milling spindle head should be cleaned before inserting the arbor.
4. Clean the spacers and the milling cutter before inserting them in the arbor. Insert the key.
5. Milling cutter should be clamped as close to the milling machine column/frame as possible.
6. Set the arbor support as close to the cutter as possible to prevent any deflection of the arbor.

EVALUATIVE QUESTIONS

1. What is the difference between cylindrical milling and face milling?
2. What is the tool material in the case of milling cutter?
3. Sketch the cross-section of the chip produced in cylindrical milling.
4. Milling results in continuous chips or powdery chip?
5. What are the most common types of milling cutter?
6. What are the advantages of the helical-toothed milling cutter with respect to the straight-toothed milling cutter?
7. What precautions should be observed when clamping or holding milling cutters on the milling machine spindle?

OBJECTIVE TYPE QUESTIONS

Choose the correct alternative

1. To produce a smooth, flat surface on a workpiece you will generally use a

   (a) lathe (b) arbor press
   (c) milling m/c (d) drill press

2. A milling machine which has a table that can be swivelled and set at any angle to the column face is called a

   (a) plain knee-and-column type milling machine
   (b) universal knee-and-column type milling machine
   (c) bed-type milling machine
   (d) drum-type milling machine

3. The operation of milling two sides of a workpiece simultaneously is called

   (a) gang milling (b) climb milling
   (c) square milling (d) straddle milling
   (e) end milling

4. A _______ must be used in gear milling operations to obtain equal spacing of teeth in gear blank.
   (a) Index plate
   (b) Gear tooth vernier
   (c) Pitch gauge
   (d) Dividing head
5. Climb milling operation should never be used when milling _______ parts because the hard surface scale on the parts will dull the cutter.
   (a) Magnesium (b) Cast Iron
   (c) Steel (d) Non-ferrous
6. When milling the teeth of a spur gear, the gear blank is mounted on _______ held between dividing head centers.
   (a) a chuck (b) a face plate
   (c) a mandrel (d) a collet
7. The position of the _______ is considered in classifying a milling machine as to horizontal or vertical type.
   (a) spindle (b) workpiece
   (c) milling cutter (d) work table or bed
8. When performing angular indexing, one complete crank turn is required to index the work through _______.
   (a) 360° (b) 180°
   (c) 90° (d) 9°
9. To obtain fine finish cuts in milling
   (a) the cutting speed should be decreased and the feed increased.
   (b) the cutting speed should be increased and the feed decreased.
   (c) both the cutting speed and feed should be decreased.
   (d) both the cutting speed and feed should be increased.

10. The rigidity and the power rating of the milling machine must be adequate when the milling operation is performed using _______.
    (a) carbide-tipped milling cutters
    (b) cast-alloy milling cutters
    (c) H.S.S. milling cutters
    (d) High carbon steel cutters
11. Helical gear milling operations can be performed only on a
    (a) Plain horizontal milling machine
    (b) Vertical milling machine
    (c) Universal milling machine
    (d) Drum type milling machine
12. To cut 20 teeth on a gear blank, the index crank of the dividing head must be turned _______, to rotate the work into position for milling each tooth.
    (a) 1/2 a turn
    (b) one complete turn
    (c) two complete turns
    (d) 1/40 of a turn (or 9 degrees)
    (e) 80 turns
13. Discontinuous or powdery chips are normally produced when milling metals such as
    (a) Mild steel (b) Tungsten
    (c) Cast Iron (CI) (d) High Speed Steel (HSS)
14. Continuous chips are normally produced when milling
    (a) Ductile materials (b) Brittle materials
    (c) Brass (d) Diamond
15. The attachment used on milling machines for indexing purposes is called a
    (a) Face plate (b) V-block
    (c) Dial indicator (d) Dividing head

16. The number of index crank turns required after each cut to mill 8 equally spaced slots in a round shaft is
    (a) 320 (b) 40
    (c) 8 (d) 5
    (e) 2½
17. A universal dividing head must be used to perform a milling operation by
    (a) Plain indexing (b) Direct indexing
    (c) Differential indexing (d) Compound indexing
18. To rotate the work mounted on a dividing head through one complete turn, the index crank must be rotated through
    (a) 40 complete turns (b) 20 complete turns
    (c) 10 complete turns (d) 1 complete turn
    (e) 1/40th of turn
19. While sharpening milling cutters by grinding, the width of the land can be reduced, by grinding a
    (a) Primary clearance angle
    (b) Secondary clearance angle
    (c) Rake angle
    (d) Suitable nose radius
20. The cutters designed for straddle milling operations are called _______ cutters.
    (a) side milling (b) face milling
    (c) plain milling (d) dovetail milling
21. The operation of a dovetail milling cutter is similar to the operation of _______.
    (a) a plain milling cutter (b) a side milling cutter
    (c) an end milling cutter (d) a shaping tool
22. A quick and accurate way to mill a perfect square on the end of a round shaft is to mount the shaft in _______ so that the shaft can be indexed.
    (a) a differential (b) a compound rest
    (c) an index plate (d) a dividing head

23. The cutting tool material capable of withstanding the highest cutting speeds is the
    (a) Stellite tipped cutters
    (b) Tungsten-carbide tipped cutters
    (c) High-speed steel cutters
    (d) High carbon steel milling cutters
24. In helical milling, the ratio of the circumference of the gear blank to the lead of the helix gives the
    (a) angle setting of the machine table
    (b) speed to use
    (c) feed and depth of cut required
    (d) no. of teeth to be cut
25. _______ of the index crank is/are required when indexing, to rotate the workpiece one complete turn.
    (a) 1/40 turn (9 degrees) (b) 1 turn
    (c) 40 turns
    (d) depends on the no. of holes in the index plate
26. One complete turn of the index crank will turn the workpiece by
    (a) 9 degrees (b) 18 degrees
    (c) 90 degrees (d) 180 degrees
    (e) 1/40th of a degree
27. The accurate spacing of teeth in a gear blank requires the use of
    (a) a dividing head
    (b) an index plate
    (c) a gear tooth vernier
    (d) a differential mechanism
28. The cutting of helical gears requires the use of a _______ milling machine.
    (a) vertical (b) horizontal
    (c) universal (d) drum-type
29. How is the feed rate expressed in the case of milling operation?

(a) mm/tooth
(b) r.p.m. of the milling cutter
(c) metres/minute
(d) strokes/minute

30. Milling cutters are mounted on a part called the _______ of the milling machine.
(a) Bracket or Brace (b) Arbor
(c) Shaft (d) Horizontal support
(e) Table

31. Milling cutters are considered as
(a) single-point cutting tools
(b) multi-point cutting tools
(c) single-edged cutting tools

32. The proper type of file to be used when filing metals such as brass or bronze is
(a) the smooth-cut file
(b) the second-cut file
(c) the coarse-cut or rough-cut file
(d) the double-cut file
(e) the single-cut file

33. The no. of teeth per inch on an hacksaw blade is called ______ of the blade.
(a) the finish (b) the module
(c) the twist (d) the pitch
(e) the cut

34. A wrench which has one solid jaw and one movable jaw is called
(a) a torque wrench (b) a power wrench
(c) a screw wrench (d) an Allen wrench
(e) an adjustable wrench

35. A power saw which employs a continuous looped blade which is driven by two wheels or pulleys is called
(a) a power hacksaw machine
(b) a circular saw machine

(c) a filing machine
(d) a band saw machine

36. The no. of teeth/inch in a band saw blade is called as
    (a) module of the blade (b) pitch of the blade
    (c) finish of the blade (d) twist of the blade
37. The cutting action of power-hacksaw blade is described as
    (a) an oscillating motion
    (b) a reciprocating motion
    (c) a continuous circular motion
    (d) a simple harmonic motion
38. Band sawing of the parts to an irregular shape or form following a layout line is called
    (a) In-line sawing (b) Out-line sawing
    (c) Contour sawing (d) Multiple sawing
39. Chipping is an operation wherein
    (a) a reciprocating motion of a file is used.
    (b) a chisel is used for metal removal.
    (c) a mark is made with a centre punch.
    (d) a scraper is used for removing the metal.

ANSWER KEY

1. (c)	2. (b)	3. (d)	4. (d)	5. (b)
6. (c)	7. (a)	8. (d)	9. (b)	10. (a)
11. (c)	12. (c)	13. (a)	14. (a)	15. (d)
16. (d)	17. (c)	18. (c)	19. (c)	20. (a)
21. (c)	22. (d)	23. (b)	24. (a)	25. (c)
26. (a)	27. (a)	28. (c)	29. (a)	30. (b)
31. (b)	32. (c)	33. (d)	34. (e)	35. (d)
36. (b)	37. (b)	38. (c)	39. (b)	

7

Boring

1. Introduction

Boring is a process of enlarging a hole by using a single point cutting tool. Holes that have been formed by casting, punching, drilling or any other process can be bored. The cutting edge may be adjusted to accommodate the different work sizes. Boring is essentially internal turning. It does not create the hole but rather enlarges the hole to a specific size. Boring process serves to accomplish the following functions:

1. To enlarge the diameter of a hole, particularly very large holes.
2. For truing a hole is to make the hole concentric and straight with respect to the workpiece axis.
3. To produce an accurate size, high quality finish holes especially in case of small production lots.
4. To machine a true hole as a pilot hole for subsequent entry of cutting tools.

Although boring can be done on most machine tools that can do turning, however boring is also possible by using a rotating tool with the workpiece remaining stationary.

2. Classification of Boring Machines

The boring machines can be classified as:

1. Vertical Boring Machine
2. Horizontal Boring Machine
3. Precision Boring Machine
4. Jig Boring Machine

Vertical Boring Machine

Vertical boring machines are turning machines that in many ways are similar to conventional lathes turned on end. In these machines, the workpiece which is mounted on a horizontal table revolves and the cutting tool is stationary. Vertical boring machine is also known as boring mill. The term 'mill' is derived from the potter's mill which rotates the work. Functionally, a vertical boring machine is similar to vertical turret lathe, except that it has two main tool heads instead of one.

Machines for boring range in size from those that handle work 30 cm in diameter to those capable of boring workpiece even 1 metre in diameter. Tool heads have both horizontal and vertical feed and so it can be used for boring and for facing. The tool heads can also be swivelled about a horizontal axis to permit boring at an angle. Fig. 1 shows the basic elements of a vertical boring machine.

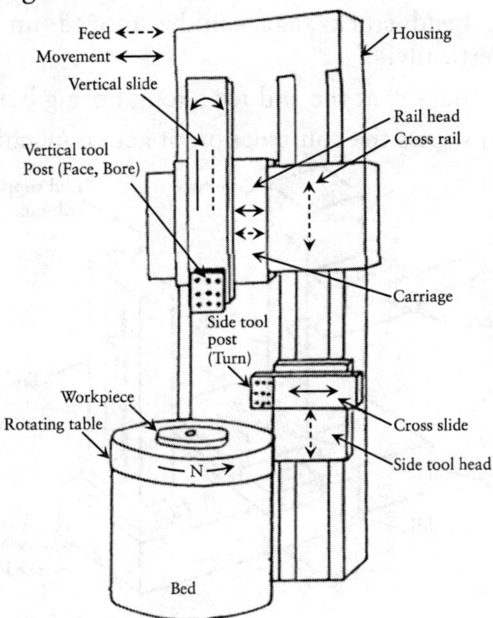

Fig. 1: Vertical boring machine

The modern boring machines are numerically controlled. These machines have the provision to make tool settings merely by setting dials and also the adjustment for a cut is preset while it is being made.

Vertical boring machines are commonly used to make round parts having short lengths in relation to their diameters. Vertical boring machines are used to make large and heavy parts.

Horizontal Boring Machine

Horizontal boring machines are constructed in such a way that the workpiece remains stationary and the tool revolves.

The basic components of these machines are shown in Fig. 2(a). The essential features of the machine are:

(i) A horizontal spindle that rotates the cutting tool.

(ii) A table that can be moved and fed in two directions in a horizontal plane.

(iii) A head stock that can be moved up and down vertically.

(iv) A support at the end for a long boring bar.

Fig. 2(b) shows the construction of actual machine.

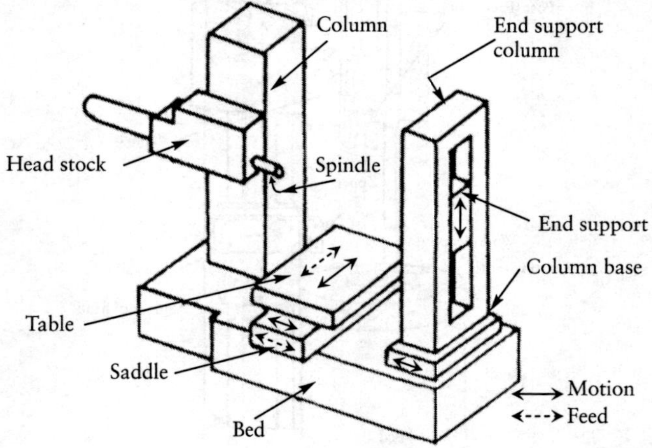

Fig. 2(a): Horizontal boring machine

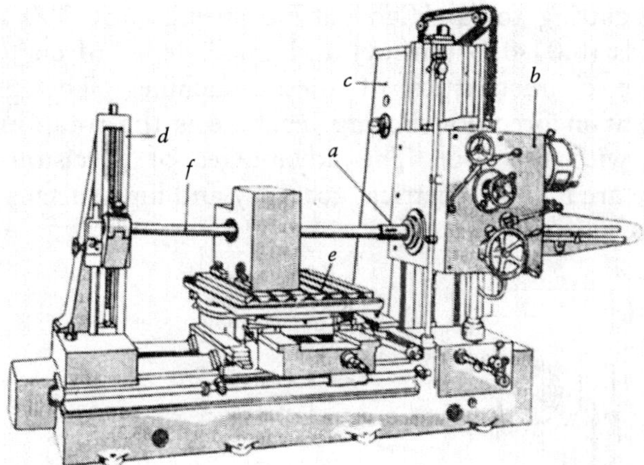

Fig. 2(b): Horizontal boring machine: (a) Main spindle, (b) Head stock, (c) Vertical column, (d) Auxiliary column, (e) Machine table, (f) Boring bar

They perform different operations at various locations on the workpiece without the need for changing the basic setup and they maintain accurate relationship between the machined surfaces. The tool is mounted either in a stub type bar held in the spindle or in a long bar whose outer end is supported in the bearings on the end support.

Horizontal boring machines are available in a wide range of capacities. The size of the machine is generally specified by the diameter of its spindle, which varies from 75 to 250 mm. Horizontal boring machines are used primarily for boring holes less than 30 cm in diameter or for long holes or for a series of in-line holes. These machines are extensively used for large, complex castings, forgings, weldments and similar workpieces.

Precision Boring Machine

These machines are designed for precision boring, facing and turning operations [Fig. 3(a)]. Precision boring machines are available in a variety of types. Configurations include single or multiple spindles arranged horizontally, vertically or at any required angle. Horizontal precision boring machine may be arranged as single-end, that is, holes are machined from one end only or it can be arranged as double-end machine as shown in Fig. 3(b). The machining is done with diamond and carbide cutting tools. High cutting speeds (150-300 m/min), fine feeds (0.01-0.1 mm/rev) and small depth of cut (0.1-0.3 mm) are characteristics of these machines. The principal movement in precision boring machine is the rotation of the spindle with the tool. The advantages of precision boring machine are high geometrical accuracy and high surface finish.

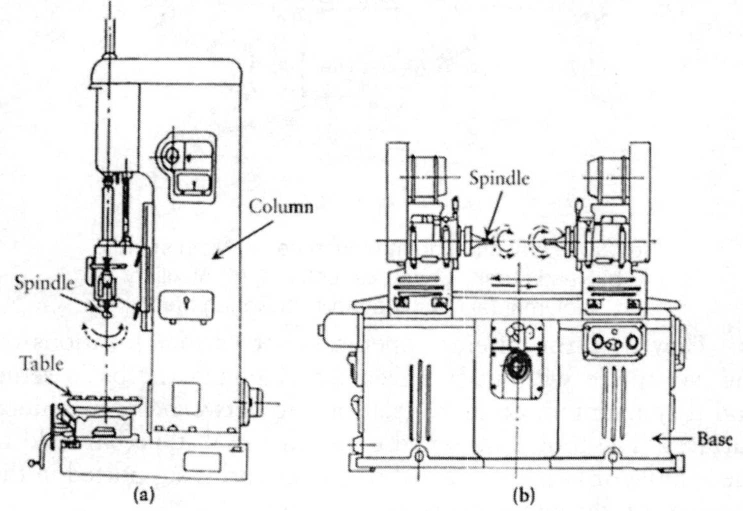

Fig. 3: Precision boring machine—(a) vertical type, (b) horizontal type

Jig Boring Machine

Jig boring machine is mostly used to machine holes to their correct position. Centre distances with tolerances as small as 1 micron can be realized. Its importance lies in the production of jigs and in tool making, where it is vital requirement for the function of the parts that certain holes are in absolute correct position with respect to each other. Jig boring machine therefore finds a wide application in tool rooms and prototype centres. Besides making holes, the jig boring machine is also used for light milling work.

The jig boring machine is somewhat similar to a vertical milling machine in construction, as shown in Fig. 4(a). The similarity lies in vertical spindle and table movements, but the jig boring machine is more rigid than milling machine. The rigidity is important to attain the required accuracy.

The jig boring machines are classified into two types—column type and cross rail type. The column type jig boring machine has a bed which carries longitudinal and transverse slide. A column, provided at the back, supports the spindle housing which can be vertically adjusted. The column type construction is used for smaller machines. Fig. 4(b) shows the construction of actual machine.

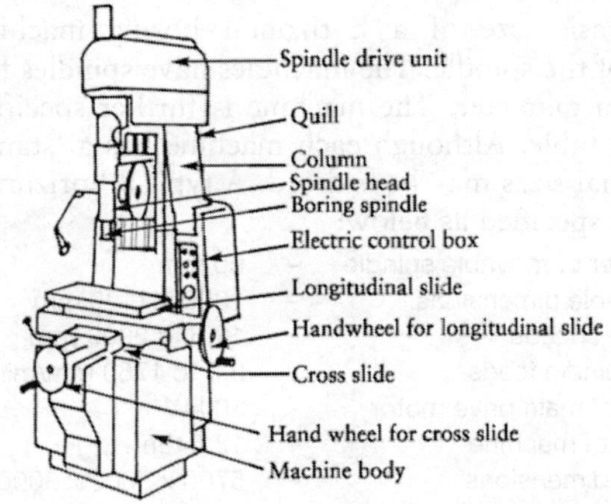

Fig. 4(a): Jig boring machine (Column type, Vertical)

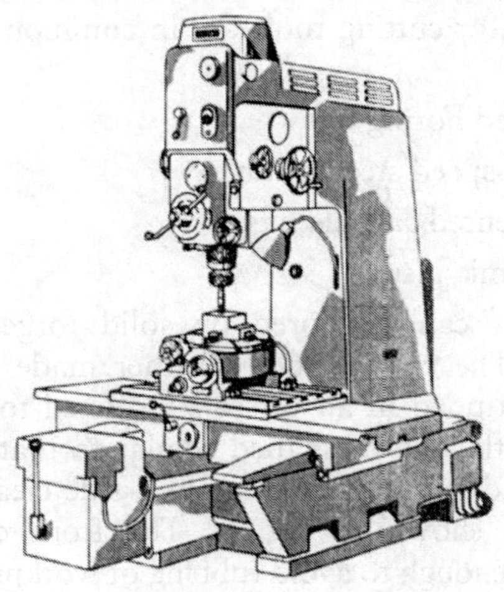

Fig. 4(b): Jig boring machine

The cross rail type is chosen for bigger jig boring machine. The bed of this machine has only longitudinal slide, which moves between two vertical columns. The cross slide slides up and down on these columns. The transverse slide is mounted on the cross rail and houses the vertical spindle unit.

3. Specifications of a Boring Machine

The size of a vertical boring machine is specified by the diameter of the revolving table. The machine is often made with table diameters from 1200 mm to 3600 mm.

The basic size of a horizontal boring machine is the diameter of the spindle. The machines have spindles from 75 to 150 mm in diameter. The machine is further specified by the size of the table. Although each machine has a 'standard' size table, special sizes may be ordered. A typical horizontal boring machine is specified as below:

Diameter of movable spindle	-	90 mm
Work table dimensions	-	1300 x 1120 mm
Spindle speeds	-	12.5 to 2000 rpm
Axial spindle feeds	-	2.2 to 1760 mm/min.
Power of main drive motor	-	10 kW
Weight of machine	-	12.5 ton
Overall dimensions	-	5700 x 3000 x 3000 mm

4. Boring Tools

Four types, of cutting tools are in common use for boring operations.

(i) Forged boring bits

(ii) High speed steel tool bits

(iii) Cemented carbide bits

(iv) Ceramic inserts

Small holes can be bored by solid forged single point cutting tools. The cutting tool is either made of HSS or the cutting end is tipped in an offset V-grooved tool holder. The cutting end of the tool is shaped similar to that of a left-hand turning tool and it has front clearance, side clearance and side rake angles as shown in Fig. 5. The front clearance angle should be large enough to avoid rubbing of workpiece at the heel.

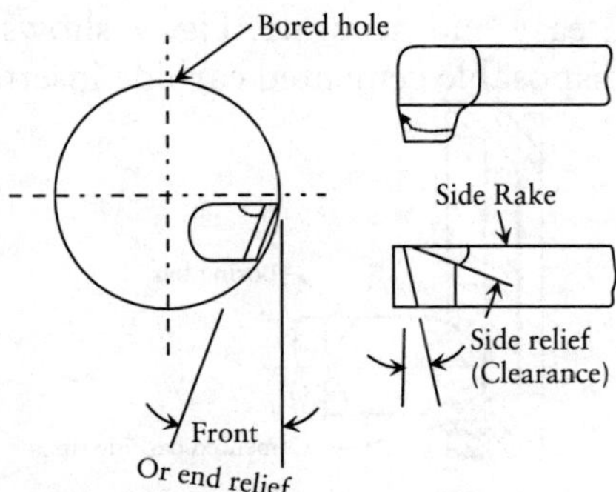

Fig. 5: Rake and relief angles of a solid, forged boring tool

The shape of cutting edge is governed by the type of operation to be performed. For shoulder turning, the forged boring tool is similar to left hand facing tool. Whereas for boring a straight hole, the cutting edge of the tool is like a left hand turning tool. For boring a hole over large length, double ended cutter is used. Fig. 6 shows sketches of boring tools.

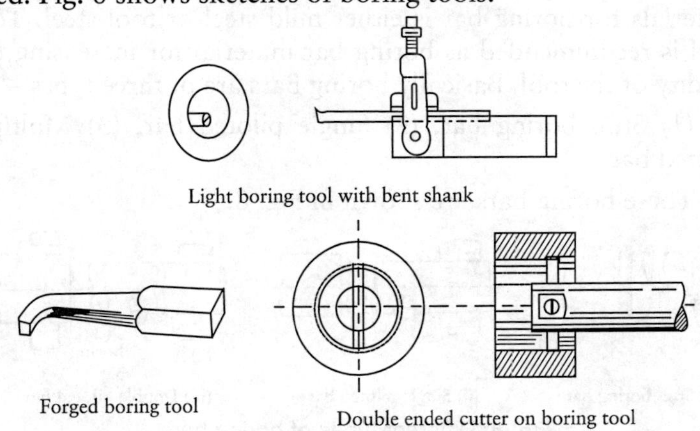

Fig. 6: Different types of boring tool

Most boring operations use a single point cutter inserted in a bar, which can enter the hole to be bored. Sometimes cemented carbide inserts are used for boring. These triangular inserts can be used for blind hole boring or boring to a square. The insert can be locked in place with a lock pin. The lock can be either tightened or loosened which makes the insert indexing quick, easy and accurate. Fig. 7 shows a boring tool holder having disposable cemented carbide inserts.

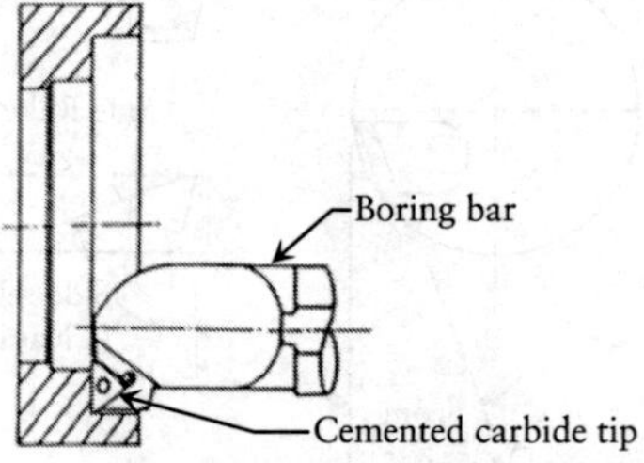

Fig. 7: Boring operation with a boring bar having disposable cemented carbide insert

5. Boring Bars

A boring bar is a round steel bar which positions and holds the tool bit. The tool bits may be secured in a boring bar at 90°, 45° and 30° angles, on either end with the help of a setscrew. Boring bars are positioned for maximum rigidity. The materials for boring bar is either mild steel or tool steel. Tool steel is recommended as boring bar material for increasing the rigidity of the tool. Basically boring bars are of three types.

(1) Stub boring bar, (2) Single piloted bar, (3) Multiple piloted bar

These boring bars are shown in Fig. 8.

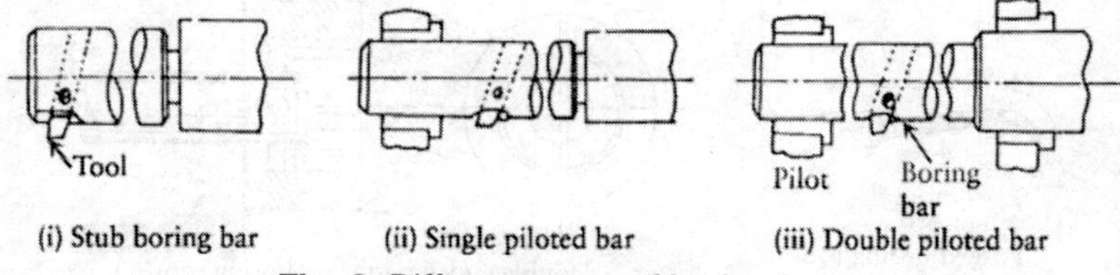

Fig. 8: Different types of boring bars

Stub boring bars are used to bore short holes. This bar acts as a cantilever beam while in action. It is difficult to secure it rigidly and may lead to chatter. Stub boring bar should be as short and as rigid as possible. For some operations stub boring bars can be made more rigid by fitting a sleeve at the supporting end. The single piloted bar is guided on its leading end. This helps the bar springing away from the work. A straight shank pilot boring bar is shown in Fig. 9. The multiple piloted bar is guided both on the leading and lagging sides. A taper shank double piloted boring bar is shown in Fig. 10. The taper shank has a standard Morse taper size.

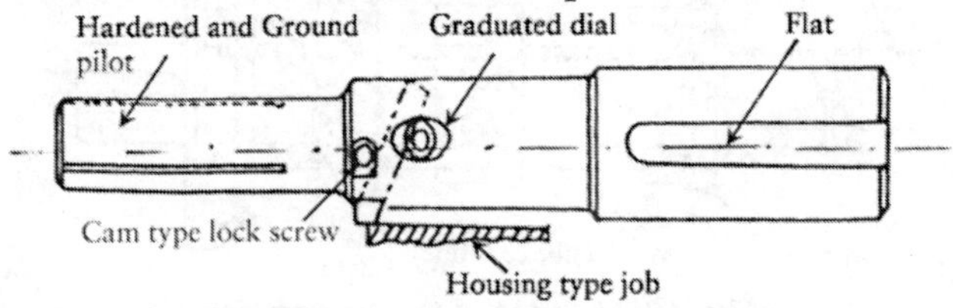

Fig. 9: Straight-shank pilot boring bar

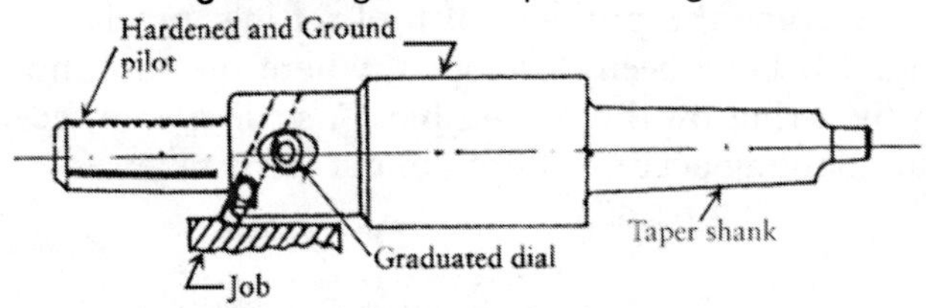

Fig. 10: Taper shank pilot boring bar

Sometimes under specific conditions, multiple cutter boring bars are also employed. These bars help in reducing the time needed for boring. There are two types of such boring bars. In the first type shown in Fig. 11, the cutters are placed around the bar. While cutting, these cutters act like a reamer. The cutters are adjustable by sliding them endwise in the tapered slots. In the second type, the cutters are placed along the bar as shown in Fig. 12.

The disadvantage of boring bar is that the working range is very limited and therefore a complete set of boring bar is required to cover the complete range of diameters. Also the tool setting is very time consuming.

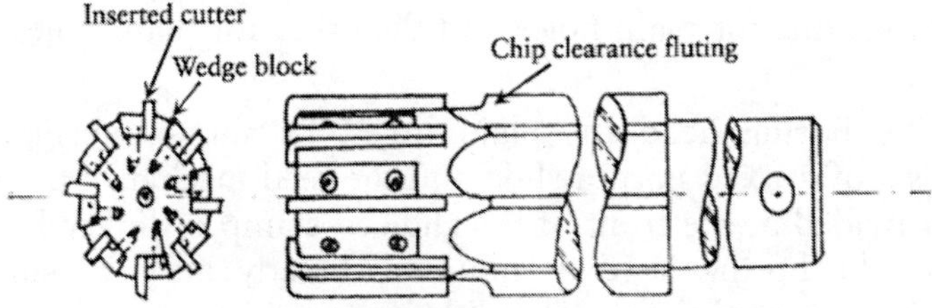

Fig. 11: Radial cutter boring bar

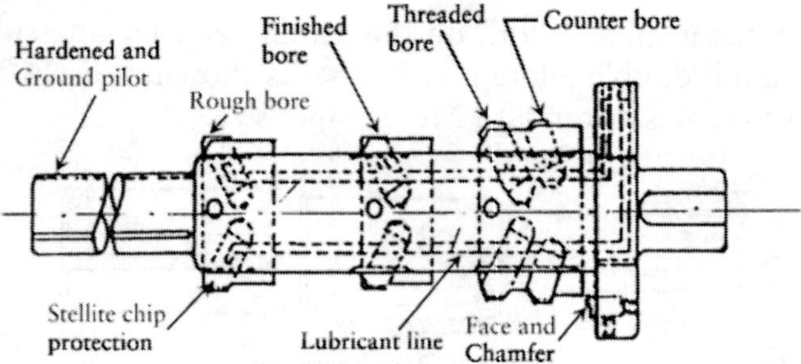

Fig. 12: Multiple cutter boring bar

To overcome the problem of tool setting, precision boring bars (Fig. 13) have been developed, where the tool movement is given by a ring on the boring bar. A scale is marked on the ring thus enabling accurate depth of cut to be taken.

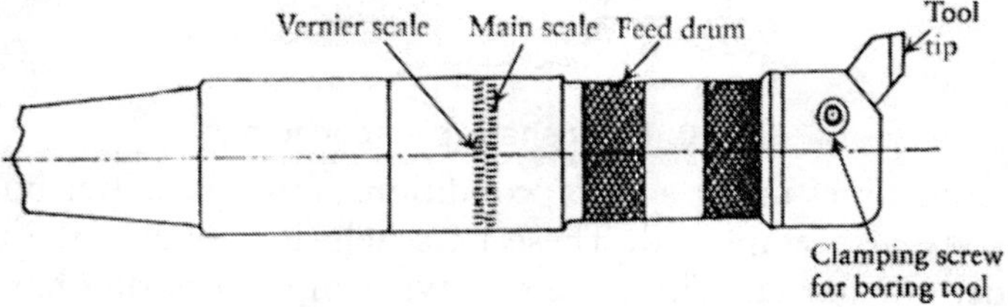

Fig. 13: Precision boring bar

6. Boring Head

This is also a boring tool holder which is more rigid as compared to boring bar. Boring head is used for boring large holes in the range of 100 to 500 mm. There is provision for precise adjustment of boring tool secured in the boring head. The boring tool can also be placed in several radial positions to suit various bore sizes. Boring heads are available in various forms and sizes. Basically two types of boring heads are available: one for plain holes and the other for plain holes and facing.

The boring head for plain holes, as shown in Fig. 14 consists of a fixed part, a slide and the feed mechanism. Holes are provided at the front of the slide to clamp ordinary boring tools. The boring head is mounted directly into the spindle nose by a tapered shank. The slide is guided in dovetail and the feed is given by a precision ground screw which moves the slide. The screw head is graduated on its circumference. This boring head is widely used on jig boring machines but the working range of this boring head is limited to small holes.

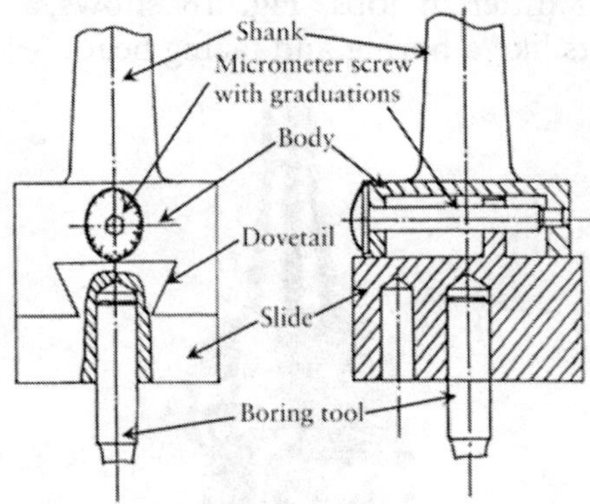

Fig. 14: Plain boring head

The boring and facing head as shown in Fig. 15 is similar to plain boring head, except that it is equipped with an additional continuous feed arrangement for the facing operation. For ordinary boring, the feed is given either by a screw or by a horizontal feed ring on the head. The motion of the feeding is then transmitted to the slide depending upon the head design. For facing operation, a separate slow motion ring is provided and a continuous feed of about 0.05 mm per revolution is achieved. If the facing is required to be done to a certain diameter

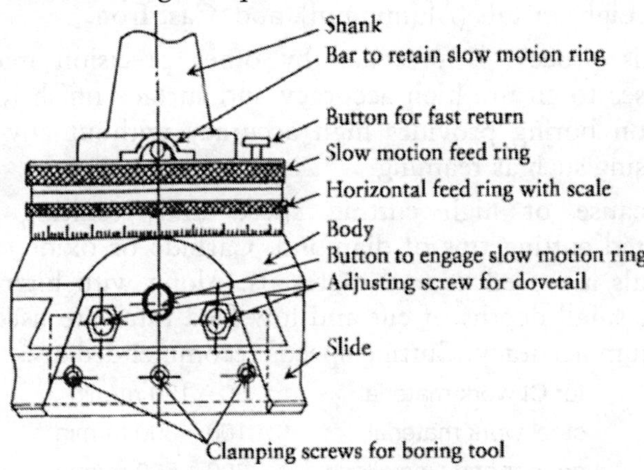

Fig. 15: Boring and facing head

only; a stop can be present which disengages the continuous feed. A choice of different boring tools and accessories has made boring head very versatile and suitable for executing a vast variety of different jobs. Fig. 16 shows a micro boring head which acts like a boring and facing head.

Fig. 16: Microboring head

Precision Boring

This machining method is primarily used for Copper alloys, Light metals (Aluminium), and Cast Iron.

This process is preceded by other precision machining processes to ensure high accuracy and surface finish (quality). Precision boring provides high accuracy without any further processing such as reaming.

Because of high cutting speed and resulting heating generated cutting tips of diamond. Carbide or oxide ceramics materials are used as tool materials. Along with high cutting speeds, small depths of cut and low feed rates are used to get maximum accuracy. Cutting speeds recommended are:

for CI work material	:	80 – 120 m/min
steel work material	:	100 – 300 m/min
copper and zinc alloys	:	300 – 500 m/min

Aluminium work material	:	300 – 800 m/min
Depth of cut	:	0.03 – 0.15 mm,
Feed rate	:	0.01 – 0.10 mm

The Boring machine must be of extremely rigid construction and boring tool should rotate without any vibrations to get the desired accuracy and surface finish.

Precision boring of engine cylinder blocks are common in manufacturing industries.

Boring Defects and Causes

Various defects that may occur during boring and the possible causes are given in the Table:

Table 7.1: Trouble Shooting Chart (Operation: Boring)

Defects		Possible Causes
1. Poor surface finish	(a)	Cutting speed too low
	(b)	Feed rate too high
	(c)	Blunt tool
	(d)	Too small nose radius
2. Chatter (vibrations)	(a)	Incorrect front clearance
	(b)	Too much back rake
	(c)	Nose radius too large
	(d)	Lack of rigidity in boring tool
3. Chipping of cutting edges	(a)	Cutting edge not sufficiently supported
	(b)	Chatter due to lack of rigidity
	(c)	Scale or inclusion in work piece material
	(d)	Interrupted cut
	(e)	Lack of rigidity in boring bar
	(f)	Over use of tools (blunt tool)
4. Cracking of edges of bored holes	(a)	Worn out or chipped cutting edge
	(b)	Improperly applied coolant
	(c)	Lack of rigidity
	(d)	Cutting speed too low
	(e)	Chatter (vibrations)

5. Excessive tool wear	(a)	Cutting speed too low
	(b)	Feed rate too low
	(c)	Incorrect tool grinding
	(d)	Poor finish on tool

REVIEW QUESTIONS

1. What is the difference between drilling and boring?
2. Name three types of boring machines.
3. Give a description of a jig boring machine.
4. Explain and list the various types to machines used in boring operation.
5. Sketch some examples of boring tools.
6. How is a boring machine specified?

OBJECTIVE TYPE QUESTIONS

Choose the correct alternative

1. In a horizontal boring machine the workpiece and tool motion respectively are
   (a) rotational and translational
   (b) stationary and rotational
   (c) stationary and rotational with translation
2. A boring tool used for boring a hole over a large length should have
   (a) one tool bit on a bar
   (b) 2 tool bits fitted on diametrically opposite sides
   (c) 4 tool bits
3. Boring is
   (a) enlargement of an existing hole
   (b) finishing of an existing hole
   (c) changing the shape of a hole
4. A Jig boring machine resembles in appearance to
   (a) vertical milling machine
   (b) vertical boring machine
   (c) horizontal boring machine

ANSWER KEY

1. (c) 2. (b) 3. (a) 4. (a)

8

Broaching

1. Introduction

Broaching is an operation in which a special tool called broach is forced across or through the workpiece, by either pushing or pulling, forming a shaped surface. The broach has a series of teeth which increase in size progressively from one end to the other. In doing so, each tooth of the tool takes a small cut through the workpiece to be broached. Broaching is an extremely rapid and accurate metal cutting process, which combines both roughing and finishing in one operation. However, no individual broach tooth performs both functions. Each successive tooth removes only a predetermined amount of stock and remains in cutting contact only for a short time. The cutting operation in broaching is performed by a single push or pull motion of the broaching tool and not by a back and forth motion. Most of the broaching operations are carried out by pull method. The broaches used in pull broaching are considerably longer than those used in push broaching. A push broach is forced through the material whereas a pull broach is drawn through the material.

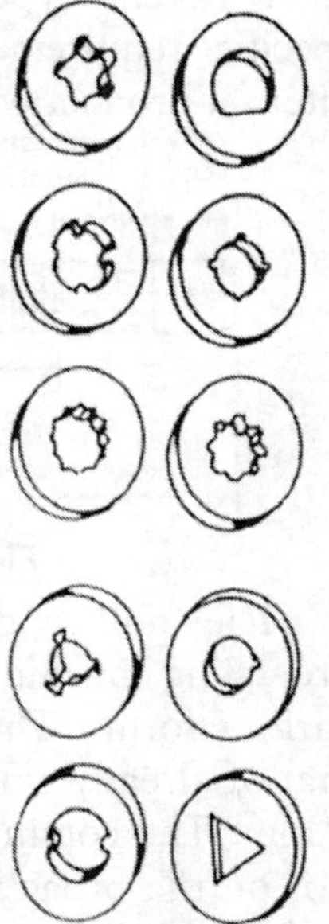

Fig. 1: Shapes of various holes that can be broached

Broaching is classified into two broad categories: internal broaching and surface broaching. Internal broaching means generating a hole of any desired shape in the workpiece already having a round hole. For example: making keyway, splines and square holes. Holes of any shape can be broached as shown in Fig. 1. In surface broaching, external surfaces on the workpieces are generated. Surface broaching machines not only produce flat surfaces but many other surfaces including those of complex contours. Examples include the external surfaces of main bearings of engine cylinder blocks, main bearing caps, connecting rods, etc.

2. Principal Parts of Broach

Different types of broaches are used depending upon specific requirements. Fig. 2 shows the main parts of an internal broach, which are described below in brief.

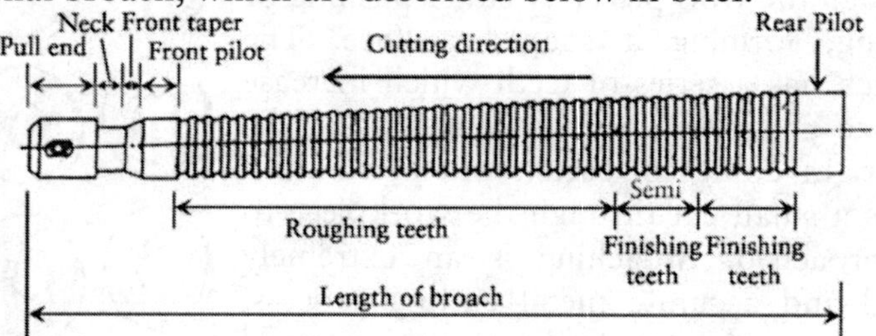

Fig. 2: Elements of internal broach

The pull end serves to engage the broach with the broaching machine through a puller head. The pull end is of various forms. The diameter of the pull end must be less than that of already existing hole in the workpiece by at least 0.5 or 1 mm. The commonly used pull end is key type pull end. The slot on the broach has a corresponding slot in the puller head and both are engaged by putting a key through the slot.

The neck is required so that the broach, when inserted into the workpiece, can be easily attached to the puller head with the latter in the position nearest to the faceplate or platen of the broaching machine. The front taper helps the workpiece to be more easily put on the front pilot of the broach. Its length ranges from 5 to 20 mm.

The front pilot aligns the broach with the workpiece and guides the broach at the beginning of the cut. The shape of front pilot must conform with the previously machined hole in the workpiece. Its length is taken equal to the length of the hole to be broached and its diameter should be equal to the minimum diameter of the previously machined hole. The cutting teeth remove the metal from the workpiece. The broach has a series of successive teeth of increasing diameter. The diameter of first cutting tooth is taken equal to that of the front pilot and the diameter of each consecutive tooth is increased, so that each successive tooth removes a small amount of material. The cut per tooth varies from 0.01 mm to 0.15 mm. The first series of cutting teeth are roughing teeth. These teeth are stepped for heavy cut and usually have chip breakers. The next series of teeth that follow roughing teeth are semi-finishing teeth. These teeth have much smaller step per tooth and are meant for progressively lighter cuts. The finishing teeth form the last group of teeth. They are four to eight teeth in number, having the same diameters. The diameter of finish teeth on a new broach is equal to that of last semifinishing tooth. Finishing teeth are reserved for finishing operation and never do any cutting. The finishing teeth also replace the last semifinishing teeth when they have been ground undersize by repeated sharpening. The rear pilot is the portion of the broach immediately following the finishing teeth. It helps in guiding the broach as it passes out of the workpiece, to maintain alignment and to avoid any damage to the broached surface by the last finishing teeth of the broach. Broach length is the total overall length of broach. This length must be less than the maximum stroke length of the broach. The broach length is limited by the rigidity of the broach. The elements of the cutting teeth on a broach are shown in Fig. 3 and are explained as below:

(i) Pitch (t) is the linear distance between successive cutting edges of the broach. Pitch is a very important design factor, as it determines the strength of the tooth and chip space. An empirical formula that may be used to find the pitch is

$$\text{Pitch} = 1.5\sqrt{\text{Length of cut}} \text{ (all in mm units)}$$

(ii) Tooth depth (*h*) is the distance from the cutting edge to the bottom of the gash.

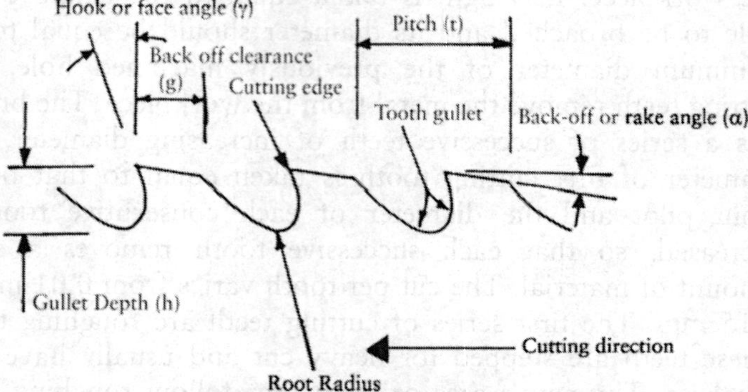

Fig. 3: Elements of cutting tooth on a broach

The pitch and tooth depth are related as t = (2.5 to 2.8) h.

(iii) Tooth thickness (*g*) is the distance across the tooth from the cutting edge to the start of next tooth radius tangent. Tooth thickness determines the number of regrinds possible before the broach cuts undersize.

(iv) Back-off angle (α). On the cutting teeth of a broach, the entire tooth thickness is relieved by a clearance angle. The value of clearance angle depends on the work material being cut. Large back-off angles lead to more rapid loss of broach size on its diameter in resharpening, causing the broach to be undersize. Therefore clearance angle (α) is made as small as possible. Clearance of 2 to 3° on the roughing teeth, 1° on the semifinishing teeth and 0.5° on the finishing teeth will give good results when broaching steel.

(v) Face angle or hook angle (γ) of the tooth depends on the material to be cut and its hardness toughness and ductility. In steel cutting, the face angle decreases with increased hardness.

Soft steel	15 to 20°
Hard steel	8 to 12°

Brittle materials call for small face angles.

Cast iron	6 to 8°
Brass	–5 to 5°

3. Broaching Machines

On a small scale, broaching can be done with the use of a hand or hydraulically operated arbor press. These presses are simple and inexpensive and are used for a variety of internal broaching operations like keyway cutting, etc. A set of broaches is used on these arbor presses. Generally, broaching machines are classified as vertical or horizontal.

i. Vertical Machines

In vertical broaching machines, the travel of the broach is vertical. The vertical machines can be further classified as: Vertical pull down machine, vertical pull up machine, and vertical surface broaching machine. Salient features of these machines are discussed below in brief:

(a) *Vertical Pull Down Machine*. This machine is recommended for high production internal broaching operations. In this machine, the broaching operation is performed on the top of the worktable. The top end of the broach is held by an upper carriage. The operation consists of passing the front pilot of the broach through the hole to be broached. This is followed by automatic attachment of the lower end of the broach to a pulling mechanism in the base of the machine and disengagement at the top. The broach is then pulled through the hole to complete the broaching operation. At the end of cutting stroke, the part is removed and the broach is moved up to engage with the upper carriage.

For higher production, it is desired to have some arrangement for rapid loading and unloading of the machine when one or more tools are used. This vertical machine can be easily adapted to the use of a shuttle table for rapid loading and unloading. Pull down broaching simplifies the problem of chip disposal, as the chips fall off easily due to the gravity. Also in pull down broaching, the supply of cutting fluid at the cutting region is easier.

(b) *Vertical Surface Broaching Machine*. In this machine (Fig. 4), the broaching is accomplished by downward stroke of a ram that carries the broach downward for the cut and then returns to top position.

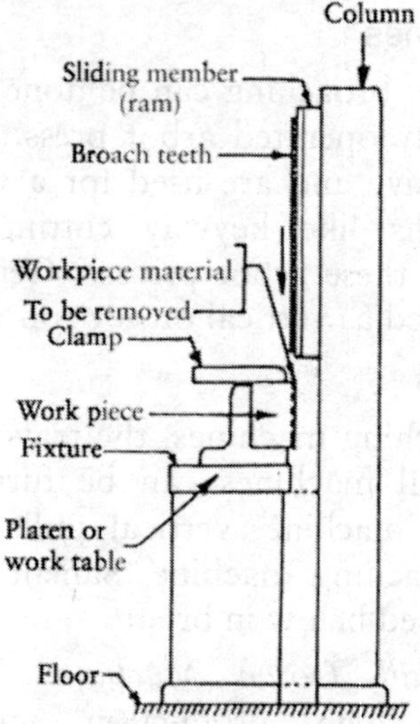

Fig. 4: Vertical broaching machine

The workpiece is held in a fixture which is fastened to a horizontal sliding table. The table recedes from the cutting position on completion of stroke and while the ram is moving upward to its starting position, the operator may remove the workpiece and reload the fixture with another workpiece. The table returns to position in time for the next cutting stroke of the ram.

A double ram machine has two ram-and-fixture sets, side by side. One ram descends with its fixture in cutting position. At the same time, the other returns while its fixture is unloaded, thus making a continuous operation.

(c) *Vertical Pull up Machines*. In pull up type machine, the workpiece is placed below the work table. The pulling mechanism is above the work table and the broach handling carriage is below it. The workpiece is placed over the front pilot of the broach. The broach is raised until it engages the puller head. As the broach is raised, the workpiece comes to rest against the underside of the work table. The broach is then pulled through the workpiece to complete the operation. The workpiece then falls free and is deflected into a container. Pull up broaching is advantageous when no work holding fixture is used, as the broached parts fall off by gravity and can be collected through a chute, thus simplifying unloading.

Most of the broaching machines used in industry are of vertical type, with their chief advantage of economy of floor space.

ii. Horizontal Machines

In horizontal broaching machines, the broach moves along a horizontal straight path. These machines can be further classified as: pull broaching machine, surface broaching machine, continuous broaching machine and rotary broaching machine. The working of these machines is explained as follows:

(a) *Horizontal Pull Broaching Machine*. In this machine, the broach is pulled through the workpiece by a drawhead which is actuated hydraulically as shown in Fig. 5. These machines are mostly used for internal broaching operation. This type of machine is preferred when stroke required is large and ceiling height is limited. The lifting of heavy broaches is avoided, but this machine occupies much more floor space than a vertical machine.

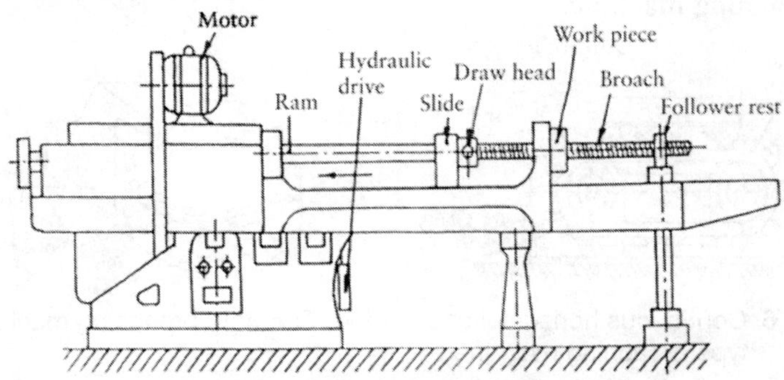

Fig. 5: Horizontal broaching machine

(b) *Horizontal Surface Broaching Machine.* Horizontal surface broaching machines are available with stroke up to 10 metres and capacities up to 100 tons. In these machines, the broach is mounted on a ram driven slide and is pulled over the workpiece surface to be broached. The workpiece is held in a fixture.

(c) *Continuous Broaching Machine.* These machines are used for surface broaching of parts that need repetitive cutting, such as gear teeth. A continuous broaching machine of chain type is shown in Fig. 6.

The machine has a continuous chain travelling in a horizontal plane over sprockets. Fixtures for locating and clamping workpieces are mounted at intervals on the chain. The broach is fixed horizontally above the chain under the bracket, as shown in Fig. 6.

The workpieces are loaded into the fixtures by clamping them manually or automatically. These workpieces pass under the broach as the chain rotates. After the workpieces are broached, they are automatically unclamped and ejected to drop into the hopper of the machine.

(d) *Rotary Broaching Machine.* In this machine, the workpieces are mounted on fixtures on a rotary table. The broaches are made in short sections for easy sharpening and are mounted on a central column and the rotary table is rotated past the broaching station. These machines are used, only for surface broaching of small parts. Fig. 7 shows a rotary broaching machine.

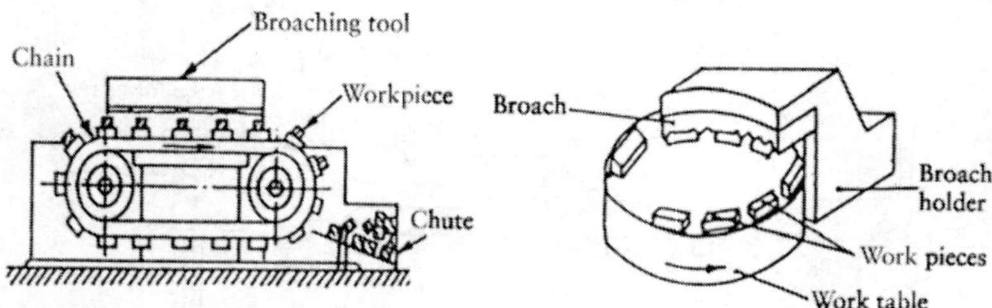

Fig. 6. Continuous horizontal chain type broaching machine

Fig. 7: Rotary broaching machine

4. Applications of Broaching

Presently the broaching is used to do many types of work, besides being used in automotive and domestic appliance industry. Typical broached parts include gear, gear-shift and steering gear levers, cylinder blocks, connecting rods, carburettors, bearings, pistons, etc. Broaching is employed equally well for both small lot jobs as well as for mass production. The main application of broaching process lies in machining irregularly shaped holes of considerable length very economically. Keyways, straight and spiral splined holes, square, hexagonal and other odd shaped holes are produced by broaching very efficiently. The broaching is useful in forming the teeth in small internal gears, in cutting suitable grooves or splines in casting, forgings and other units intended to fit splined shafts. The temperature of the workpiece remains fairly constant as the workpiece is broached, thus avoiding the occurrence of error as a consequence of temperature rise in the workpiece.

5. Advantages of Broaching

1. Broaching is faster than other machining operations, resulting in higher rate of production with better finish and more accuracy.
2. As each tooth of the broach takes a small cut only once in one operation, the broach has a longer life.
3. The broach performs both roughing and finishing operation.
4. As the machining cycle is quite simple, the broaching operation does not need a highly skilled operator.
5. The cutting force of the broach serves to clamp the workpiece and hold it firmly in position.

6.Limitations of Broaching

1. Initial cost of a broach is very high.
2. The broaching machine is a very costly machine tool. Therefore the operation is justified only for mass production.

3. Broach sharpening is an expensive and difficult process and requires a separate sharpening procedure.
4. A surface having an obstruction in the way of broach travel cannot be machined.
5. Delicate and very light jobs are difficult to broach.
6. One broach is used to produce only one type of surface. Therefore for getting different shapes and sizes, different broaches are required.
7. Blind holes cannot be easily produced through broaching.
8. Variations in hardness or inclusions in work material are not desirable.

7. Specifications of Broaching Machine

The capacity of a broaching machine is the maximum pulling force developed by the slide and its length of stroke. Vertical broaching machines are made with single, double or multiple rams. The stroke of the broach can be up to 2500 mm and the weight of the machine can be up to 50 tons. Horizontal broaching machines are made to handle broaches up to 2300 mm long and 300 mm diameter. They have capacities from 5 to 50 tons. The continuous broaching machines are standard machines, but their chain size, handling equipment, broach support, etc., can be modified to fit a particular job. A typical horizontal broaching machine is specified as below:

Maximum pulling force	-	5000 kg
Maximum stroke	-	1320 mm
Cutting speed (variable)	-	1 to 10 m/min
Return speed	-	30 m/min
Maximum workpiece diameter	-	340 mm
Maximum broach diameter	-	60 mm
Total power required	-	6.5 kW
Weight of machine	-	2450 kg

8. Broaching Tools

All broaching tools are of special design because each one of them is made for a single use and a specific machining operation. Depending on the broaching techniques, broaching tools may be classified as internal broaches and external broaches.

(a) *Internal Broaches*

Majority of internal broaches are based on the pull method. Internal push broaches are used only for sizing operations. Broaching of round holes is the simplest of all internal broaching operations. If a hole already exists in the workpart and production requirements are high, then internal broaching is more economical than other processes. Other broaches used for internal broaching are as follows:

(i) *Rotary Cut Broaches.* These broaches are designed heavier to withstand heavy cutting loads and are spaced in staggered fashion along its axis to generate the entire circumference of the hole. These broaching tools are primarily used for taking deep cuts underneath a poor-quality surface for example for rough forgings, malleable iron casting and sand casting. Once the top surface is penetrated, the underlying metal is semifinished and finished by the rest of the broaching tool. Rotary cut broaches are of three types—hexagonal type, radial type and spline cut type.

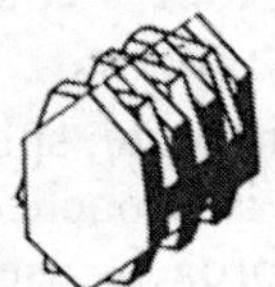

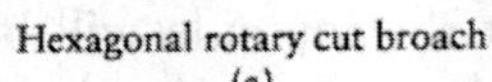

Hexagonal rotary cut broach
(a)

Radial rotary cut broach
(b)

Spline rotary cut broach
(c)

Fig. 8: Different types of rotary cut broaches

Hexagonal rotary cut broach [Fig. 8(a)] removes comparatively small stock and is used for small-diameter holes. The radial rotary cut broach [Fig. 8(b)] removes greater amount of material as compared to hexagonal type, because the cutting teeth are connected by arcs rather than by flats. Spline type broach [Fig. 8(c)] permits maximum metal removal and gives the operator a greater degree of flexibility compared to the other two.

(ii) *Keyway Broaches*. This type of broaching tool is in the form of a narrow, flat bar with cutting teeth spaced along one surface as shown in Fig. 9. Almost all keyways in machine tools and work parts can be made by using keyway broach. Both internal and external keyways can be cut. If several keyways are required to be cut on a part, then either a single keyway broach can be used to cut all the keyways by indexing the part, or a multiple-spline broach can be used which is equivalent to a series of keyway broaches combined in one tool, and the cutting teeth spaced around the periphery of the tool. Multiple spline broach leads to higher production as compared to a single keyway broach.

(iii) *Polygonal Broaches*. Polygonal holes are broached by square, hexagonal, rectangular and other broaches. One main feature of their design is the provision of several groups of teeth along their length with different cuts per tooth.

(iv) *Spline Broach*. Broaching a helical spline requires a broaching tool on which the teeth are ground in a helical path around the broach axis with the helix angle corresponding to that required on the work part. This is depicted in Fig. 10. The broach, instead of being pulled straight through the workpiece, is rotated during the cutting stroke, so that each tooth operates through a helical track. For this, the broaching tool is coupled to ball bearing puller that leaves the tool free to revolve.

For broaching straight-sided splines or involute splines, spline broaches are used. Straight-sided spline is broached by the usual keyway broaching method. The broach used is similar to the keyway broach, except that the top side will be radiused corresponding to that of outside diameter of the shaft.

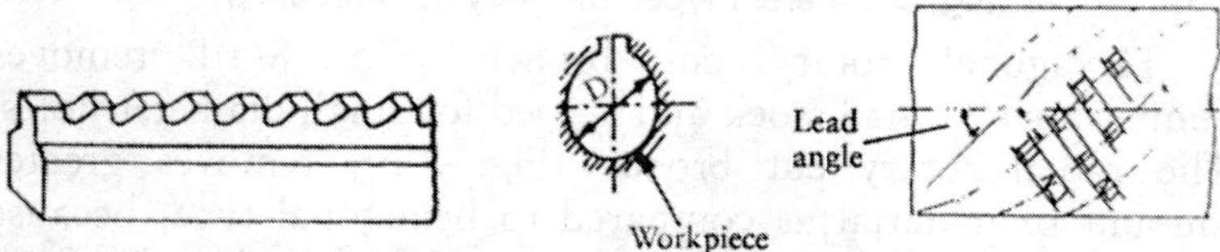

Fig. 9: Keyway broach **D:** Dia. of hole **Fig. 10:** Helical broach

(v) *Burnishers*. This broaching tool is designed to polish rather than cut the material. The surface is polished by cold working rather than by cutting. Generally, these tools are used on nonferrous workparts. Burnishing tools are relatively short and are used when surface finish and accuracy are of vital importance. Sometimes, burnishing buttons are included behind the finishing teeth on the conventional broach tool. Fig. 11 shows a burnisher.

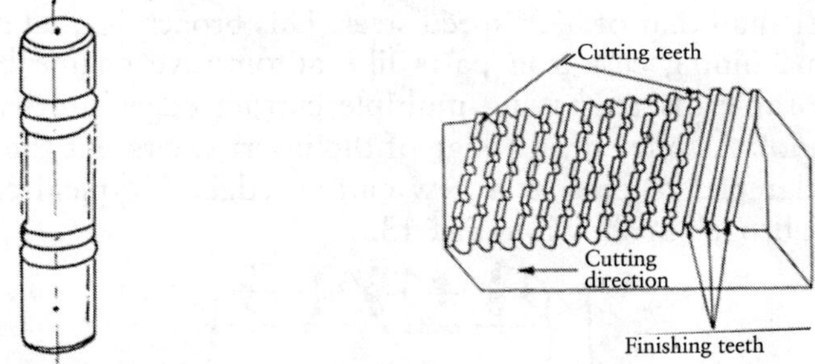

Fig. 11: Round burnishing broach **Fig. 12:** Surface broach

(b) *External Broaches*

These broaches are used for broaching the surfaces. The conditions under which components are broached externally differ somewhat from those of internal broaching. The operating cycle in external broaching is much more rapid and the working face of the component is overhung, in the form of a cantilever, in many instances. So component must be massive enough to withstand the broaching forces. The broaches used for broaching surfaces are discussed below:

(i) *Surface Broach*. These broaches, as shown in Fig. 12, are used for machining external surfaces. The surface broach is different from internal broaches, as surface broach is built up in the form of short sections. For designing a surface broach, the surface of the workpiece to be machined is broken down into separate elements of such a shape that the broach required to machine each element is simple to manufacture and convenient in operation. These requirements are met by a number of broach sections, which are mounted in a common holder and may operate in parallel, consecutively or in some combination. The broach sections are clamped with screws and wedges in the holders.

(ii) *Rotary Broaches*. These are special types of surface broaches used for producing workparts having external radial forms. Mostly, the broach is mounted on a rotating faceplate and makes one revolution to give the desired shape.

In recent years, cemented carbide inserts are being used in broach design, as the life of broaches with carbide teeth is longer than that of high speed steel. This broach is used mainly for machining cast iron parts like automotive engine blocks. Disposable inserts having multiple cutting edges are used on this tool. As one cutting edge of the insert wears out the insert is reclamped to present a new cutting edge. A typical carbide insert broach in shown in Fig. 13.

Fig. 13: Carbide insert broach

REVIEW QUESTIONS

1. What is meant by broaching?
2. How are broaching tools classified?
3. What is the pitch as applied to a broaching tool?
4. What is the function of burnishing teeth on broaches?
5. What advantages does a vertical pull-down machine have over a vertical pull-up machine?
6. List various advantages of broaching.
7. Describe the operation of broach.
8. Briefly describe the important elements of a broaching tool.
9. Compare the main features of two types of broaching machines.

OBJECTIVE TYPE QUESTIONS

Tick the most correct response

1. The surface finish obtained in broaching operation is of the order of

   (a) 8 microns (b) 0.8 micron
   (c) 0.08 micron (d) 0.008 micron

2. The maximum depth of cut which the tooth of a broach cuts is

   (a) 1 mm (b) 0.15 mm (c) 0.01 mm (d) 0.5 mm

3. A broach has

   (a) roughing teeth, semifinishing teeth and finishing teeth
   (b) roughing teeth and finishing teeth only
   (c) only finishing teeth
   (d) normally 30 teeth

4. The pitch for teeth of internal broaches is given by the relationship

   (a) 0.35 times depth of chip space
   (b) 1.25 to 1.5 times depth of chip space
   (c) 2.5 times depth of chip space
   (d) None of the above

5. For proper broaching, the minimum number of teeth that should be engaged in the work at a time is

   (a) 1 (b) 2 (c) 3 (d) 4

6. In broaching

   (a) the job is completed in one stroke of the machine
   (b) the tooling cost is low
   (c) the rate of production is very slow

7. The finishing teeth of a broaching tool are provided with

   (a) large amount of land
   (b) smaller amount of land
   (c) no land

8. The range of hardness of the material which can be broached is
   (a) 10 to 20 Rockwell C
   (b) 25 to 40 Rockwell C
   (c) 60 to 80 Rockwell C
9. Broaching is primarily done for
   (a) better finish (b) cylindrical jobs
   (c) mass production (d) hard materials
10. The broaching operation in which the tool moves past the stationary work is
    (a) push broaching (b) pull broaching
    (c) continuous broaching (d) surface broaching
11. Broaching operation is frequently used in automobile industry as
    (a) it is an automatic machine
    (b) it is a mass production machine
    (c) operation is completed in one stroke
12. The front teeth of a broach
    (a) remove minimum metal
    (b) remove maximum metal
    (c) remove no metal
13. ______ are harder and more wear resistant than tungsten carbide tools but are weaker in tension.
    (a) Low carbon steel tools
    (b) High carbon steel tools
    (c) H.S.S. tools
    (d) Ceramic tools
14. Stellite tools are formed by ______ process.
    (a) Sintering (b) Forging
    (c) Machining (d) Cold rolling
    (e) Casting
15. Cermic tips are prepared from ______ powder.
    (a) Tungsten (b) Carbon
    (c) Aluminium Oxide (d) Silicon carbide
    (e) any metal

16. ______ are cast alloys containing tungsten and chromium carbides in a matrix of cobalt and chromium.
    (a) Mechanites (b) Tungsten Carbides
    (c) Bakelite (d) Stellites
17. Carbide tips are fixed to the shanks of cutting tools by
    (a) forging (b) sintering
    (c) welding (d) soldering
    (e) brazing
18. The most wear-resistant grade of carbide used for cutting tools is the straight ______.
    (a) Iron carbide (b) Calcium carbide
    (c) Tungsten carbide (d) Nickel carbide
19. Which one of the following must be considered while selecting the proper grade of carbide tool to use?
    (a) The type of operation to be done
    (b) The type of work material to be machined
    (c) The finish requirements of the machined part
    (d) The speeds to be used in turning the part
20. During grinding of carbide tool bits, ______ wheel is generally used.
    (a) Aluminium Oxide (b) Silicon Carbide
    (c) Diamond (d) Cobalt
    (e) High Speed Steel (HSS)
21. Powder Metallurgy (PM) techniques are used in the production of
    (a) High carbon steel tools
    (b) HSS tools
    (c) Tungsten carbide tools
    (d) Twist drills.
22. What is the material of the cutting tool generally used on a lathe?
    (a) CI
    (b) Mild steel

(c) Low-carbon steel
(d) High Speed Steel (H.S.S.)
(e) Brass

23. Among the following abrasives, the hardest one is ______.
(a) Al_2O_3 (Aluminium Oxide)
(b) SiC (Silicon Carbide)
(c) B_4C (Boron Carbide)
(d) CBN (Cubic Boron Nitride)
(e) Diamond

24. The hardness of carbon tool steels can be increased when alloyed with ______ and vanadium.
(a) Tungsten (b) Chromium
(c) Silicon (d) Manganese
(e) Silver

25. The type of wear which occurs on the rake face of tool bits is referred to as
(a) Flank wear (b) Crater wear
(c) Tip wear (d) End wear
(e) Radius wear

26. What is the approximate % range of Carbon content in the High Speed Steel (H.S.S.) tool material?
(a) 0.6 – 1.0% (b) 4 – 6%
(c) 10 – 18% (d) 18 – 42%

27. What is the approximate % content of Vanadium in H.S.S. (High Speed Steel) cutting tool material?
(a) 16 (b) 4
(c) 0.1 (d) 1

28. The usual % of Tungsten content in the High Speed Steel (H.S.S.) cutting tool material is
(a) 18% (b) 4%
(c) 1% (d) 0.6%

29. The approximate percentage content of Chromium in H.S.S. (High Speed Steel) cutting tool material is

(a) 1% (b) 4%
(c) 18% (d) 0.6%

30. In metal cutting operations, when the cutting speed is increased, the life of the cutting tool
    (a) remains the same (b) decreases
    (c) increases (d) gets doubled

ANSWER KEY

1. (b)	2. (b)	3. (a)	4. (c)	5. (c)
6. (a)	7. (b)	8. (b)	9. (c)	10. (d)
11. (b)	12. (b)	13. (d)	14. (e)	15. (c)
16. (d)	17. (e)	18. (c)	19. (b)	20. (c)
21. (c)	22. (d)	23. (e)	24. (b)	25. (b)
26. (a)	27. (d)	28. (a)	29. (b)	30. (b)

9

Grinding

1. Introduction

Grinding is a process of removing metal, but in a smaller volume. To 'Grind' means 'to abrade', to wear away by friction. In grinding, the material is removed by means of a rotating abrasive wheel. The action of grinding wheel is very similar to that of a milling cutter. The wheel is made up of a large number of cutting tips made of projecting abrasive particles in the grinding wheel.

Grinding is done on surfaces of almost all shapes and materials of all kinds and it is also able to produce accurate and fine surfaces.

Grinding possesses the following advantages over other cutting processes:

(i) Grinding is the only method of cutting hardened steel, etc.

(ii) Extremely smooth finish desirable at contact and bearing surfaces can be produced due to large number of cutting edges on the grinding wheel.

(iii) No marks as a result of feeding are there, as the wheel has considerable width.

(iv) Very accurate dimensions can be achieved in a very short time.

(v) Very little force is required in this process, so permitting its use on very light work that would otherwise tend to spring away from the tool.

Far greater advances in design, construction and use have been taking place during the past few years in the grinding machine than in any other machine tool. Until recently the grinding machine was regarded as a tool room machine, particularly useful only for finishing hardened steel. However, now, it is considered as one of the most important machine tools for manufacturing purposes. It is owing to the remarkable development of the machine itself and also of the abrasive wheels used, that grinding is used today as the means of producing very accurate and fine finished surfaces accurately.

The function of the grinding machine is, like every other machine tool, the removal of metal by means of a suitable cutting tool. There are various types of grinding machines.

2. Types of Grinding Machine

Grinding machines may be of roughing type (such as grind stones, bench grinders, flexible shaft grinders, cut-off wheels) or they may be of precision type.

Precision type grinding machines can be classified as:

1. Cylindrical grinder
   (a) Centre type (b) Centreless
2. Internal grinder
3. Surface grinder
   (a) Planer type (b) Rotary type
4. Tool and cutter grinder
5. Special grinding machines
6. Jig grinding machine
7. Newly developed grinding machines

Cylindrical Grinders

(a) *Centre Type Cylindrical Grinders*: In these grinders, the workpiece is rotated between centers. These grinders are primarily used for grinding contoured cylinders, tapers, faces and shoulder, fillets and even cams and camshafts. There are two types of grinding operations where the workpiece is held between fixed centres. (i) Traverse Grinding and (ii) Plunge Grinding.

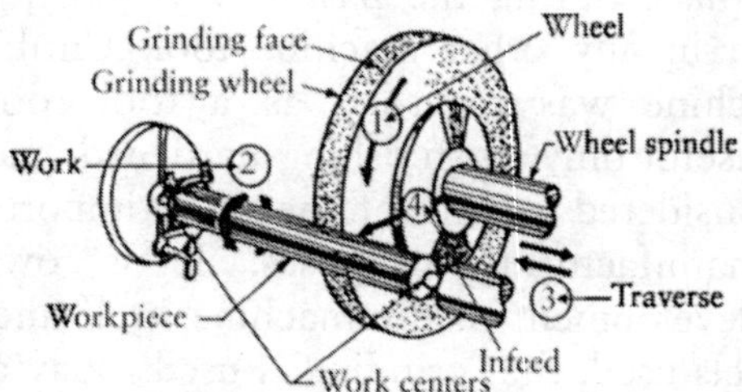

Fig. 1: The movements of an external cylindrical grinding machine

In traverse grinding, the work reciprocates past the grinding wheel. As the workpiece passes, the wheel removes fixed amount of metal from the diameter of the workpiece. At the end of the pass, the wheel is advanced another increment of distance (just as in lathe) for removing further metal layer (Fig. 1). This grinding is employed when the workpiece is longer than the maximum width of the wheel.

In cylindrical grinding, the workpiece is mounted between headstock and tail stock centres. The grinder is designed to make four movements.

(a) Rotating the grinding wheel about its own axis
(b) Rotating the workpiece on its axis between centres
(c) Reciprocating the workpiece horizontally
(d) Moving the wheel into and away from the work.

In plunge grinding, as shown in Fig. 2, a wheel is used as wide as the surface to be ground. The wheel is fed into the revolving workpiece, which is mounted between the centres and does not reciprocate past the wheel. The infeed is continuous rather than intermittent.

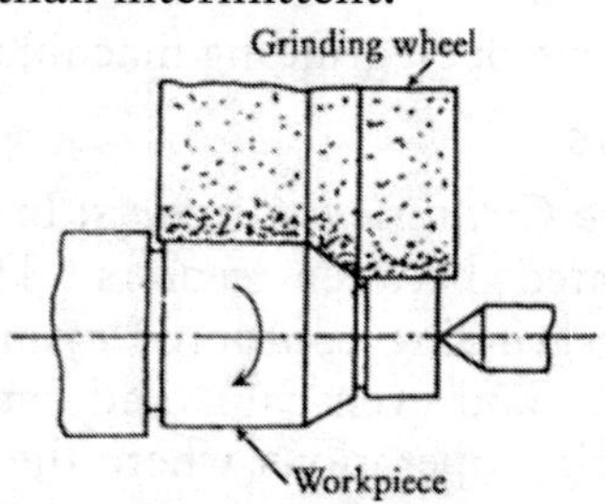

Fig. 2: Plunge grinding

(b) *Centreless Cylindrical Grinding*: It is accomplished by grinding cylindrical surfaces without rotating the workpiece between fixed centres. The workpiece is supported between three components as can be seen in Fig. 3(a). The three components are (i) Grinding wheel, (ii) Regulating wheel, and (iii) the work rest. Photograph of a centreless grinder is also shown in Fig. 3(b).

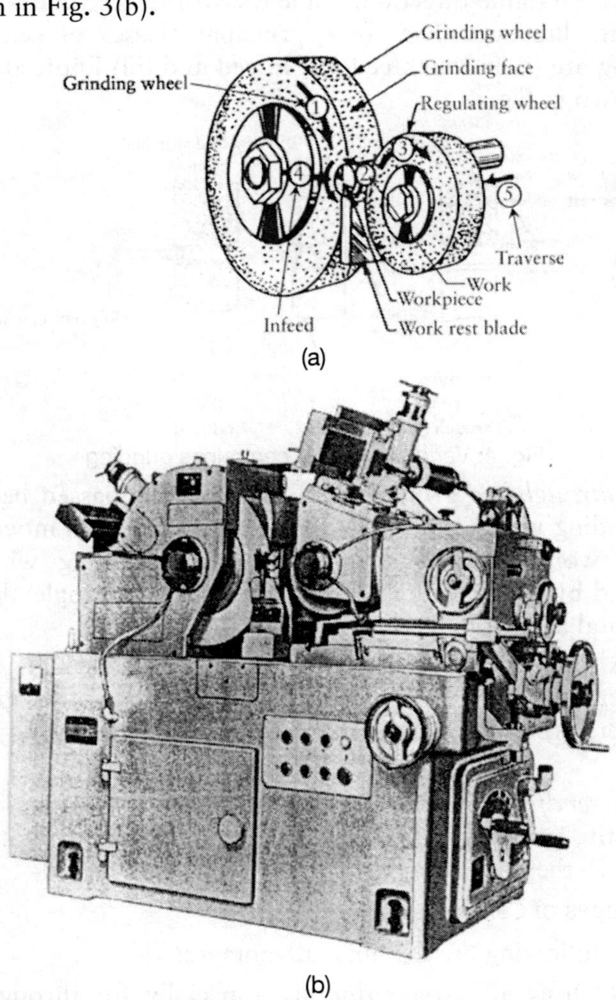

Fig. 3: (a) External cylindrical centreless grinding machine, (b) Centreless grinder

The grinding wheel does the actual grinding while the work-rest positions the workpiece for grinding. The functions of the regulating wheel are: to govern the speed of the rotation of the workpiece, to govern the sizing of the workpiece and to govern the rate of workpiece travel through the grinder in through feed grinding. The grinding and regulating wheels rotate in the same direction, while the workpiece rotates in the opposite direction. The three principal classes of centreless grinding are (i) Throughfeed, (ii) Infeed and (iii) Endfeed. They are shown in Fig. 4.

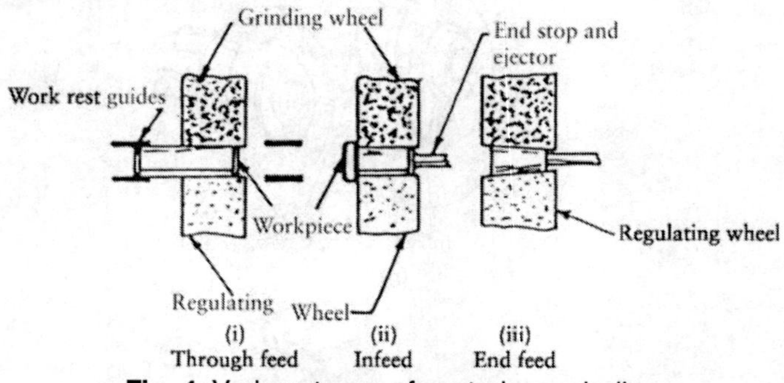

Fig. 4: Various types of centreless grinding

In throughfeed grinding, the workpiece is passed between the grinding wheel and regulating wheel. The actual movement of the workpiece past the face of the grinding wheel is imparted by tilting the regulating wheel at slight angle about a horizontal axis from 0° to 7-10°.

The infeed method is used on the work which has a shoulder, head or some portion larger than the ground diameter, so that it does not pass completely through the wheels.

The endfeed method is used primarily on taper work. Either the grinding wheel or regulating wheel, or both are dressed to the desired taper.

Advantages of Centreless Grinding

The following are the main advantages:

1. It is almost continuous, especially for throughfeed grinding, so it is used for production work.

2. The work is supported throughout its entire length, so there is no chatter or deflection.
3. In the operation of the machine, even less degree of skill will serve the purpose.
4. The size of the work is easily controlled.

Internal Grinder

These machines are used mainly for finishing round holes. The application of these machines is extensive. Range of type of parts and hole sizes are only limited by the capacity of the machine. Holes (straight or tapered) may be finished in such parts as gears, bushings, cutters, gauges, etc. In general small wheels are used. Frequent truing is necessary for maintaining the accuracy of ordinary wheels. There are three types of internal grinders.

(a) Chucking, in which the workpiece is held in a chuck and rotated. The revolving wheel spindle moves in and out of the hole during the operation.

(b) Centreless, in which the workpiece is rotated by the outside diameter between the rolls. The outside diameter of the workpiece generates a path for grinding the inside diameter. The inside diameter is ground concentric to the outside diameter (Fig. 5).

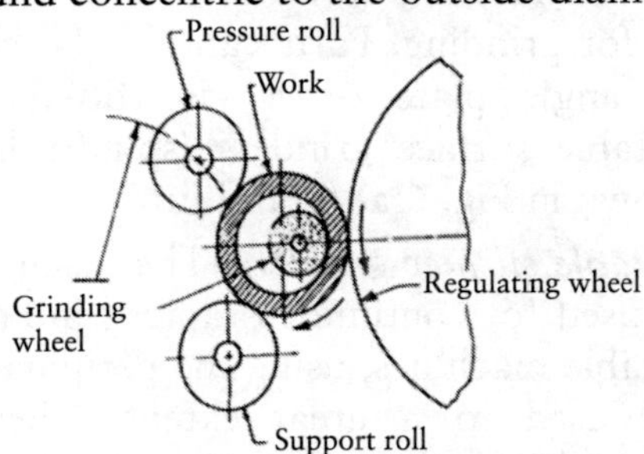

Fig. 5: Internal centreless grinding

(c) Planetary, in this the workpiece is held stationary and mounted on the table of the machine. The wheel not only rotates on its axis but also travels in a circular path of the desired diameter. This machine is used only for bore grinding of heavy or irregular shaped work, which is difficult to rotate.

Surface Grinder

Surface grinding is a term generally used to describe the grinding of flat surfaces. Many parts that were formerly machined on milling machines and shapers are machined on surface grinders. Surface grinders can be classified as:

(i) Planer type (reciprocating table)

(ii) Rotary type (revolving table)

In both the above types, the spindle can be either horizontal or vertical.

(i) *Reciprocating table surface grinder*: The horizontal spindle, reciprocating table surface grinder using the periphery of a straight wheel, is the most common surface grinder. They are built in a wide range of sizes for flat grinding. The size of a surface grinder is determined by the size of the surface to be ground. The machine with reciprocating table is particularly suited to pieces that are long or have stepped profiles at right angles to the direction of motion.

For most of the operations on surface grinders, the workpiece is held on a magnetic chuck. The magnetic chuck is widely used, as it is fast to apply and will hold a large number of small pieces for grinding. Parts can also be held in fixtures like V-blocks, angle plate or a vise fitted to the table. Reciprocating table surface grinders (spindle horizontal and vertical) are shown in Fig. 6(a) (i) and (ii).

(ii) *Rotary table surface grinders*: The machine with rotary table is widely used for continuous rapid grinding. Horizontal spindle rotary table machines, using the periphery of a straight wheel are not used to a great extent. They are usually constructed in small sizes.

Vertical spindle rotary table machines, using cylinder wheels or segments are built with one to five spindles mounted on a central column. These machines are used so that roughing and finishing cuts can be taken on large parts with only one pass through the machines.

Rotary table surface grinders (spindle horizontal and vertical) are shown in Fig. 6(a) (iii) and (iv).

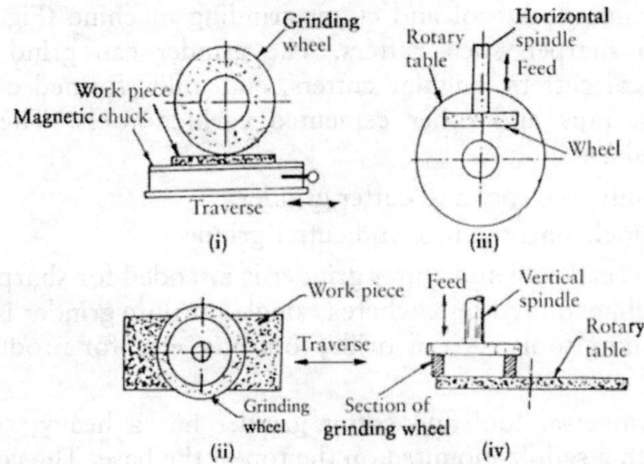

Fig. 6(a): Types of surface grinders—(i) horizontal spindle reciprocating table, (ii) vertical spindle reciprocating table, (iii) horizontal spindle rotating table, (iv) vertical spindle rotating table

The photograph in Fig. 6(b) shows a horizontal table reciprocating table surface grinder.

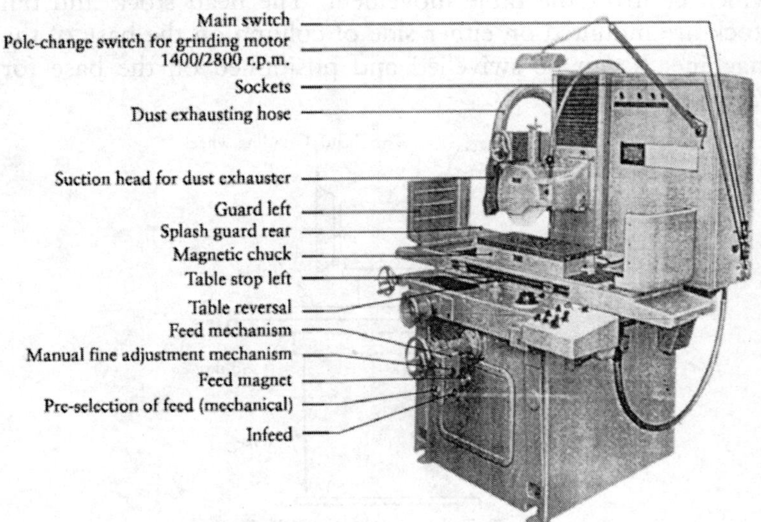

Fig. 6(b): Reciprocating table surface grinder

Tool and Cutter Grinder

When milling cutters, reamers or drills become dull, they need grinding. A tool and cutter-grinding machine (Fig. 7) is used to sharpen such cutters. The grinder can grind plain cylindrical cutters, angular cutters, end mills, formed cutters, reamers, taps and other cemented carbide tools. They are classified as:

1. Universal tool and cutter grinder
2. Single purpose tool and cutter grinder

Universal tool and cutter grinder is intended for sharpening of miscellaneous cutters, whereas single purpose grinder is used for grinding tools such as drills, tool bits, etc., for production plants.

A universal tool and cutter grinder has a heavy, rugged base with a saddle mounted on the top of the base. The column supporting the wheel head is mounted on the saddle and it can be moved up and down and swivelled.

The table moves on a top base which is mounted over the saddle. The top of the base contains the gears and mechanism which controls the table movement. The head stock and tail stock are mounted on either side of column on the base of the machine. It can be swivelled and positioned on the base for varied set ups.

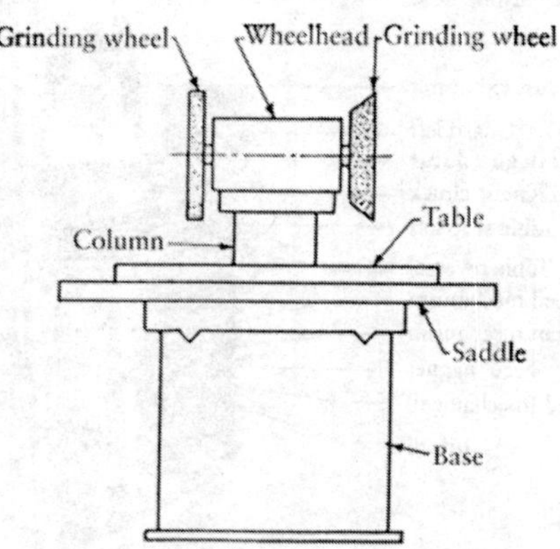

Fig. 7: Tool and cutter grinder

Special Grinding Machines

In special grinding machines, the form grinder, thread grinder and gear grinder are very important.

(a) *Form grinder*: In form grinding, a formed wheel feeds towards the revolving work, which does not traverse. The wheel has the exact shape and form of the profile to be ground. Figure 8 shows the form grinding.

(b) *Thread grinder*: In thread grinding, the profile of the grinding wheel depends on the shape of the thread to be ground. The wheel is so formed as to have the exact shape of the thread. Figure 9 shows thread grinding. Thread of any form can be ground on these machines, either from a blank or as finishing operation after milling.

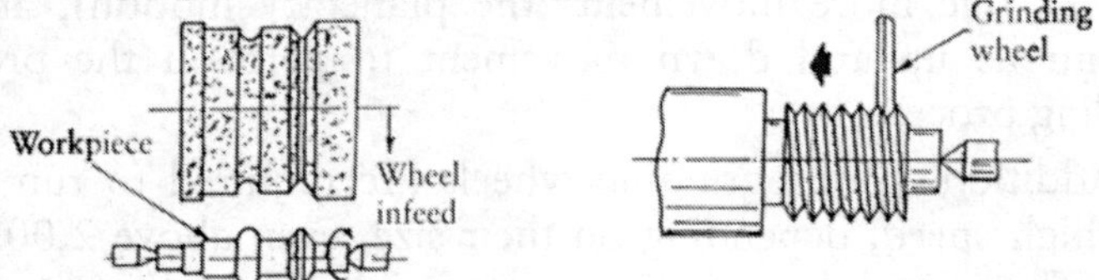

Fig. 8: Form grinding **Fig. 9:** Thread grinding

(c) *Gear grinder*: As shown in Fig. 10, this grinder makes use of a formed wheel, shaped exactly to the profile of the shape between two adjacent teeth. This curved profile varies for each change in gear diameter, or for each change in number of teeth. Such a profile is difficult to form and maintain. A rough cut gear is mounted on an arbor and the arbor is held in an indexing and locking work head. With the gear locked in position, it is reciprocated slowly under the wheel. At proper intervals, indexing head shifts the gear from one tooth position to the next.

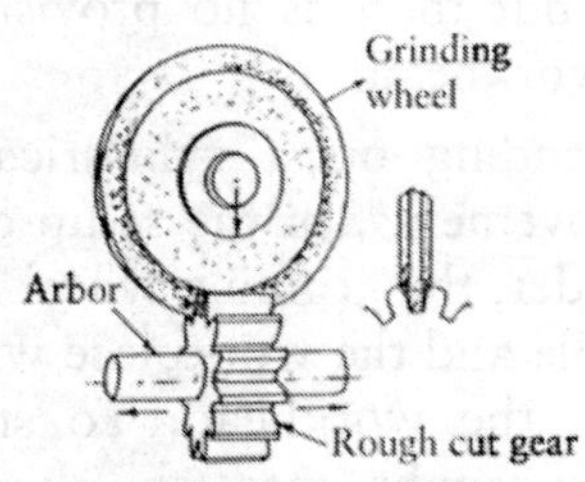

Fig. 10: Gear grinding

Jig Grinding Machine

Jig grinding machines were originally developed to locate accurately and grind holes to size in hardened materials. They utilize the same principle of rectilinear positioning employed in jig borers.

Jig grinder can be used for work other than cylindrical hole grinding also. Most of the machines are equipped with a taper grinding attachment, which is especially used in die making for the piercing holes. In addition to round holes, it is also possible to grind radii (internal and external), segments of a circle, tangents, flats and angles. Certain machines are equipped with a slot grinding attachment, which makes them still more universal in profile grinding. The jig grinding operation involves one more movement (the planetary motion), and a continuous up and down movement to perform the proper grinding process.

Additionally the grinding wheels are required to run at a very high speed, depending on their size, even above 2,00,000 rpm. This certainly requires also special drive units to reach this much high speed. The grinding wheel determines to a great extent the surface quality, the cylindricity and the roundness of a ground hole. A jig grinding machine is shown in Fig. 11.

The jig grinding process does not only involve the movements of the ordinary boring process, like the rotation of the main spindle and a slow vertical feed of a few hundredths to a few tenths of a millimeter, there are other movements also. For a better understanding of the necessary movements, it shall first be referred to a very similar grinding process that is the internal grinding on a cylindrical grinder. This operation is basically the same, but there is no provision to set different coordinates on the workhead.

The internal grinding on a cylindrical grinder involves three continuous movements. In the setup of internal grinding on a cylindrical grinder, the grinding wheel is fixed on the high speed grinding spindle and the workpiece with its permachined bore is clamped on the workhead. To start the work, the grinding wheel is brought into the bore, the horizontal movement of the workhead is set in a way that the slide travels in the range of the bore. Then the grinding wheel and the horizontal movement is started. Additionally the workhead must also rotate the job. With these three continuous movements, the internal grinding process can be performed. Exactly the same three movements must also be available in the jig grinding process. The basic differences of the two processes are that in jig grinding the working direction is vertical and the job is clamped onto a fixed table, thus not moving at all. The rotational movement which is given by the workhead on internal grinding must now be given to the grinding wheel itself, and the horizontal movement of the slide becomes now a vertical movement of the quill. In other words, all the three movements are carried out by the grinding wheel, that is the up and down movement, the circular path movements of the grinding wheel or the planetary motion as it is normally called, and the rotation of the grinding wheel itself. As there are three different movements, there are also three different speeds to be maintained. The grinding speed of the wheel depends on the diameter and the material of the grinding wheel. The circumference speed for conventional grinding wheels is about 20 to 25 m per sec, for diamond and "Borazon" wheels it is about 10 m per sec.

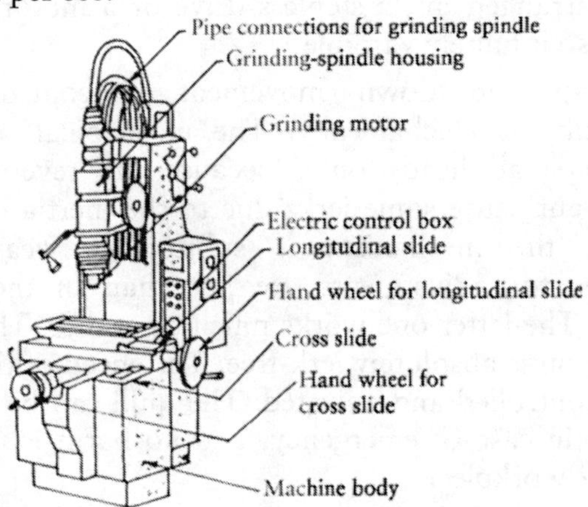

Fig. 11: Jig grinder

The planetary motion gives the surface speed of the job. Since the grinding wheel and therefore also the relatively heavy motor describe an eccentric path, this speed is kept low, because of the centrifugal force. The speed depends on the eccentricity of a dc-motor. Normally, separate charts are supplied by the manufacturer of the machine for selection of different speeds.

The selection of the third speed, for the up and down movement, is not critical. It has to be adjusted according to the planetary motion, in such a way that the grinding wheel describes a new path at every pass. The infeed (depth of cut) is given by a separate handwheel mounted on the grinding head.

The grinding head in jig grinder looks even more bulky than boring spindle head. One explanation therefore is that, a grinding head has to be still more rigid and accurate. This is not because of high cutting forces, it is because of the very high speeds of the grinding wheels and its incident vibrations. Additionally, the grinding motor describes an eccentric path. To reduce the effect of the resulting centrifugal force to a tolerable amount, the construction of the complete quill has to be strong enough.

The planetary motion is simply given by a small motor on the grinding head. The speed selection is either by a belt changing arrangement, a stepless drive or a motor, where the speed is also infinitely variable.

The up and down movement is controlled either hydraulically or mechanically. The mechanical operation is limited to small heads only, because the reversing of the motion might cause some jerks due to the inertia of the quill. Moreover, the mechanism is subject to wear and the possibilities for adjustments are less than in the hydraulic operation. The latter one works much smoother. The reversing of the motion is absolutely jerk-free. The speed of the quill can easily be controlled and adjusted. The quill can be stopped by hand only in case of emergency, and no harm is done to the machine or workpiece.

In order to grind the different materials efficiently, different kinds of grinding wheels are available. The grinding wheels may be grouped into categories according to their material. There are aluminium oxide, silicon carbide, "Borazon" and diamond wheels.

The aluminium oxide wheels are recognized by their colour (pink, brown, blue). The pink and blue ones are slightly superior to the brown ones, in view of cutting capacity, whereas the obtainable surface finish and wear characteristics are better on the brown ones. The silicon carbide wheels with its characteristic green colour are slightly harder than aluminium oxide wheels.

The "Broazon" wheels are a new development. Basically, it is a cubic boron nitride of almost the same hardness as diamond. It is produced in a similar way as the synthetic diamonds. Its use in all the grinding operations becomes more and more important and in jig grinding it plays an important role.

"Borazon" wheels are available in diameters as small as 0.5 mm, they are best suited to grind extremely small holes, as well as high alloyed tool steel and high speed steel. It is also possible to grind any other hardened steel very effectively with these wheels. Unhardened and soft steel may cause some problem because of clogging. The obtainable surface quality is as good as with conventional wheels. Since they are removing the material very effectively, the heat creation in the workpiece is less and therefore a better dimensional accuracy is possible. "Borazon" is not recommended to grind cemented carbide. Therefore diamond wheels are mainly reserved for this purpose only. They are available in extremely small diameters.

Newly Developed Grinding Machines

Automation has been associated with advancement in technology. In the process of automation for small batch production and mass production, hydraulic tracer controlled machine tool and programmed operating cycle machine tools have been evolved. However, this requires templates, cams, stops, electrical trip dogs, timers, etc.

Keeping automation in mind different types of machines have been developed for the operations which were being done before, using some attachments or fixtures on cylindrical, internal, surface and centreless grinding machines. Thus required jobs are being done achieving better results in minimum time, without fatigue.

(i) *Centreless grinding machine with profile dressing*: Both the grinding and the regulating wheel heads are provided with hydraulically operated wheel dressers having stepless variable dressing speeds and are controlled independently. Profile dressing is possible with the interchangeable copy template. Top slide of dressing units is mounted on preloaded antifriction ball guideways. A spring loaded stylus against a template achieves high copying accuracy.

(ii) *Roll camber grinding machine*: This machine is suitable for grinding rolls with either concave or convex cambering. It can also be used for normal cylindrical grinding.

The cambering on the roll is produced by copying it from a master template fitted on rear side of the bottom table. Based on the profile of the template, a torque is transmitted to ball screw and nut through lever arrangement for small infeed movement of grinding wheel. The infeed slide of the machine, i.e. wheel head is on antifriction ways. The grinding operation is achieved with micron precision, as the transfer from template to wheel head takes place through ball screw and nut arrangement. The screw and nuts are hydraulically preloaded. By adjusting the lever length (Fig. 12) different values of camber with the same template can be achieved. To adjust the required camber, it is necessary to move the follower arm to the predetermined position on the lever arm.

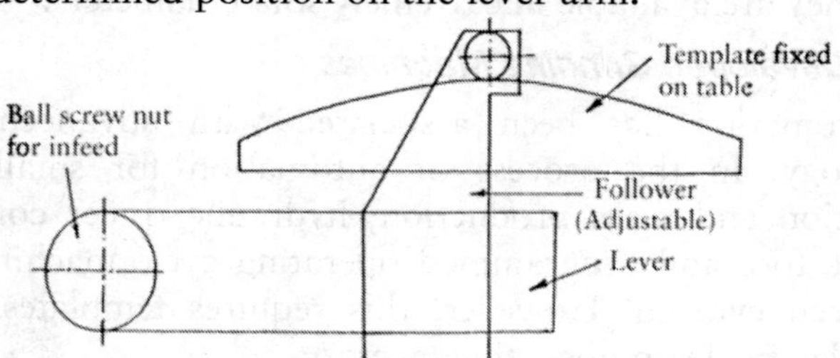

Fig. 12: Principle of roll camber grinding machine

(iii) *Duplex grinder*: Duplex grinder is a surface grinding machine used to generate two parallel surfaces at high production rate. The machine has two grinding heads and utilises the faces of bonded grinding wheels fastened to steel plates which are in turn attached to spindle face plate.

The two grinding wheels (Fig. 13) are running in opposite direction and workpieces are fed between them to grind two parallel faces simultaneously to close tolerance and fine finish. The production rate is quite high in this machine. Piston rings can be produced at a rate of about 1500 pieces/hour attaining flatness and parallelism within 0.01 mm.

The finish and flatness obtained by this process is better than any other grinding process using periphery of grinding wheel.

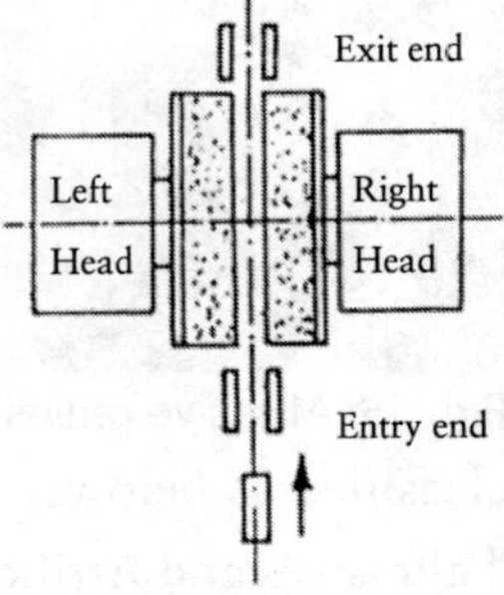

Fig. 13: Duplex grinder

(iv) *Angular wheel head grinder with profile dressing*: For grinding multiple diameters, shoulders and profiles in one plunge cut the wheel is trued with a hydraulic copy truing device mounted on the wheel head.

In this machine, wheel head is swivelled and fixed at 30° to the work axis (Fig. 14). By doing so, wheel width is increased and facing on job becomes easier.

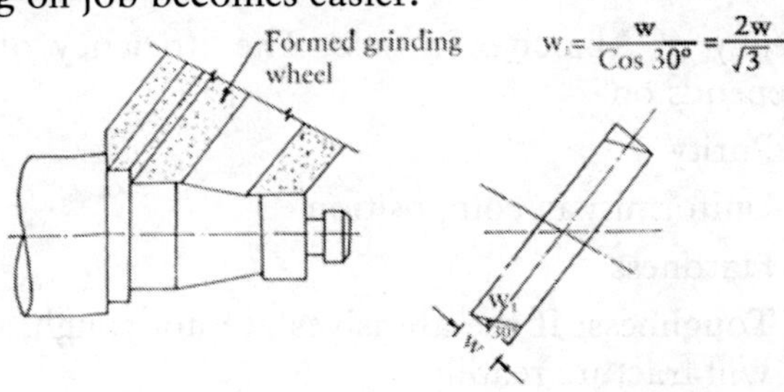

Fig.14: Angular wheel head grinder

3. Various Terms Related to Grinding Wheel

Various terms related to specification of grinding wheel like abrasives, bond, grainsize, grade and structure are explained below.

Abrasives

Grinding wheels are made of abrasive particles bonded together by suitable bond. An abrasive is a hard material which can be used to cut or wear away other materials. Abrasive grains of silicon carbide are shown in Fig. 15.

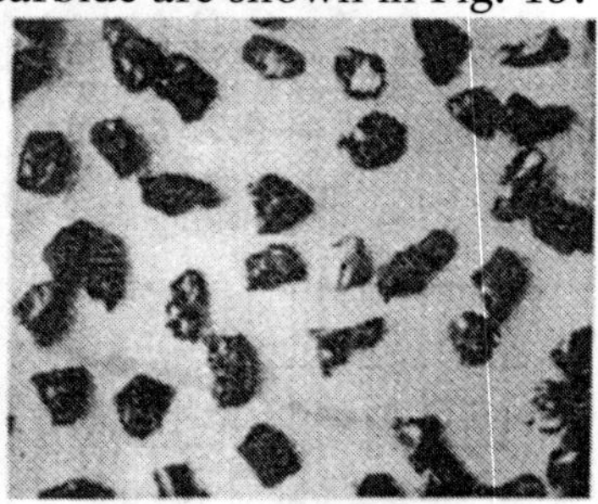

Fig. 15: Abrasive grains

Abrasives can be classified as below:

Abrasives: Natural abrasives and Artificial abrasives

Natural abrasives	*Artificial abrasives*
(a) Sand stone or solid quartz	(a) Silicon carbide
(b) Emery	(b) Aluminium oxide
(c) Natural corundum	(c) Boron carbide
(d) Diamonds	(d) Zirconium oxide
(e) Garnet	(e) Electro corundum
	(f) Synthetic diamonds

Efficiency of Abrasive Particle: The efficiency of abrasive particle depends on

(i) Purity

(ii) Uniformity in composition

(iii) Hardness

(iv) Toughness: If the abrasives are not tough, the wheel will fracture readily.

(v) Sharpness of fracture: Better cutting action is obtained by sharp edged abrasives. Natural abrasives give rounded edges and are not efficient in cutting.

Manufacture of Abrasives

(a) *Sandstone*: Sandstone wheels are rarely used. Although they are cut from high grade quartz or sandstone, yet they do not wear evenly because of variation in the natural bond.

(b) *Corundum and Emery*: Both are crystal aluminium oxide in combination with iron oxide. Corundum is obtained by melting bauxite or clay (Aluminum oxide) in an electric furnace.

Corundums are particularly suitable for grinding steels of high strength and toughness, cast steel and malleable iron.

(c) *Diamond wheels*: They are made with resinoid bond and are useful in sharpening cemented carbide tools. Due to their rapid cutting ability, slow wear, free cutting action they are quite economical. Very little heat is generated with their use. Diamond wheels are used for grinding carbide tips.

(d) *Silicon carbide*: It was discovered during an attempt to manufacture precious gems in an electric furnace. The hardness of this material according to Moh's scale is slightly over 9.5, which approaches the hardness of a diamond. Raw materials consisting of silicon sand, petroleum, coke, saw dust and salt are heated in an electric furnace to around 2300°C and held there for a considerable period of time. The product consists of a mass of crystals surrounded by partially unconverted raw material. After cooling, the material is broken up, ground and crushed to grain size. Silicon carbide crystals are very sharp and extremely hard. Silicon carbide is suitable for grinding soft materials and hard metals, CI and glass.

(e) *Aluminium oxide*: It is made from the clay-like mineral 'bauxite' which is the main source of aluminium. Aluminium oxide is slightly softer than silicon carbide, but is much tougher. Most manufactured wheels are made from aluminium oxide.

Bonding Material

A bond is a material that holds the abrasive grains together enabling the mixture to be kept in desired shape. Various types of bonds are designed as below:

(i) Vitrified bond – V
(ii) Silicate bond – S
(iii) Shellac bond – E
(iv) Rubber bond – R
(v) Resinoid bond (Bakelite) – B
(vi) Resinoid reinforced – BF

The bonding material should have the following properties:

(i) The bonding material should withstand the grinding temperature.
(ii) It should be able to withstand the spindle speed without disintegrating.
(iii) The bonding material should be capable of retaining the abrasive grains during cutting action.
(iv) It should be friable enough to dislodge the worn grits.
(v) The bond should be rigid to enable the grit to penetrate and cut the work.

Various types of bonds are as follows:

(a) *Vitrified bond*: The abrasive grains are mixed with clay-like ingredients that are changed to glass-like material upon being burnt at high temperature. Addition of water is required and the wheels are shaped in metal moulds under an hydraulic press. The time for burning, varies with wheel size, being anywhere from 1 to 14 days. These wheels are porous, strong, unaffected by water, acid, oils and climatic or temperature conditions. Vitrified bonds are used for most grinding operations. The maximum standard cutting speed generally used for vitrified bond wheels is 30 m/sec.

(b) *Silicate bond*: Silicate bonds require much lower temperature to harden the wheel than do vitrified bonds. In this sodium silicate is mixed with abrasive grains and mixture is tapped in metal moulds. After drying for several hours, the wheels are baked at 260°C from 1 to 3 days. The process is recommended for large wheels, as they do not crack or warp in the baking process.

Silicate wheels release the grains more readily giving a milder (or cooler) grinding action and are suitable for grinding edges of tools.

(c) *Shellac bond*: Shellac bond, initially used for cut-off operations, are also used for hardened materials. The abrasive grains are first coated with shellac mixed in a steam-heated mixer. The material is then placed in heated moulds and rolled or pressed. The wheels are finally baked for few hours at 150°C. This bond is used for thin wheels as it is strong and elastic. Alkaline cutting fluids react with this bond.

(d) *Rubber bond*: Pure rubber with sulphur as a vulcanizing agent is mixed with abrasives by feeding the material between heated mixing rolls. After it is rolled to thickness, the wheels are cut out with dies and vulcanized under pressure.

- ❖ Very thin wheels can be made by this process
- ❖ Rubber bonded wheels are used for high speed grinding (45-80 m/s)

As they afford rapid removal of the stock, these wheels are used for cut-off operations. When subjected to heat, the bond softens and releases the grains. Thus, cutting fluid should normally be used with this bond.

(e) *Bakelite or resinoid bond*: Abrasive grains in this process are first mixed with thermosetting synthetic resin powder and liquid solvent, then moulded and baked. The bond is very hard and strong and can be operated at cutting speed of 45-80 m/s. It is used for general purpose grinding and for snagging operations.

(f) *Oxychloride bond*: This bond contains oxides and chlorides of magnesium. Oxychloride is a cold setting cement requiring an aging period. As most cutting fluids affect this bond, it is normally used for dry grinding.

Grain Size

Abrasives must be refined and made to standard sizes before they are manufactured into wheels. For this, the abrasives are crushed and screened, that is they are passed through screens containing meshes of standard sizes. These sizes are measured by the number of meshes per inch through which the grains will pass. For example, a grain that passes a screen having 20 mesh opening per inch, is called a 20 grit size and measures roughly 1/20 inch across. The grain size is indicated by grit number as shown below:

Coarse Grain – 6, 8, 10, 12, 14, 16, 20, 24

Medium Grain – 30, 36, 46, 54, 60

Fine Grain – 70, 80, 100, 120, 150, 180

Very Fine Grain – 220, 240, 280, 320, 400, 500, 600, 800, 1000, 1200

Grade

Grade of a wheel is a measure of how strongly the grains are held by the bond. The bonding material in a wheel surrounds the individual grains and links them together. Different types of grades are represented by English letters from A to Z, as shown below.

A to H – Soft Grade

I to P – Medium Grade

Q to Z – Hard Grade

Different types of wheel grade are shown in Fig. 16.

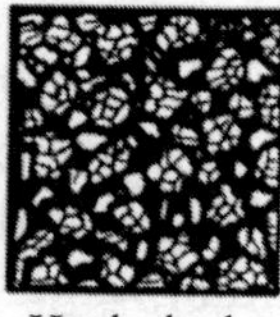

Hard wheel

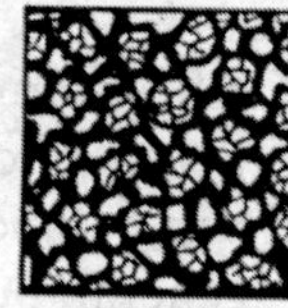

Medium wheel

Soft wheel

Fig. 16: Wheel grades

By "hardness" of the grinding wheel, we do not mean the hardness of the abrasives but the strength with which the abrasives are held by the bonding medium.

Structure

The size and no. of pores (voids) present given an idea of the structure. The structure of grinding wheel refers to relationship of abrasive grain and bonding material and of these two elements to the void between them. Different types of structures are represented by range of numbers.

1 to 8 – Dense structure (less voids)

9 to 16 – Open structure (more voids)

Numbers 5 to 8 represents medium structures and there are most widely used structures.

Grinding Ratio

It is defined as the ratio of the volume of material removed to the volume of wheel wear. The grinding ratio depends on parameters such as the type of wheel, wheel speed, cross-feed, down-feed, and the grinding fluid used. Values ranging from a low of 2 to over 200 have been observed in practice. A high grinding ratio, however, may not necessarily result in the best surface integrity of the part.

Grinding Wheel Shapes

Straight wheel, wheel recessed on one side and wheel recessed both sides are classified as straight wheel types and are used for cylindrical grinding, surface grinding, off-hand grinding and snag grinding. The recess is provided to give clearance for mounting flanges. Tapered wheel is used for snag-grinding operations. Cylinder wheel type is used on either horizontal or vertical spindle surface grinding machine. Either the peripheral surface or the face of the wheel is used as the grinding surface. A straight cup wheel is used on horizontal or vertical spindle surface grinding machine. It is also used for off-hand grinding. The wheel face may be plain or bevelled. Saucer wheel is used primarily for resharpening saws. Flaring cup and dish wheel are used for tool and cutter grinding. The flaring cup shape, which may have plain or bevelled face is also used in conjunction with a resinoid bond as a snag-grinding wheel. Various grinding wheel shapes are shown in Fig. 17.

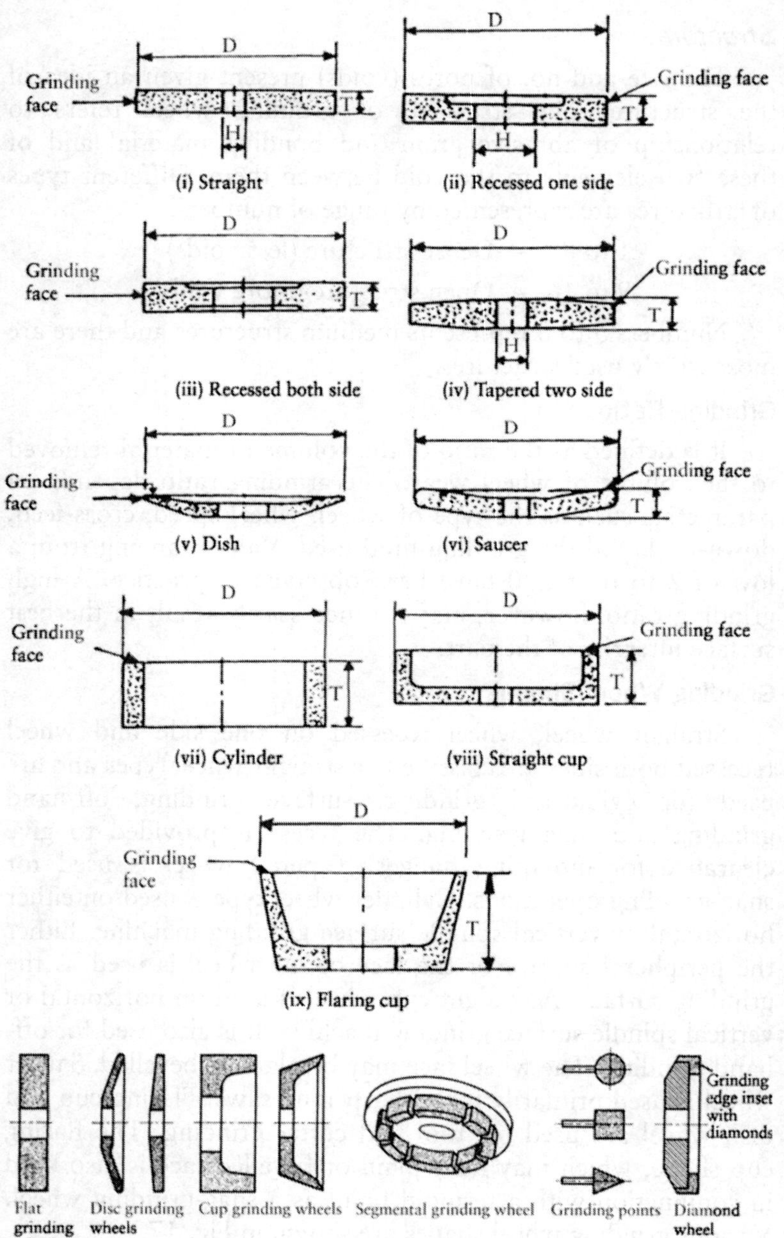

Fig. 17: Different types of grinding wheels and shapes

4. Factors in the Selection of Grinding Wheels

The important factors are the hardness and shape of the workpiece to be ground. However, the following factors must be considered while selecting a grinding wheel.

Constant Factors

The most important factor to be considered is the material to be ground, which influences the selection of the abrasive, grain size, grade, structure and bond. Abrasive should be aluminium oxide for steel and its alloys and silicon carbide for cast iron and non-ferrous metal. Grain size: a fine grain for hard and brittle materials and coarse grain for soft, ductile materials; Grade of wheel: hard wheel for soft materials, and soft wheel for hard materials; Structure: dense structure for hard and brittle materials and open structure for soft and ductile materials. At times, the bond selection is affected by the material to be ground, but more often it is influenced by the variable factors explained later.

The next constant factor considered is the accuracy and finish needed. It affects the selection of grain size and the bond. The grain size should be coarse for rough cutting and fine for fine finishing. The bond used should be vitrified for roughing and resinoid, rubber and shellac for high finish.

The third factor is the area of contact. It influences the grain size, grade and structure. It is recommended that grain size should be fine for small areas of contact and coarse for large areas of contact. The grade recommended is—the smaller the area of contact, harder the wheel. For structure, dense structure is recommended for small area of contact and open structure for large area of contact.

The fourth constant factor considered is the nature of the grinding operation. It affects the bond used. Precision grinding calls for vitrified bonded wheels. If an exceptionally fine finish is desired, one of the organic bond (resinoid. rubber or shellac) can be used.

Variable Factors

There are four variable factors to be considered. The first variable factor is wheel speed. It influences the grade of the wheel and bond of the wheel to be used. Higher the wheel speed softer should be the wheel. Resinoid bond wheels are preferred for higher speeds.

The second variable factor is the feed rate. It affects the grade of the wheel to be used. Higher the feed rate, harder should be the wheel.

The third variable factor is condition of the grinding machine. Spindles that are loose in their bearings and have shaky foundations necessitate use of harder wheel than would be needed if the machine were in better condition.

The fourth factor is the operating characteristics, i.e. production rate expected.

Standard Codification of a Grinding Wheel

Due to the diversity of methods employed by the various grinding wheel manufacturers for the making of grinding wheels, an universal marking system has been accepted as standard. It is divided into six parts, with the markings placed in the following sequence:

Position 1	**Position 2**	**Position 3**	**Position 4**	**Position 5**	**Position 6**
Kind of abrasive and manufacturer's prefix	Grain size	Grade	Structure	Bond type	(Manufacturer/ company's own marking system)

Figure 18 shows how a specific wheel is marked and how it should be ordered from the manufacturer. The chart shows that a grinding wheel marked 51 A-36-L-7-V-23 has been chosen. The letters and numbers contained in this identification are explained in the figure. This codification is as per Bureau of Indian Standard Code IS 551: 1989.

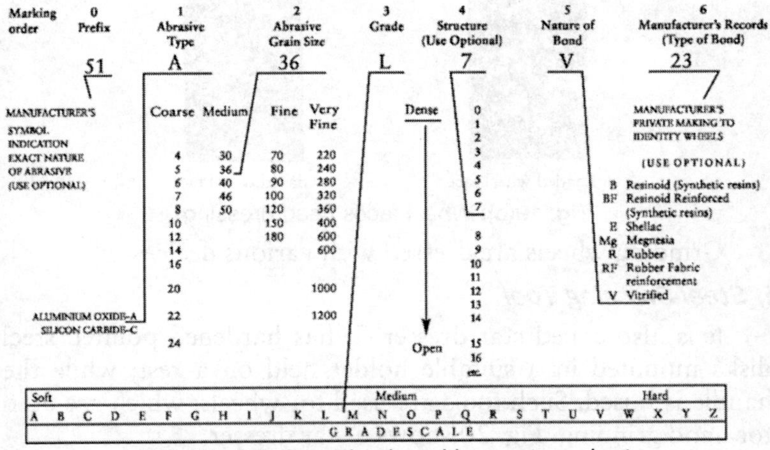

Fig. 18: Grinding wheel marking system chart

6. Wheel Dressing

Accurate running of grinding wheels is achieved by wheel dressing. Wheel dressing improves cutting action of the grinding wheel. Correct method of dressing a grinding wheel is shown in Fig. 19(a).

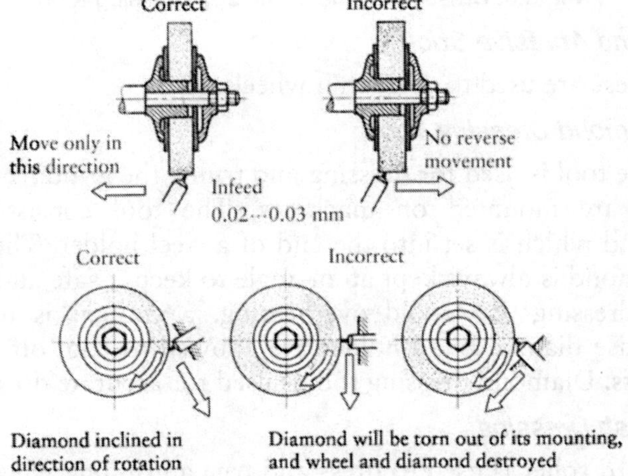

Fig. 19(a): Correct method of dressing a grinding wheel

Whenever the wheel face is loaded or glazed (voids filled with work material debris), it needs to be dressed as depicted in Fig. 19(b).

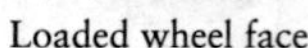

Loaded wheel face

Glazed wheel face

Fig. 19(b): Wheel faces need dressing

Grinding wheels are dressed with various decices.

i. *Steel-Dressing Tool*

It is also called star dresser. It has hardened pointed steel disks mounted in a suitable holder held on a rest; while the handle is raised. Such tools are used for wheels, which are used for hand grinding. Fig. 20 shows a star dresser.

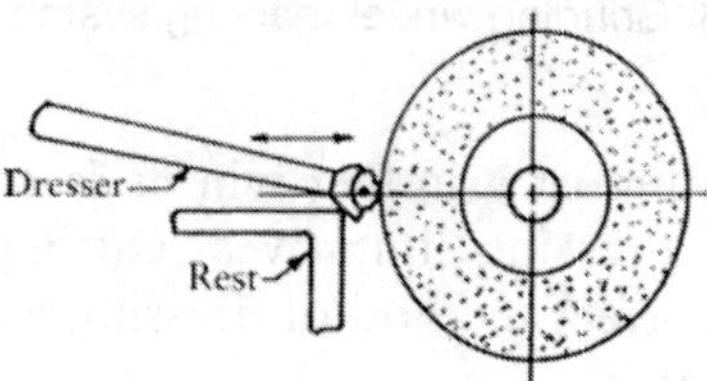

Fig. 20: Dressing a wheel with star dressing tool

ii. *Round Abrasive Stick*

These are used to dress thin wheels.

iii. *Diamond Dressing Tool*

The tool is used for dressing and truing the grinding wheels which are mounted on machines. The tool consists of a diamond which is set into the end of a steel holder. The point of diamond is always kept at an angle to keep it safe and to get good dressing. To avoid overheating, a coolant is used or otherwise diamond and holder are allowed to cool off after a few cuts. Diamond dressing tool is used for accurate dressing.

iv. *Crush Dressing*

Hard roller is used to dress and true a grinding wheel. The roller, sometimes diamond impregnated, is pressed against revolving grinding wheel. The reverse of the form needed on the wheel is given to the roller; which displaces and crushes the grains and imprints the form on the wheel. For intricate forms, crush dressing is very economical.

v. *Reactive Dressing*

It is a practice followed in grinding operations to restore cutting efficiency of wheel. When the wheel becomes dull and needs more power; the work speed is increased automatically. It causes the wheel face to break down and resharpen itself, due to increase in grain depth of cut.

Wheel dressing has the effect of resharpening a blunted wheel and regaining its original shape.

7. Wheel Truing

Truing is required to correct uneven wear on the face of the wheel. Truing is done to create a true and concentric surface on a wheel and this truing is attained by dressing. It also includes the altering of cutting face to produce special contours. As soon as fresh wheel is fitted, it becomes necessary to true its face. Fig. 21 shows wheel faces out of shape thus needing to be trued.

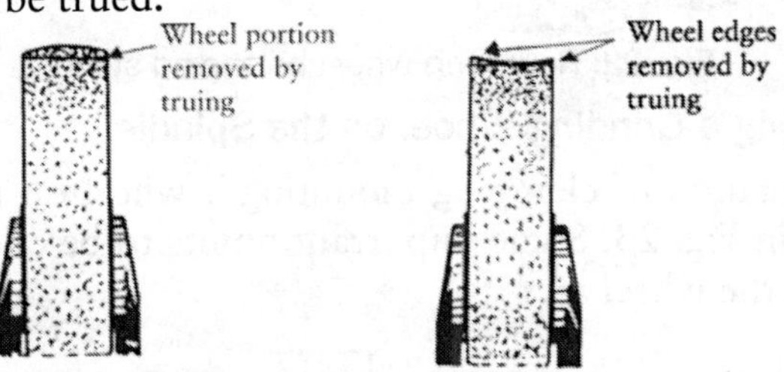

Fig. 21: Wheel faces worn out of shape, require truing

8. Balancing of Grinding Wheel

Balancing is necessary to make the wheel run smoothly and thus obtain a smooth ground surface. From the point of view of wheel life and surface finish on the job, vibration in grinding operation is critical. The machine should be rigid and bearings in good condition. Vibrations to a large extent are caused by out-of-balance and out-of-round wheels. Since the grinding wheel speeds are high, slight out-of-balance condition may give rise to large centrifugal forces. This may result in excessive vibrations, poor surface finish, faster wheel breakdown and may even be dangerous to the operator. Therefore, particular attention should be paid to the balancing of wheels. Generally balancing weights are provided on the mounting flange of the grinding wheels.

By mounting the wheel on a static balance stand equipped with two rollers or pairs of over-lapping discs (Fig. 22), the wheel is brought to a static balance by adjusting the position of the balancing weights provided on the mounting flange. Then it is mounted on the grinding machine and dressed to concentricity. It is once again removed from the machine and rebalanced. It can now be used for grinding. For getting better results, wheel can be dynamically balanced while it is running on the machine.

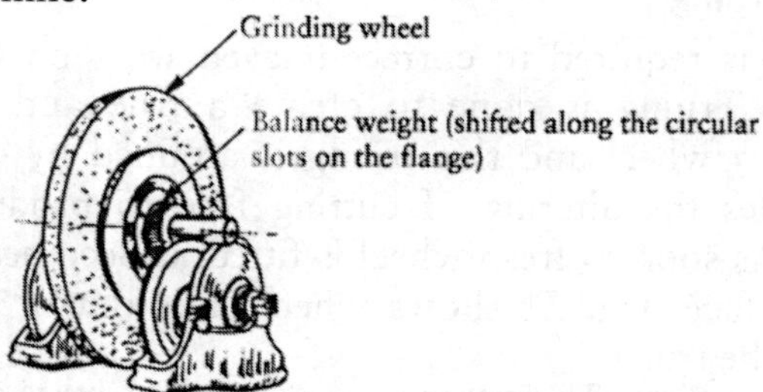

Fig. 22: Revolving wheel balancing stand

9. Mounting a Grinding Wheel on the Spindle

The method of clamping mounting a wheel on the spindle is shown in Fig. 23. Some important points to be considered in mounting the wheel are:

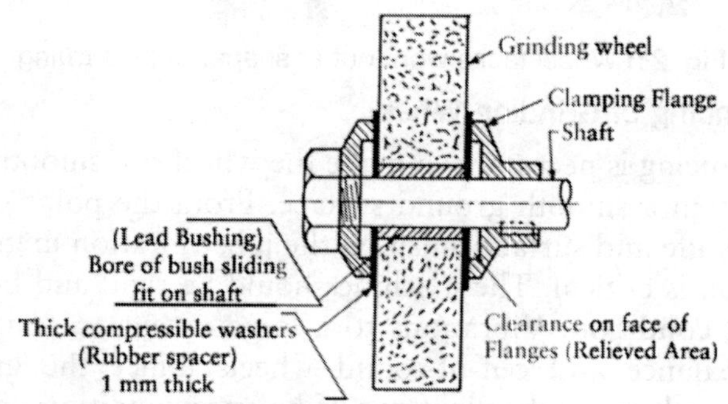

Fig. 23: Mounting of a grinding wheel

(i) Make sure that the wheel is not cracked. Although the cracks are not visible; yet the crack in the wheel can be found. To test for cracks, hold the wheel through its hole and lightly tap it with an object like a wrench handle. A wheel in good condition will give a very clear ringing sound, while a cracked one will give a dull or flat sound. Damaged wheels must not be used.

(ii) Ensure that the wheel has proper packing on both sides around the hole.

(iii) The wheel should slide into the spindle or arbor having clearance up to 0.1 mm. If hole is small, it should be reamed.

(iv) The wheel flanges should be of safety type. Wheel flanges should extend upto 50% of radius of wheel. In case of big stones, it is very important. Washers made of thick, soft pulp paper or rubber should be placed between flanges and wheel sides.

(v) Wheel should never be excessively tightened.

(vi) Wheel should never be operated without guard. Also ensure that wheel moves freely in its guard and housing by rotating the wheel with hand.

(vii) See that wheel is balanced and dressed properly.

(viii) While starting the machine keep yourself away to one side for a minute or so.

(ix) Whenever grinding wheel is attached to the machine it must be tested for smooth running for 5 minutes at full operating speed.

10. Wheel Speed and its Impact on Grinding

If the wheel speed is increased, at a constant longitudinal or rotary feed rate, the size of the chip removed by a single abrasive grain is reduced. This reduces the wear of the wheel.

If the wheel speed is reduced, the wear is increased. From this, it is clear that from the view point of wear, it is better to operate at higher wheel speeds. However, it is limited by the allowable speeds at which the wheel can be worked, as well as the power and rigidity of grinding machine. Normally the grinding wheel speed ranges from 20 to 40 m/sec. The wheel speed also depends on the type of grinding operation and the bond of grinding wheel, e.g. resinoid bonded wheels can be generally used at higher peripheral speeds than vitrified bond wheels.

Table 1 gives various wheel speeds for various grinding operations and types of bonds.

Table 1

Type of grinding	Wheel speed (surface speed) m/sec.
Rough grinding wheel (Vitrified bond)	25
Rough grinding wheel (Resinoid bond)	45
Surface grinder (Vitrified bond)	20-25
Internal grinder (Vitrified bond)	20-35
Centreless grinding wheel (Vitrified bond)	30-80
Cylindrical grinding wheel (Vitrified bond)	30-35
Cutting off/Parting off wheels with resinoid bond or rubber bond	45-80
Hand grinding of wheels	20-25
Automatic grinding of tools	25-35
Hand grinding of carbide tools	18-25

11. Work Speed and its Effect on Grinding

Work speed can be defined as the speed at which the workpiece traverses across the wheel face or rotates around between centres. If work speed is high; wheel wear is increased, but the heat produced is decreased. On the other hand, if the work speed is low the wheel wear decreases, but heat produced is more. The ratio of wheel speed to work speed is of importance and should be maintained at proper value. Low work speeds result in local overheating and bring about deformation of hardened workpiece. It affects the mechanical properties of the workpiece and very often micro cracks will appear on the workpiece. The increase in work speed is limited by premature wheel wear and vibrations induced by wear. The approximate work speed in m/sec. for various materials are given in Table 2.

Recommended Surface speeds for different work materials (m/sec.)

Table 2

Material	Cylindrical grinding Roughing	Finishing	Internal grinding	Surface grinding
Soft steel	11-15	6-8	15-20	
Hard steel	14-16	6-10	18-22	
CI	12-15	6-10	18-22	
Brass	18-20	14-16	28-32	8-15
Aluminium	50-70	30-40	32-35	

Cooling: Hardened work materials are ground with a coolant. Coolants generally used are water with soda, soap and oil additives.

12. Common Troubles, Causes and Their Remedies in Grinding Process

Common troubles, causes and their remedies in grinding process are presented in Table 3.

Table 3

Common Troubles	Causes	Remedies
1. The wheel becomes glazed	Grade too hard, grain too small, wheel speed too high	Wheel needs dressing
2. The wheel becomes loaded or clogged with work particles	Grain too small; grade too hard, wheel speed too high	Wheel needs dressing
3. The wheel fails to cut; excessive pressure required	Grade too hard and wheel glazed	Wheel needs dressing
4. The wheel wears too rapidly	Grade too soft; grain too large; wheel speed too low	Give less feed and less traverse
5. Wheel or work chatters or vibrates	Wheel loaded; grade too hard; wheel not balanced	Use centre rest or back rest
6. Work surface gets overheated	Wheel glazed or loaded	Use coolant
7. Work surface is ground Inaccurate	Lack of coolant; work out of balance, or failure to use centre rest or back rest	Remove all defects

13. Safety in Grinding Machines

Safety during grinding operation can be maintained only if the operator is aware of the hazards present in grinding.

Injuries resulting from grinding machine operations are generally severe/serious.

The principal hazards of grinding can be:

1. Injury to the eyes from flying particles and spark
2. Inhalation of dust, generated by the grinding process
3. The danger due to bursting of wheel
4. Contact with the revolving wheel

All the above hazards can be eliminated or minimized by mechanical guarding machine guards or by the use of personal protective equipment (i.e. goggles) or both.

The general safety precautions to be observed while operating a grinding machine are the same as that intended for other machines. But, certain precautions are to be observed exclusively in the case of grinding machine operations. Maximum permissible speeds must never be exceeded.

Safety Devices in Grinding Operations

Listed below are the various safety devices which are used during grinding:

1. Safety goggles and face shields must be worn by the operator while working.
2. Permanent eye shields or screens fitted on the machine.
3. Interlocked grinder guard with trip switches.

QUESTIONS

1. What are natural and artificial abrasives? Why are the latter preferred over the former?
2. What is meant by Grain, Grit, Structure and Grade of a grinding wheel?
3. Why are truing and dressing necessary in grinding wheel?

4. Why is balancing done? Suggest some methods for balancing a grinding wheel.
5. Explain with suitable sketches, the relative work, wheel and table movements on different types of surface grinders.
6. What are different types of internal grinders? Describe them.
7. Explain the principle of centreless grinding. How do 'Through feed' and 'Infeed' methods differ in centreless grinding?
8. Explain the factors to be kept in mind while selecting a grinding wheel.

REVISION QUESTIONS

1. What are the principal artificial grinding media?
2. Why are soft grinding wheels used for hard materials, and hard wheels for soft materials?
3. How is the hardness grade of a grinding wheel indicated?
4. What does grain "100" mean?
5. Why must grinding wheels be balanced?
6. Why must soft packing material be placed between the grinding wheel and the flange?
7. Why should the workpiece support point be as close as possible to the grinding wheel?
8. What precautions must be taken when dressing a grinding wheel with a diamond tool?

OBJECTIVE TYPE QUESTIONS

Choose the correct alternative:

1. In grinding operation, the actual cutting action is done by

   (a) resin (b) abrasives

   (c) rubber (d) void spaces

   (e) steel

2. The grinder which is used to produce flat ground surfaces is called a
   (a) flexible shaft grinder
   (b) surface grinding machine
   (c) cylindrical grinding machine
   (d) centreless grinding machine
   (e) internal grinding machine
3. The depth of cut on a surface grinder is regulated by raising or lowering the _______.
   (a) work table (b) wheel head
   (c) workpiece (d) magnetic base
   (e) Diamond dresser
4. In centreless grinding operation the regulating wheel and the grinding wheel
   (a) must always rotate at the same speed as the work speed
   (b) must always rotate at different speeds, but in the same direction
   (c) must always rotate at the same speed, but in opposite direction
   (d) must always rotate at different speeds, and in opposite direction
5. Grinding machines equipped with rotating work tables are classified as
   (a) tool and cutter grinders
   (b) cylindrical grinders
   (c) surface grinders
   (d) centreless grinding machines
   (e) internal grinding machines
6. The operation of sharpening a grinding wheel is called
   (a) truing (b) dressing
   (c) aligning (d) wheel balancing
7. Surface grinders are usually equipped with a _______ for holding the work to be ground.

(a) pneumatic chuck
(b) fluid chuck
(c) mechanical chuck (three-jaw chuck)
(d) magnetic chuck

8. In centreless grinding operation, the workpiece is made to rotate
   (a) by the grinding wheel
   (b) by the regulating wheel
   (c) by the work rest
   (d) directly by the electric motor
9. Among the following abrasives, the hardest one is
   (a) aluminium oxide (b) silicon carbide
   (c) diamond (d) cubic boron nitride
10. When grinding wheels become loaded or glazed, they may be dressed, to restore sharpness, with a
    (a) pumice stone dresser (b) tool steel dresser
    (c) diamond dresser (d) sharp edged H.S.S. cutter
11. The grinding wheels employed in surface grinding operations are classified as
    (a) plain grinding wheels
    (b) cup wheels
    (c) Dished or saucer wheels
    (d) grinding points
12. In centre-type cylindrical grinding operations when the work is mounted between centres, the work is made to rotate by
    (a) using a live or rotating centre in the head stock spindle.
    (b) the frictional drive of a regulating wheel.
    (c) the same general method used to rotate work mounted between centres on a lathe.
    (d) the movement of the grinding wheel itself.
13. When grinding wheels become loaded or glazed, they must be

(a) balanced properly (b) aligned accurately
(c) trued (d) dressed

14. A centre-type cylindrical grinder that permits swivelling the wheel head and head stock at an angle to the table ways is called a
    (a) Tool and cutter grinder
    (b) Surface grinder
    (c) Internal grinding machine
    (d) Bench grinder
    (e) Universal grinder
15. In cylindrical grinding operations, the work is always rotated
    (a) at a much slower speed than that of the grinding wheel
    (b) at a much faster speed than that of the grinding wheel
    (c) at the same speed as the grinding wheel
    (d) at 30 r.p.m.
16. The peripheral speed of a 150 mm diameter grinding wheel rotating at 300 r.p.m. is
    (a) 20 m/min
    (b) 141.5 m/min
    (c) 450 m/min
    (d) π (150) (3000)/60 m/min
17. A slight taper is to be ground on the full length of a long shaft mounted between centres on a universal cylindrical grinder. This can be accomplished by
    (a) Off-setting the tail stock
    (b) Swivelling the table on its base
    (c) Swivelling the wheel head
    (d) Swivelling the workpiece
18. To grind the 60° on the end of a machine centre point on a universal cylindrical grinder
    (a) the table is swivelled to 30°
    (b) the headstock is swivelled to 60°

(c) the headstock is swivelled to 30°
(d) the workpiece is swivelled to 30°
(e) the headstock is swivelled to 120°

19. Plunge cut grinding is accomplished on a cylindrical grinder by:
(a) Traversing the table
(b) Using the table cross feed
(c) Feeding the grinding wheel into the work while the table remains stationary
(d) Feeding the workpiece into the grinding wheel while the grinding wheel remains stationary

20. The correct setting of the height of the work in _______ operations is obtained by the setting of the work rest blade.
(a) cylindrical grinding (b) surface grinding
(c) centreless grinding (d) internal grinding
(e) ball grinding

21. In centreless grinding operations, the work is rotated and fed past the face of the grinding wheel by the _______.
(a) work rest (b) regulating wheel
(c) table (d) electric motor directly

22. Holes in parts which have been hardened by heat treatment can be finished to accurate size only by
(a) Drilling (b) Boring
(c) Internal grinding (d) Reaming

23. The surface speed of the smaller grinding wheels used for internal grinding operations ranges from
(a) 15–30 meters/min
(b) 300–1000 metres/min
(c) 1200–2000 metres/min
(d) 10,000–20,000 metres/min

24. The approximate cutting speed for surface grinding operations is generally

(a) 15 metres/min. (b) 100 metres/min.
(c) 600 metres/min. (d) 1800 metres/min.
(e) 20,000 metres/min.

25. The grit size of the abrasives used in the grinding wheels is usually specified by
    (a) the Hardness number
    (b) the size of the wheel
    (c) the softness or hardness of the abrasive
    (d) the Mesh number or Grit No.

26. In grinding practice, the term "grade of the wheel" refers to
    (a) the fineness of the abrasives used
    (b) the strength of the bond of the wheel
    (c) the finish that can be obtained
    (d) the hardness of the abrasives used in the manufacturing of the wheel
    (e) the structure of the wheel

27. In the standard marking system used for specifying the grinding wheel, the grade of the wheel is usually represented by
    (a) a Number
    (b) a Letter of the English Alphabet
    (c) the Hardness of the abrasives
    (d) the finish of the wheel

28. In grinding practice, the term "Hardness of the wheel" refers to
    (a) the hardness of the abrasives used
    (b) the strength of the bond of the wheel
    (c) the finish of the wheel
    (d) the hardness of the workpiece that can be machined with the grinding wheel

29. In Centreless grinding operation, the regulating wheel is usually
    (a) bigger than the grinding wheel
    (b) smaller than the grinding wheel

(c) of the same size as the grinding wheel
(d) smaller than the diameter of the workpiece.

30. Lapping, honing and superfinishing are basically
    (a) slow speed abrasive machining processes
    (b) high speed abrasive machining processes
    (c) grinding processes
    (d) material removal processes wherein large quantity of material is machined

31. Which one of the following processes removes maximum amount material out of a workpiece?
    (a) Honing (b) Lapping
    (c) Grinding (d) Superfinishing

32. For achieving extremely smooth surface finish at bearing surfaces, the process of machining generally employed is
    (a) milling (b) shaping
    (c) drilling (d) grinding

33. Grinding operation is used for
    (a) forming (b) shaping
    (c) dressing (d) finishing

34. For grinding shafts and spindles use
    (a) tool and cutter grinding
    (b) cylindrical grinding
    (c) thread grinding
    (d) surface grinding

35. For grinding flat surfaces use
    (a) internal grinding (b) thread grinding
    (c) cylindrical grinding (d) surface grinding

36. The highest cutting speed is used in
    (a) surface grinding (b) centreless grinding
    (c) internal grinding (d) cylindrical grinding

37. Grinding ratio may be defined as
    (a) Wear of grinding wheel/volume of metal removed

(b) Volume of metal removed/Wear of grinding wheel
(c) Wear of grinding wheel × Volume of metal removed

38. Surface speed (in m/min) of the grinding wheel in centreless grinding is
(a) 15–16 (b) 100–500
(c) 1000–1500 (d) 1500–1800

39. Workpiece is supported as follows in centreless grinding
(a) on magnetic chuck (b) in universal chuck
(c) in collet chuck (d) none of the above

40. The workpiece advances in centreless grinding due to
(a) Machine drive
(b) Effort applied by operator
(c) Force exerted by regulating wheel
(d) Force exerted by grinding wheel

41. Artificial abrasives are
(a) Sandstone, Emery, Diamond, Quartz
(b) Silicon carbide, Aluminium oxide, Boron carbide
(c) Garnet
(d) Corundum

42. For grinding work materials like bronze, brass, copper, aluminium, etc. the abrasive used is
(a) Silicon carbide (b) Garnet
(c) Aluminium oxide (d) Corundum

43. For softer materials, the grain of abrasives used is
(a) Coarse grain (b) Fine grain
(c) Medium grain (d) Both (b) and (c)

44. If condition of machine is such that it produces some vibrations, then use
(a) Hard grade wheel
(b) Soft grade wheel

(c) Both (a) and (b)
(d) Grinding wheel with fine grain abrasives

45. For harder materials, the grain of abrasives used is
(a) Coarse grain (b) Fine grain
(c) Medium grain (d) Both (a) and (c)

46. Majority of the grinding wheels use the following type of bond
(a) Resinoid (b) Silicate
(c) Rubber (d) Vitrified

47. In finish grinding, the grinding ratio is kept as
(a) 5–10 (b) 10–25
(c) 25–50 (d) 50–100

48. Grade of a wheel, i.e. strength of a bond is designated from A to Z, where
(a) A is very soft (b) Z is very hard
(c) Z is very soft (d) Both (a) and (b) are true

49. If a grinding wheel is designated as 30 A 36 H 6 VB then the letter H indicates
(a) Grade of bond (b) Type of bond
(c) Structure (d) Abrasive

50. In the above question, the letter 'V' indicates
(a) Type of abrasive (b) Type of bond
(c) Type of structure (d) Grade of bond

51. A grinding wheel is said to be glazed
(a) when it becomes unbalanced
(b) when the abrasive grains have been replaced
(c) when the abrasive grains become dull and stop cutting
(d) when some lubricant is added to the abrasive grains

52. Truing of grinding wheel is done by
(a) Glazing the wheel (b) Dressing the wheel
(c) Loading the wheel (d) Balancing the wheel

53. Grinding wheels are balanced _______ by shifting weights on one flange of the wheel mount.
   (a) Statically
   (b) Dynamically
   (c) Both (a) and (b)
54. Which operation should be performed in the last in connection with grinding wheel?
   (a) Balancing (b) Dressing
   (c) Truing (d) Glazing
55. Which of the following grinding processes has highest possible speeds?
   (a) Internal grinding (b) Surface grinding
   (c) Cylindrical grinding (d) Cutting off operations

ANSWER KEY

1. (b)	2. (b)	3. (b)	4. (b)	5. (c)
6. (b)	7. (d)	8. (b)	9. (c)	10. (c)
11. (a)	12. (c)	13. (d)	14. (e)	15. (a)
16. (b)	17. (c)	18. (c)	19. (c)	20. (c)
21. (b)	22. (c)	23. (c)	24. (d)	25. (d)
26. (b)	27. (b)	28. (b)	29. (b)	30. (a)
31. (c)	32. (d)	33. (d)	34 (b)	35. (d)
36. (d)	37. (b)	38. (d)	39. (d)	40. (c)
41. (b)	42. (a)	43. (a)	44. (a)	45. (b)
46. (d)	47. (c)	48. (d)	49. (a)	50. (b)
51. (c)	52. (b)	53. (a)	54. (a)	55. (d)

10

Jigs and Fixtures

1. Introduction

There are different views on what jigs and fixtures are. Some of these are as follows:

(i) Most shop floor people (engineers and supervisors) use the term jig to describe work holding devices (other than chuck/collets).

(ii) A jig or fixture is any device that guides drill or other tools, so as to produce work that is interchangeable within the tolerance set by manufacturing requirements.

(iii) The terms jigs and fixtures are also used for devices or frames that hold pieces in their proper position while being welded.

(iv) A distinguishing definition for jigs and fixtures that seems to be generally holding is as follows. A jig is a work holding device, which is not fastened to the machine on which it is used. A fixture is also a work holding device, but one that is bolted or otherwise fastened to the machine. That is, the jig may be moved around on the table of a drill press to bring each bushing under the drill spindle. A fixture on the other hand is fastened to the table or base of a machine, and either the tool is moved to the point of operation, as in the case of a radial drill, or table is moved under the cutting tool as in a milling machine.

2. Uses of Jigs and Fixtures

The jigs and fixtures are useful in the following ways:

1. In increasing the production and production rate.
2. In ensuring high accuracy of parts.
3. For providing inter-changeability.
4. In reducing expenditures involved in quality control.
5. In reducing the cost of production.
6. In saving labour.
7. In increasing the versatility of the machine tool or in a way, it widens the technological capacity of machine tools.

The jigs and fixtures should possess the following components to fulfil their basic functions:

(i) Locating elements
(ii) Clamping elements
(iii) Tool guiding elements (for jigs) or tool setting elements (for fixtures).
(iv) Base plate for holding together the entire asembly.

Locating elements are generally in the form of locating pins. Locating pins are the pins which are inserted in the body of jig and fixture. The workpiece is pushed against this pin to establish relation between workpiece and the jigs or fixture.

The purpose of clamping element is to exert a force to press a workpiece against the locating elements and hold it there against the action of cutting forces.

3. Difference Between a Jig and Fixture

Jig	Fixture
1. A Jig holds and positions the work and locates or guides the cutting tool with respect to the workpiece.	1. A fixture also holds and positions the work but does not guide or locate the cutting tool.
2. Usually, a jig is not fixed to the machine table.	2. A fixture is bolted or clamped to the machine table.
3. It is usually lighter in construction and moved around on the table.	3. It is usually heavy in construction and fixed to the table.

4. Jigs are used on drilling, reaming, tapping and counter boring operations.	4. Fixtures are used for milling, grinding, shaping, planning, boring and welding operations.

4. Principles of Location

Assume a workpiece in a space as shown in Fig. 1. In a free state, it can move in either of the two opposite directions along XX, YY or ZZ axes. These six movements are called 'movements of translation'. The workpiece can also rotate in either of two opposite directions around each axis. These six movements are called 'rotational movements'. It means we have twelve degrees of freedom of a workpiece in space. To confine the workpiece accurately, all these degrees of freedom must be restricted.

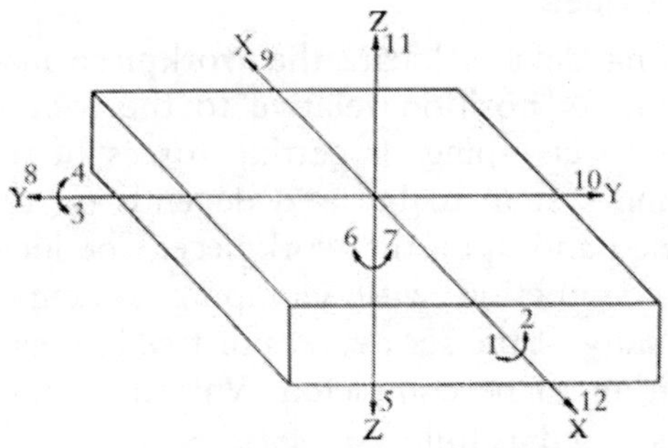

Fig. 1: Workpiece in space

(a) The workpiece is made to rest on three pins *A*, *B* and *C* inserted in the base of the body a shown in Fig. 2. This restricts the rotational movement along XX, YY and also downward movement, i.e. 1, 2, 3, 4, 5 shown in Fig. 1 have been arrested.

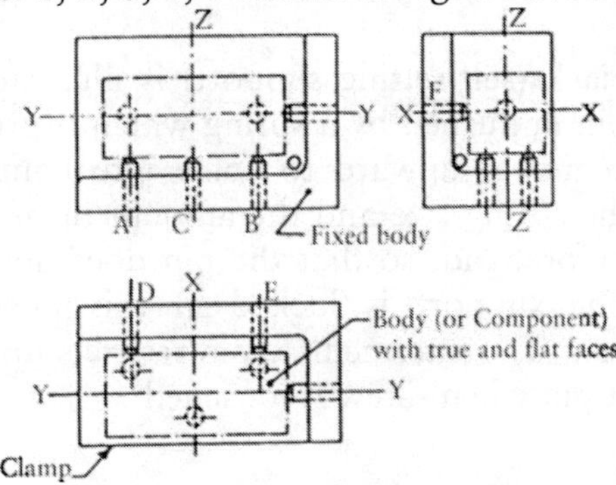

Fig. 2: Six point location principle

(b) Insertion of two more pins *D* and *E* restricts the rotational movement about Z-axis and movement towards left, i.e. 6, 7, 8 have been arrested.

(c) Insertion of one pin *F* stops the movement 9.

In this way, *six locating pins restrict* nine degrees of freedom. Three degrees of freedom are still free and for that three pins are needed, but this will lead to a complete enclosure of the workpiece making its loading and unloading difficult; so a clamping device is used to restrict the rest three degrees of freedom. The method is called '3-2-1' or 'six point location' principle. 3 restraints in the base plane, 2 restraints in the perpendicular plane and 1 restraint in the other 3rd plane.

5. Locating Devices

The locating devices locate the workpiece inside a jig or fixture, so that its position relative to the tool may not be disturbed due to clamping or cutting forces of the tool. The type of locating system to be used depends on the operation being performed and upon the workpiece. The locating system must be in conjunction with clamping system and should eliminate as many of the six degrees of freedom as is necessary for the operation to be completed. Various types of locating devices are discussed as follows:

Location from Rough Surfaces

Rest buttons and adjustable supports are used for location from rough surfaces. Floating pads are used for locating flat machines surfaces (Fig. 3). Spring jacks are also used for locating.

A spring jack (self-setting support) is illustrated in Fig. 4. The vertical pin in pushed by a spring which is precompressed. Thus, the pin moves upward to come into contact with the workpiece. The spring size and the amount of precompression are selected beforehand, so that the pin does not displace the workpiece. The support is locked in this position by the tightening screw. The intermediate pin restricts upward motion of the support pin when screw is loosened.

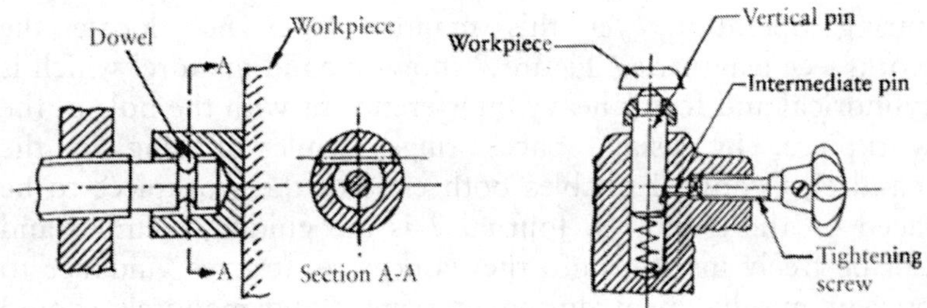

Fig. 3: Floating pad **Fig. 4:** Spring jack

Location from External Cylindrical Surfaces

A workpiece is located from external cylindrical surfaces by a Vee-block. Wide V-block are used for locating finish-machined datum surfaces and narrow ones for rough surfaces. There are two types of V-blocks—fixed and sliding. A sliding V-block is secured to the jig body by cap head screw and dowel pins. The sliding V-block may be actuated by hand nut or cam (Fig. 5). All sliding parts should be made of steel and hardened and ground. The bottom of the vee should be slotted to remove the sharp corner.

Vee blocks are made of straight carbon steel grade 45 or case hardening steels grade 08 to 20. Very large vee blocks are made of grey CI with hardened steel pads secured by screws to working surfaces. Rigid V-blocks with an angle of 90° are most commonly used in jigs and fixtures.

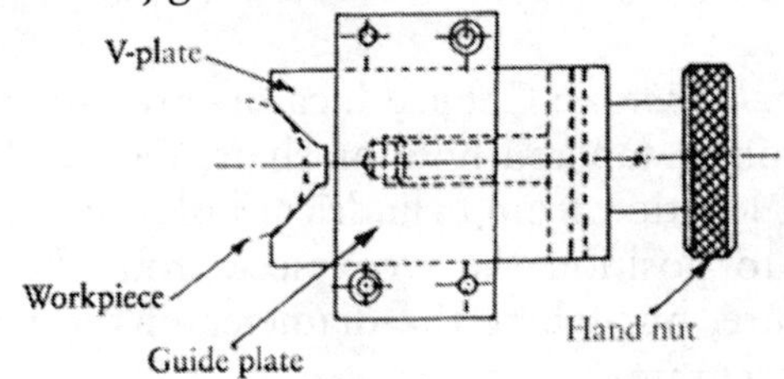

Fig. 5: Sliding V-block

Location from Holes

Workpieces are located from holes by pins or mandrels. A plain mandrel with a very small taper (1:2000) is shown in Fig. 6. This is forced into workpiece through a hand press (sometimes by wooden mallet). The interference between mandrel and workpiece prevents the workpiece from slipping during operation, yet this mandrel does not locate the workpiece lengthwise. Figure 7 shows a solid mandrel which is cylindrical and forms heavy interference fit with the hole of the workpiece. By using spacer rings, while pressing in the mandrel, Groove 1 enables both ends of the workpiece to be faced on the mandrels. Journal 2 is the guiding member and can be freely inserted into the workpiece. It gives guidance to prevent misalignment during pressing. Solid mandrels should be made of steel, heat treated (hardened) and precisely ground.

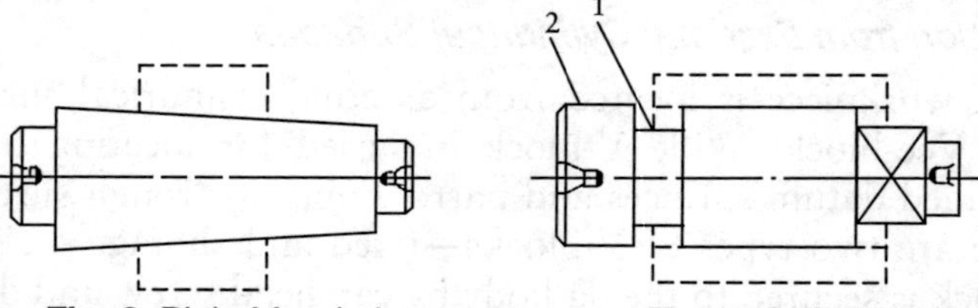

Fig. 6: Plain Mandrel **Fig. 7:** Solid Mandrel

Various types of pins made of hardened steel are the most widely used locating devices that are used to locate a component in a jig or a fixture. The pins may be either of the following:

(1) Locating pins and (2) Support pins

1. Locating Pins: Locating pins may be used if reamed or finely finished holes are available for positioning purposes. Depending upon their form, locating pins are classified as below:

(a) *Conical locators*: Conical locators are used to locate the workpiece from a tapered hole or shaft. Figures 8(a) and (b) show conical location from cylindrical holes or shafts where it is necessary to position the workpiece from the axis of the location feature, but where the diameter of the feature is not particularly accurate.

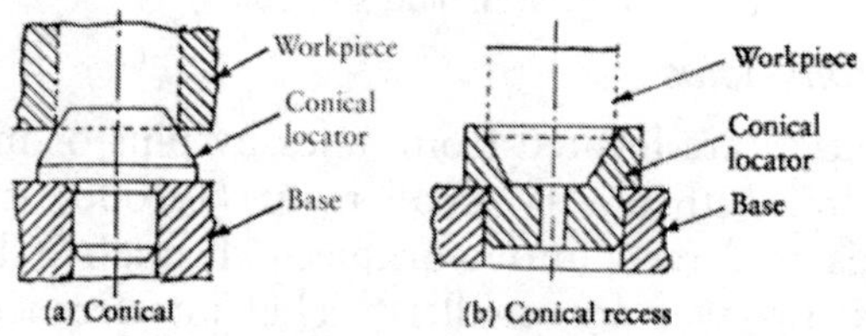

Fig. 8: Fixed conical locators

(b) *Cylindrical locators*: Figure 9(a) shows a short cylindrical locator which will constrain all degrees of freedom with the help of clamping force, except that of rotation about its own axis YY, for which a second locator must be used for complete location. Locators must be accurately positioned relative to the base and kept as short as possible. If a long locator must be employed to give greater support to a weak workpiece, location must only take place at extreme ends of the locator and so the post must be relieved as shown in Fig. 9(b).

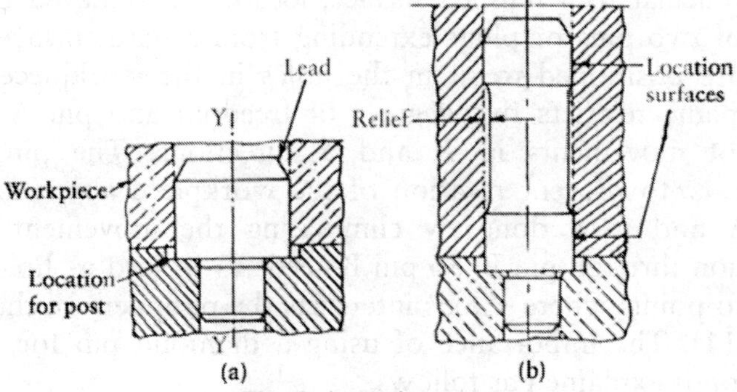

Fig. 9: Location post

Location posts should be given a generous lead to facilitate loading and should sit in a recess in the base, so that dirt will not prevent the workplace from being correctly seated.

2. Support Pins: Support pins enable the component to be supported at convenient points. Support pins are of two types: fixed and adjustable. These are shown in Figs. 10(a) and (b) respectively.

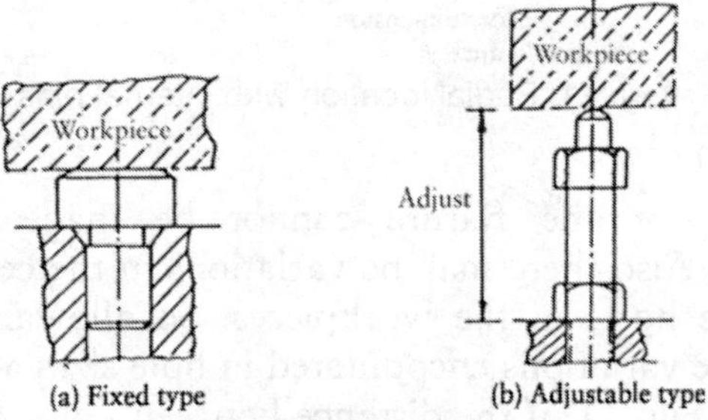

Fig. 10: Support pins

Bush Location

Shaft type components can be accommodated in hardened steel bushes. Small bushes may be pressed or driven into the jig, or fixture body. Large bushes may be a push fit in the body and locked with the help of screws. In all bushes, a good entrance chamfer or radius is provided to facilitate loading of the workpiece.

Radial Location

When the workpiece contains accurate holes that are perpendicular to a finished surface; location is achieved by the help of two pins or plugs extending from a base surface. The pins are positioned to fit in the holes in the workpiece. The base plane restricts five degrees of freedom and pin A takes care of movements in *X* and *Y* directions. The pin B is necessary to prevent rotation of the workpiece about axis of pin A and it is done by eliminating the movement in *Y* direction through pin B. So pin B needs to fit into its hole only on two points where *Y* axis intersects the periphery of the hole (Fig. 11). The importance of using a diamond pin for radial location is explained as follows:

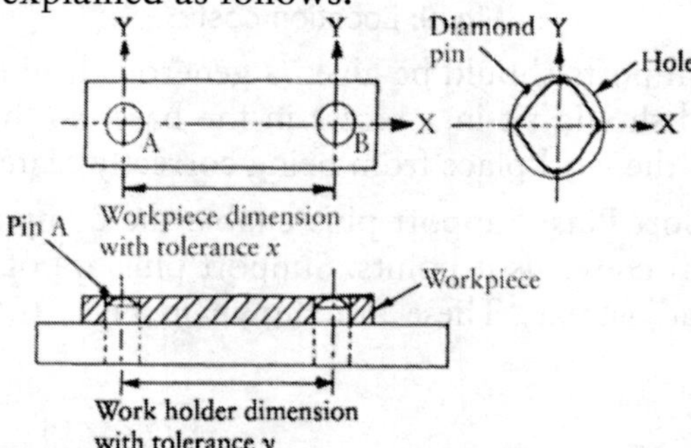

Fig. 11: Radial location with internal pins

Diamond Pin

The jig or the fixture cannot be made to accurate dimensions. Also there may be variations in the centre distance between the holes in the workpieces. So allowances must be made for the variations encountered in hole sizes and locations. Refering to Fig. 11, if the distance between holes A and B vary to the extent of tolerance *X* and distance between pins A and B has a tolerance Y, then if the workpiece is effectively located on pin A, the allowance between pin B and hole B will be (X + Y). Extreme cases occur when both hole and pin centre-to-centre dimensions are at maximum or minimum conditions. Due to this, there will be large allowance between the hole and the pin at B in Y-direction. It will lead to undesired radial rotation around the pin A and so the ultimate purpose, for which pin B is put, is lost.

To get better radial location, pin B is made of diamond shape and therefore it is called a diamond pin. It is relieved on two sides to allow for variation in X-direction, but it will provide cylindrical locating surfaces in Y-direction.

Location from Outer Contour

For locating a workpiece by taking its outer contour as reference, locating nests are used. Locating nest is actually a cavity or pocket which accomodates the workpiece. Different types of locating nests are—cavity nest, pocket nest and partial pocket nest.

- (i) *Cavity Nest*: In this type of nest, the shape of the workpiece is sunk into the nest plate. Generally the cavity is not made all around the contour to the exact size, but only a few points determine the workpiece location as shown in Fig. 12(a). A provision is also made for easy removal of workpiece.
- (ii) *Pocket Nest*: In this type, a pierced plate mounted on the rest plate locates the workpiece as shown in Fig. 12(b). In this case also, only a few points of the workpiece are located.
- (iii) *Partial Pocket Nest*: It is very similar to pocket nest. It consists of two or more splits which are independently mounted on the support (rest) plate [Fig. 12(c)]. In this case, removal of workpiece is easier.

When position pins or buttons are put along the contour of the workpiece, they are called nest pins [Fig. 12(d)]. Nest pins should be made foolproof and should be located on extreme positions.

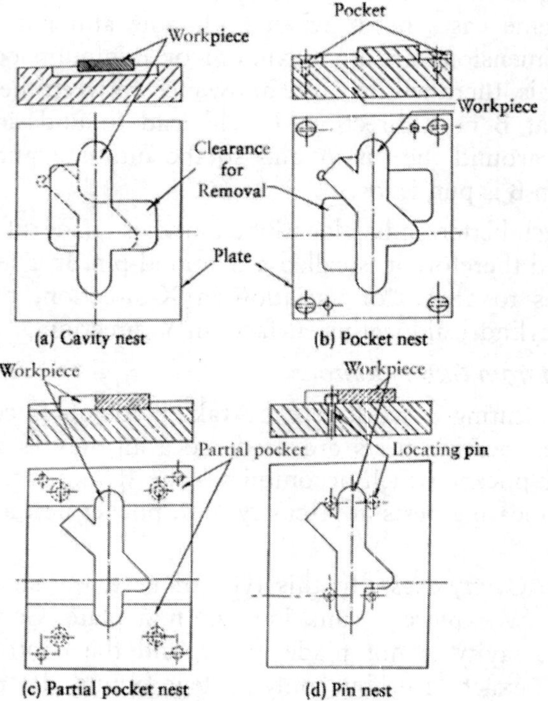

Fig. 12: Locating nests

Centralizers

By this method the physical middle line is automatically brought into the middle of the location. This means that the cutting depth is same all round and the centre of gravity is located properly. Since the machining allowance is equally distributed, the cutting forces are the same (Fig. 13). Three types of centralizer are: (a) Angular blocks, (b) Linkage controlled, and (c) Self-centering chucks.

Angular blocks are very popular among the group. Most commonly used is the concave type with two flat surfaces at a certain angle. The angular block can be fixed or adjustable type as shown in Fig. 14. In linkage controlled type, the most commonly used design is the Scissor type (Fig. 15). Self-centering chucks and collets are part of the accessories of more or less every lathe and it should be clear how they work. Mainly the collet chuck is used in jig and fixture design.

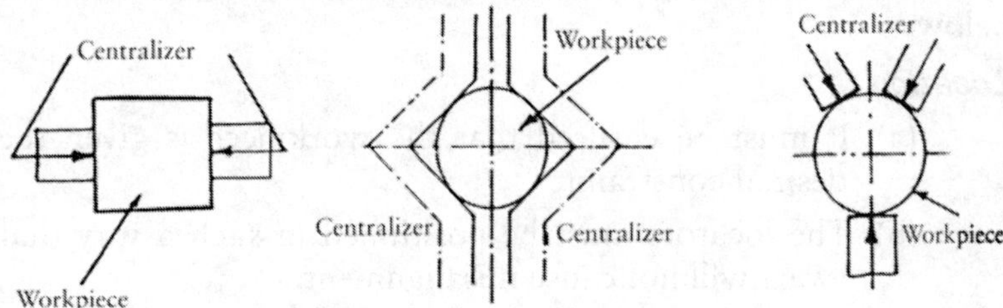

Fig. 13: Centralizers

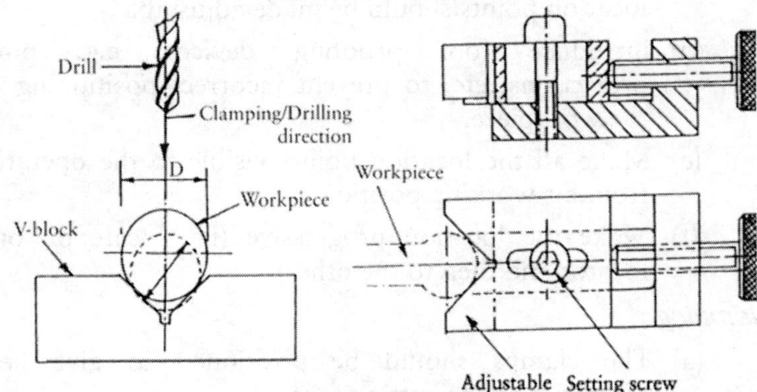

Fig. 14: Angular blocks

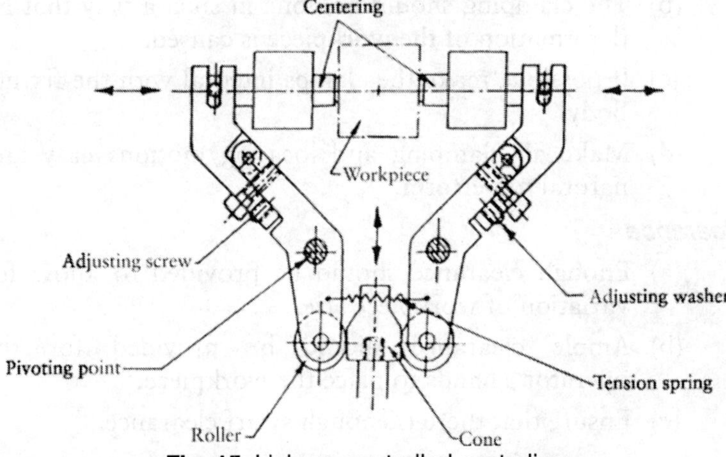

Fig. 15: Linkage controlled centralizer

6. Principles of Jig and Fixture Design

The basic principles of jig and fixture design are listed below:

Location

(a) It must be ensured that the workpiece is given the desired constraint.

(b) The locators must be positioned in such a way that swarf will not cause misalignment.

(c) If a rough casting or forging is being machined, the location points should be made adjustable.

(d) Introduce fool proofing devices, e.g. pins, projections, etc. to prevent incorrect positioning of the workpiece.

(e) Make all the location points visible to the operator from his working position.

(f) Make the location progressive (i.e. locate on one locator and then to the other).

Clamping

(a) The clamps should be positioned to give best resistance to the cutting forces.

(b) The clamping should be done in such a way that no deformation of the workpiece is caused.

(c) If possible, make the clamps integral with the fixture body.

(d) Make all clamping and locating motions easy and natural to perform.

Clearance

(a) Enough clearance should be provided to allow for variation of workpiece size.

(b) Ample clearance should be provided for the operator's hands to place the workpiece.

(c) Ensure that there is enough swarf clearance.

Stability and Rigidity

(a) Provide four feet so that uneven seating will be avoided.

(b) Make the equipment as rigid as is necessary for the operation.

(c) Provide means of positioning and bolting the equipment to the machine table or spindle, if required.

Handling

Make the equipment light and easy to handle. It must be ensured that no sharp corners are present. If it is heavy, provide lifting points.

General

(a) Keep the design simple to minimise cost.

(b) Use standard parts as much as possible.

(c) Method of location and clamping should be such that idle time is minimum.

(d) Design for safety.

7. Purpose of Clamping Elements

The main function of clamping element is to ensure dependable contact between the workpiece and locating element and to prevent displacement and vibration of the workpiece during operation.

With the help of a supplementary clamping device, rigidity of the system is increased and it leads to higher accuracy and better finish.

In many cases, clamping devices are employed to ensure proper location and centering of the workpiece. In this way they serve as locating cum clamping devices like in 3-jaw universal chucks.

If a very heavy workpiece is to be machined, whose weight is greater compared to the forces developed in the cutting process, then no clamping device is used. Clamping devices are also useless if the loaded workpiece is deprived of all its degrees of freedom.

The point of application of clamping force is selected from the conditions of maximum rigidity and stability in clamping and minimum deformation of workpiece. For a very high accuracy, the additional requirement is constancy of the clamping force.

For small batches of work, simple clamps may be quite suitable but for high rates of production, quick acting clamps are essential.

8. Types of Clamp

Various types of clamp are discussed below in brief:

Lever Clamps

Figure 16 shows a solid heel clamp. In this, the heel is solid and is one of piece with the clamp. By rotating the clamp, the workpiece is loaded into the jig or fixture or removed from these. It will be seen that tee and heel are shaped to ensure adequate clamping over a range of workpiece height. The clamp is prevented from rotating during clamping by the pin at the heel end.

In a loose guided heel clamp as shown in Fig. 17, a lever clamp guided by a loose heel is driven into the fixture body. The short stem turned on the end of the heel fits loosely into a keyway in the clamp bar. The clamp is simply reciprocated by hand, for loading and unloading the workpiece.

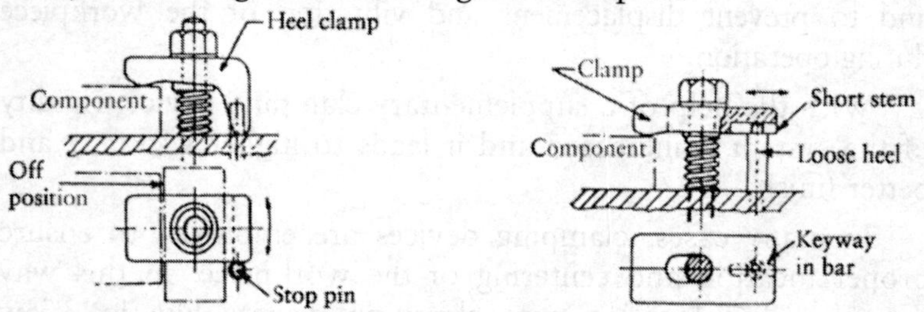

Fig. 16: Solid heel clamp **Fig. 17:** Loose guided heel clamp

Figure 18 shows a more elaborate system in which a slotted clamp plate is used, so that the workpiece can be released without clamp rotation. Simple reciprocatory motion loads and unloads the workpiece.

Figure 19 shows the use of an adjustable heel pin, which is quite often used at early machining operations, where the raw material workpiece height is likely to vary considerably.

When a fixed heel pin is used, variation of workpiece height may cause insecure clamping by the nuts. This can be offset by using a pair of spherical washers (Fig. 20). Spherical washers are sold by manufacturers of standard jig and fixture parts.

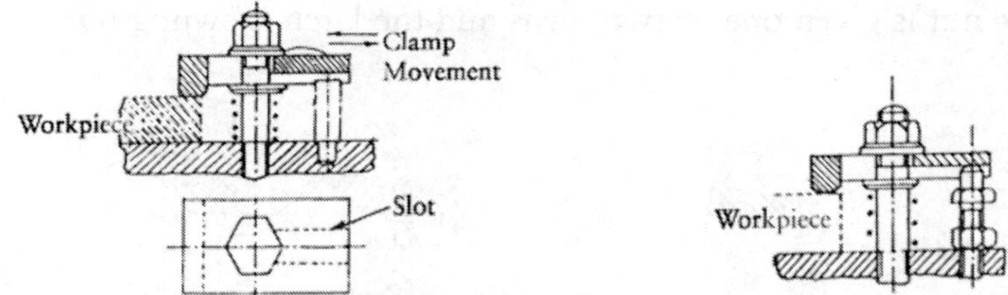

Fig. 18: Sliding clamp with heel **Fig. 19:** Clamp with adjustable heel pin

C-Clamps

'C' clamps are shown in Fig. 21(a) and (b). In these clamps, once the clamp is removed or swung away, the component can pass freely over the nut. It means locking nut needs about one turn to lock or unlock the clamp. So it is quick acting. The plain washer can be secured to the fixture by a small chain to prevent it from being lost.

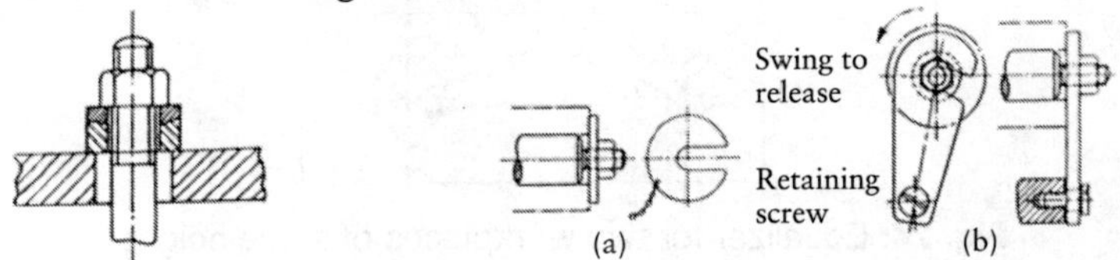

Fig. 20: Spherical washer **Fig. 21:** 'C' clamps

Quick Acting Nut

A simple and effective clamp is a nut with interrupted threads (Fig. 22). The length of nut is about 2 to 3 times the thread diameter and diameter of clearance is bigger than the outside diameter of the threads. Also, the hole axis is inclined at angle of 3° to 7° with the axis of the nut. When the nut is assembled over the male thread, it is inclined to the axis of clearance hole. When the nut engages, it is dropped on to screw threads and is tightly locked by giving it about half a turn.

Latch Type Clamp

Figure 23 shows a latch type clamp. This type is very easy to operate. Latch and stud movements are controlled. The latch is in the fully open position when the faces indicated by X are in contact. A hand nut clamps the workpiece. To unload, the nut is given one or two turns and the latch is swung out.

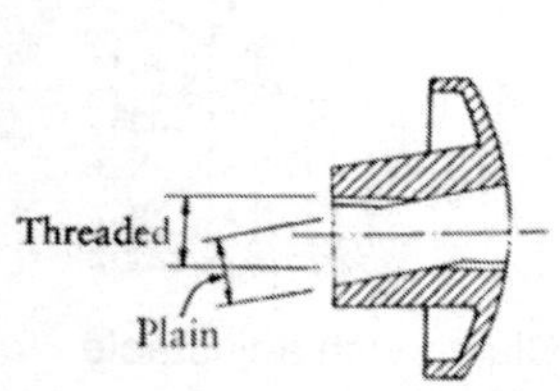

Fig. 22: Quick acting nut

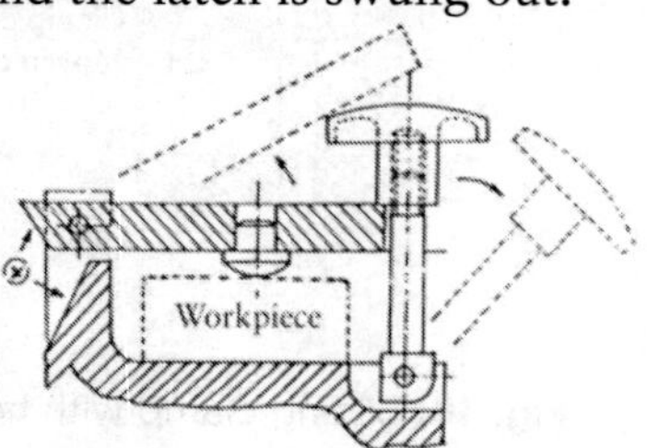

Fig. 23: Latch type clamp

Equalizer

The equalizer is meant to distribute a single force to several points on the workpiece. When two workpieces of the same height are clamped by a strap and center screw, then this type is a simple equalizer and clamp at the same time (Fig. 24).

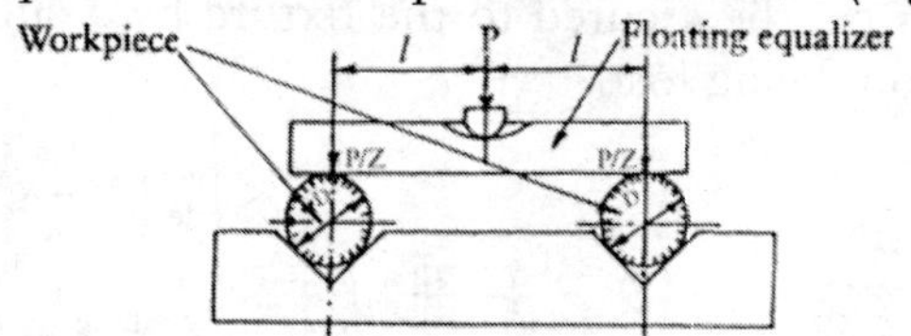

Fig. 24: Equalizer for two workpieces of same height

Many centralizers are also serving as equalizers. For example an angular V-block which holds a cylinder is centralizing it and equalizes the clamping force. Basically centralizer forces the workpiece into a certain desired position. It controls the motion and position of locating points. The equalizer is forced into a certain balanced position through the workpiece by clamping force. Self-adjustment is possible because the equalizer is made floating (Fig. 25).

Equalizers are used for the following situations:

- To distribute one clamping force equally to two or more points on a workpiece.
- To distribute one point clamping force evenly over a workpiece.
- To clamp on rough surface or on surface of different heights.
- To center and clamp one of several workpieces.

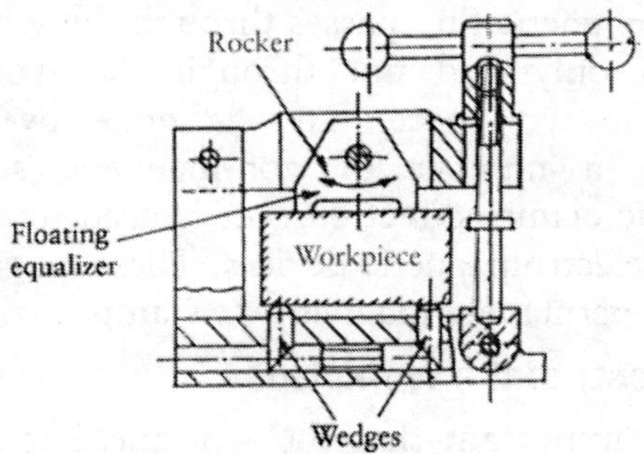

Fig. 25: Equalizer

Other Types of Clamping Systems

A number of other clamping devices are discussed below in brief:

(i) *Pneumatic clamps*: In pneumatic clamps/pressurised air is used for clamping the workpiece. The compressed air generally used in the production shops has a pressure from 5-6 kg/cm^2. This type of clamp ensures quick, uniform and constant clamping force but this clamp occupies comparatively large space.

(ii) *Hydraulic clamps*: These type of clamps make use of fluid under pressure. Compared to pneumatic pressure, these clamps operate at higher pressures (up to 80 kg/cm^2).

(iii) *Vacuum clamps*: It is quite difficult to hold thin flat sheets by conventional clamps as the surface of sheets may get distorted. For this type of situation, Vacuum clamping can be done to hold the sheet effectively.

(iv) *Magnetic clamps*: Magnetic clamps utilize the attractive force of the magnets for clamping purpose. The magnets can be either permanent or electromagnetic.

In permanent magnet type clamp, the workpiece to be clamped is placed on the work surface of the clamp. Below the working surface, there are a number of permanent magnets. A lever is used to operate the clamp. When the lever is in 'ON' position, the magnetic flux passes through the workpiece to complete the magnetic circuit. When the lever is in 'OFF' position, the magnetic flux passes through the working surface of the clamp only and not through the workpiece, thus unclamping the workpiece. This is done by aligning the magnets with a number of non-magnetic separators. In electromagnetic clamp, direct current is used for clamping the workpiece on electromagnetic devices. These magnets are more powerful than permanent magnet type clamps.

9. Support (Rest) of the Workpiece

It is quite important that the workpiece to be machined rests well on some base plate or support. If the workpiece is supported on a plain plate, then some dirt grooves should be made on the plate as shown in Fig. 26. This helps a lot to keep the plate clean, because the dirt (mixed with oil, etc.) can be wiped and absorbed in these grooves. The contact area between the workpiece and the plate should be kept as minimum as possible. A good alternative to the dirt groove is inserted small rails, as shown in Fig. 27. For rough parts, it is recommended to give three-point support by fixed pins, buttons or adjustable screws.

10. Drill Bushes and Jigs

Drilling jigs are used during drilling of holes in products. In drilling jigs actual guide for the drill is in the form of an accurately located and precise hole in a hardened steel piece called drill bush. Drill bushes are used to locate and guide drills, reamers, etc., to get positional accuracy of holes. The drill bushes are usually press fitted into a mild steel plate. The portion of the jig into which these bushes are fitted is called bush plate.

Drilling jigs are either clamped on the workpiece or the workpiece can be housed in the Jig. If the number of holes to be produced is more than one, the drill jig is moved on the machine table by hand and is brought to position under the drill. If the torque is not high, the jig is held by hand, otherwise stop is used.

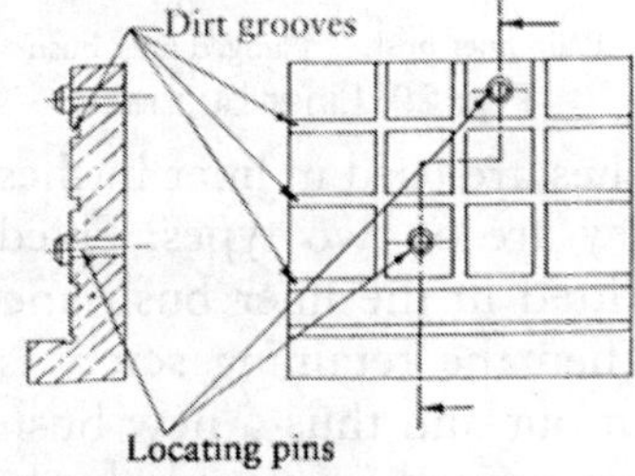

Fig. 26: Grooved plain plate

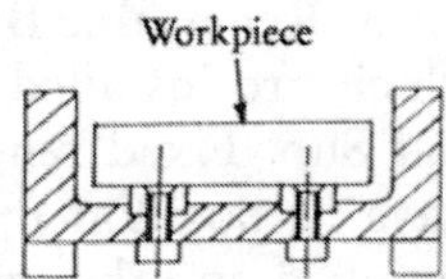

Fig. 27: Rail support

Drill Bushes

Drill bushes may be classified as follows:

1. Fixed bushes: (a) Plain, (b) Headed (or flanged)
2. Liner bushes: (a) Plain, (b) Headed
3. Renewable bushes
4. Clamping bushes

1. Fixed Bushes: A plain fixed bush is mainly used when a hole is produced with only one tool. These are employed when centre distances of holes are too close to permit the fitting of liners and renewable bushes. A flanged bush permits an adequate entry radius to be employed. It also enables the length of the bush to be increased relative to the depth of the bush plate, if the latter has to be restricted. Finally a head enables the bush to be used as a stop for the tool. Both types of fixed bushes are shown in Fig. 28. The outer diameter is generally press fitted into the jig plate, i.e. H7/m6 fitting. The inner diameter of the bush should be in the F7 tolerance, but not exceeding 0.1 mm.

2. Liner Bushes: These bushes serve two purposes. Their main job is to act as a hardened guide for renewable bushes. They may also be used to guide tools. They are similar in form to fixed bushes. These bushes can be with or without heads. A liner bush is always used in conjunction with a renewable bush. Figure 29 shows the plain and headed liner bushes.

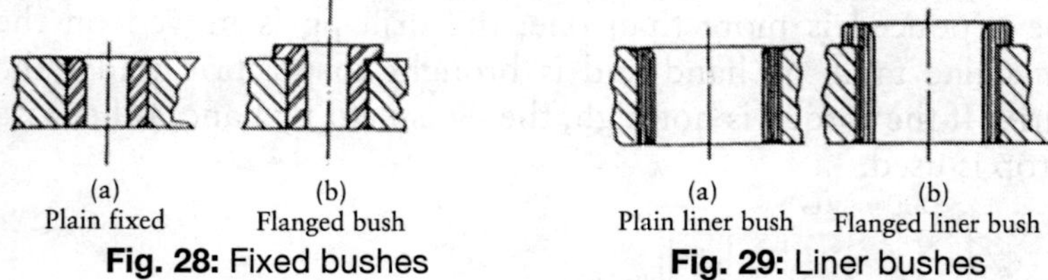

Fig. 28: Fixed bushes **Fig. 29:** Liner bushes

3. Renewable Bushes: These bushes are used in liner bushes which are installed in jig plate. They are of two types: Fixed and Slip. Fixed renewable bush is fitted in the liner bush and stays in position until worn out. Then the retaining screw is removed and the worn bush is taken out and thus a new bush is easily substituted. Figure 30 shows a fixed renewable bush. When any hole requires two or more tools to be used, then it is necessary to supply a slip bush for each tool. Renewable bushes are interchangeable in a given size of liner.

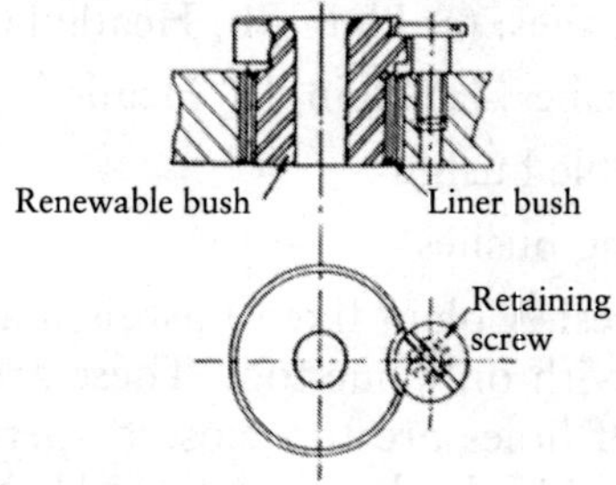

Fig. 30: Fixed renewable bush

For easy removal, its head is knurled. The renewable bushes are prevented from rotation or getting lifted with the drill by the help of screw head as shown in Fig. 31.

4. Clamping Bushes: If a bush is provided with a thread on its outside diameter, it may be used for clamping and locating a component. The clamping bush is shown in Fig. 32. The eccentricity between the thread of the bushing and the thread in the jig plate may result in hole being drilled out of centre. Again rapid wear of the threads in the plate will cause inaccuracy in the exact location of the hole to be drilled.

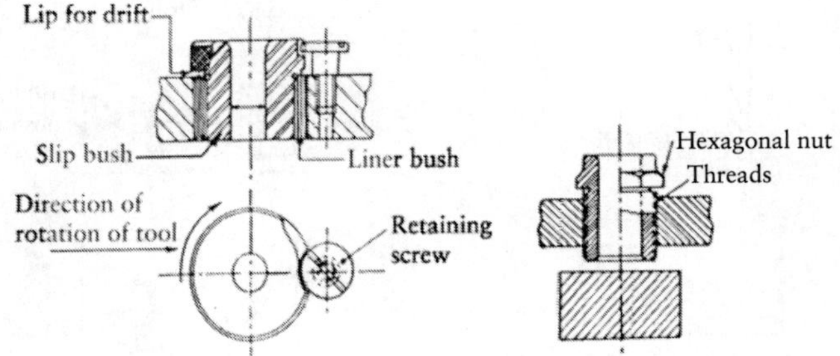

Fig. 31: Slip bush

Fig. 32: Clamping bush

Clamping bushes have the following advantages:

- They are easy to machine accurately because they are round.
- Centre-distance in the mild steel plates can be made accurately, either by jig boring or on the milling machine. Since no heat treatment is done on the plate, the centre distance remains unchanged.
- In case the bush is worn out it can be replaced quickly at low cost.

A drilling jig might be a simple plate with one or several bushes. But it can also be a very intricate, complex device offering to machine the component in one setting from all the sides. Economical aspect has to be considered. Sometimes it is more economical to produce 3 simple jig plates than one complicated jig. This all depends upon the quality and number of the components as well as equipment available to make the jig.

The drilling jigs can be classified as follows:

(i) *Template Jig*: When plain holes are to be drilled a simple plate jig having required number of holes made in it, is used. Especially when the holes are closely spaced, the plate is used directly for guiding the tools. The plate is hardened in this case. This jig is used if only a few parts are to be made. A template jig is shown in Fig. 33.

(ii) *Plate Jig*: It is different from the template jig in the sense that in this case, drill bushes are incorporated in the holes of the plate as shown in Fig. 34. These bushes when worn out can be replaced.

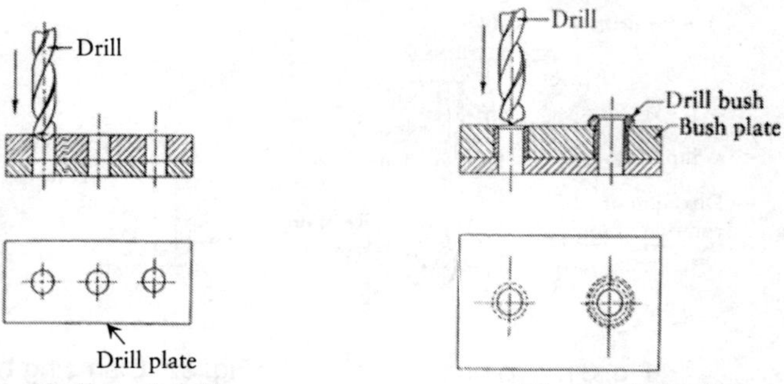

Fig. 33: Template jig **Fig. 34:** Plate jig

(iii) *Swinging Latch Jig*: A drill Jig which is fitted with a pivoted plate (latch or lid) in which one or more drill bushes are fitted is called a swinging latch jig. The plate is arranged to swing about the pivoted point and it eventually provides means of entry for loading and unloading the component.

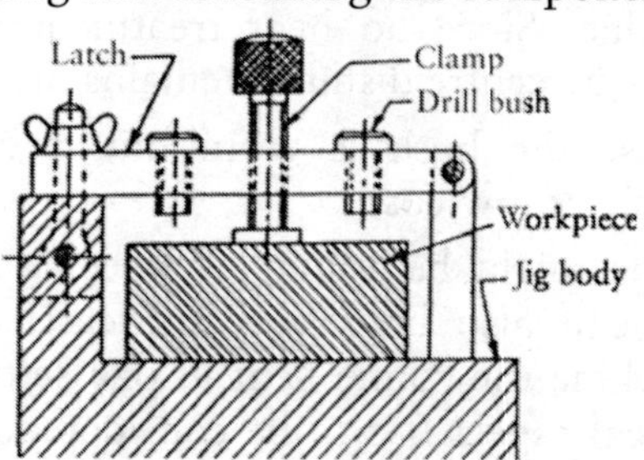

Fig. 35: Swinging latch jig

(iv) *Box Jig*: When drilling is to be done in more than one plane in a component, the jig is provided with same number of drill bush plates as the number of planes. It means a box type jig is needed as shown in Fig. 36. Each bush plate has got a foot opposite to it. One side of the box is fitted with a latch to provide opening for loading and unloading of the component.

While designing a box jig, it should be kept in mind that the body of the jig should be as light as possible, as any extra weight of the jig will lead to unwanted fatigue along with reduction in productive efficiency. Also special care should be given for chip removal in this jig.

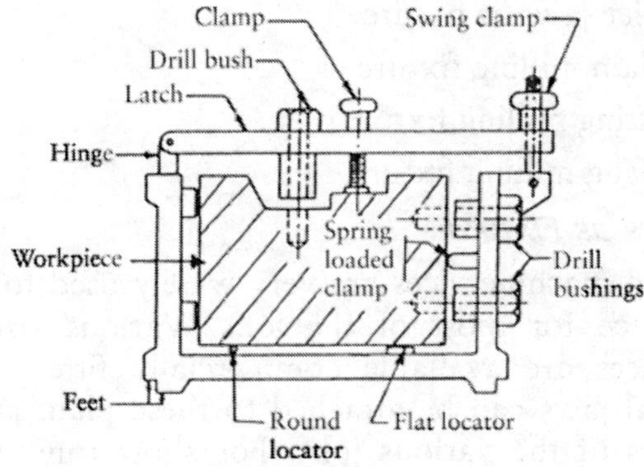

Fig. 36: Box jig

(v) *Pot Jig*: Figure 37 illustrates a pot jig. This jig is suitable for use with circular components, which have both an external and internal diameter. The body of the jig is in the form of a pot and it carries the workpiece. The bush plate is carried by the workpiece. The body is fitted with liner bush, which locates the component.

The drill bush is located on a post. The workpiece is supported to the point of drilling and swarf clearances are provided. The drill plate is located to line up with the swarf clearance grooves.

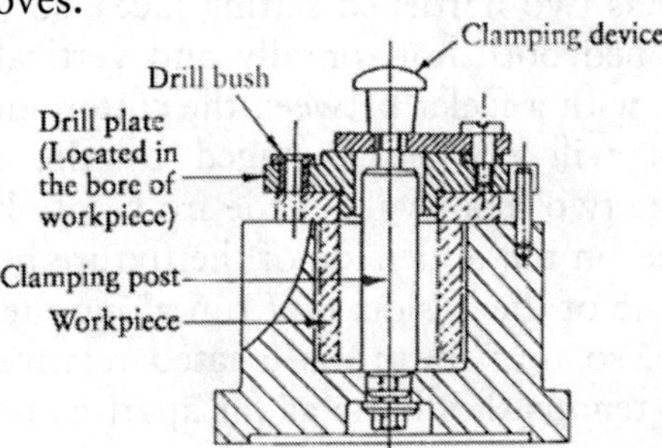

Fig. 37: Pot jig

11. Fixtures for Milling Operations

Milling is one of the most important production processes and introduction of form relieved cutters has made it possible to use milling machine for mass production. Fixtures used to hold the components during milling can be classified as:

1. Vice jaws as fixture
2. Plain milling fixture
3. String milling fixture
4. Gang milling fixture

1. *Vice Jaws as Fixture*

Standard machine vices are very widely used for clamping the workpiece for most of the jobs. Various sizes of cam operated vices are available commercially fitted with plain jaws. Special jaws can be attached to these plain jaws to suit the contours of the various jobs. For short runs, the special jaws may be made of suitable sections of aluminium alloy or soft mild steel. For long runs, all the wearing parts of the special jaws should be hardened and should be of renewable type. Vice jaws have their limitations, the criterion usually being the size and the complexity of the workpiece.

2. *Plain Milling Fixture*

It is a specially designed fixture used to hold a single component. Figure 38(a) shows the arrangement for a simple fixture used to hold the workpiece. The setting block is located relative to the fixture location system and retained in position by screws. It has two hardened setting faces, so that the table can be positioned both horizontally and vertically. The table setting is done with a feeler between the cutter and setting face, so that block will not be damaged by the cutter during machining. The two tenons are made from case hardened steel and are located on the underside of the fixture box. These two tenons sit in one of the T-slots that run along the length of the machine table, so that fixture is located relative to the table feed. The two tenons should be as far apart as possible to give maximum accuracy. The Fixture is boled to the table by two or four tee bolts that are placed in the T-slots.

3. *String Milling Fixture*

When a number of component are kept one behind each other, in a line, the fixture is called a string or line milling fixture. Fig. 39 shows a string milling fixture for milling six pins at a time. The pins are strung behind each other. The fixture moves relative to the cutters, so two factors have to be considered. First: how many components should be accomodated in the row, second: how the components can be brought as close as possible to avoid cutting air idle time during machining operation. The problem is discussed for circular components.

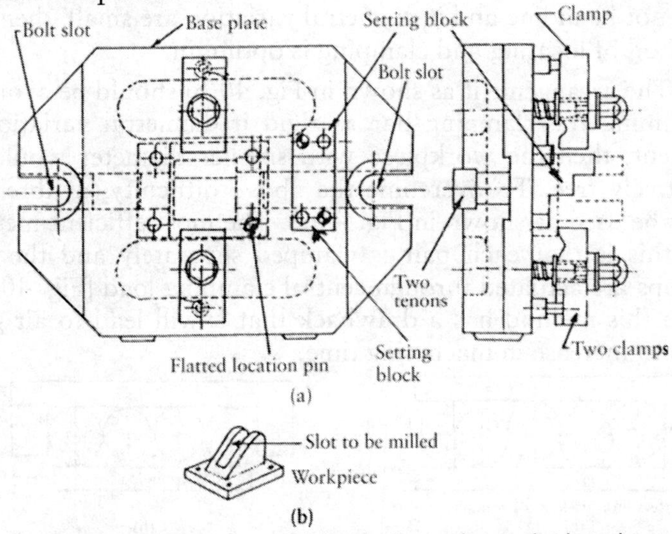

Fig. 38: (a) Plain milling fixture, (b) Workpiece whose slot is to be milled

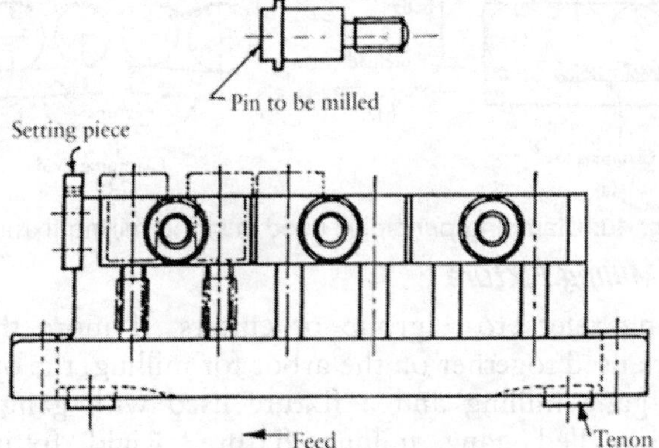

Fig. 39: String milling fixture for pin

Figure 40 shows the conditions affecting the clamping together of a number of pieces (of circular cross-section) when diametral variation is present. If the workpieces are clamped in a channel, Fig. 40(a), they will be fully constrained when each component has linear contacts at three points. If the diametral variations are present, then centre line of various components will not be in line and if diametral variation are small, then this method of locating and clamping is optimum.

The arrangement as shown in Fig. 40(b) should be avoided. Assuming the clamping bar as rigid if diametral variation is present, then the workpiece with smaller diameter would be relatively free. To overcome the above difficulty, a fibre pad may be used as shown in Fig. 40(c). The most efficient method for this is that each pair is clamped separately and the two clamps are actuated through central clamping load [Fig. 40(d)]. Even this method has a drawback that it will lead to air gaps thus to increase in machining time.

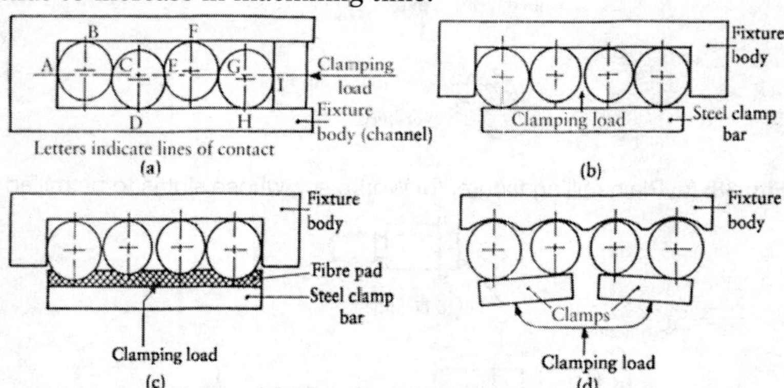

Fig. 40: Clamping principles used in string milling fixture

4. *Gang Milling Fixture*

A gang refers to a group or cutters. If more than two cutters are used together on the arbor for milling, the operation is called gang milling and a fixture used with gang milling cutters is called gang milling fixture. Rigid fixtures are designed to absorb forces due to heavy cutting load.

Figure 41 shows a gang milling fixture for milling pins. It is not necessary to try to keep the components as close together as possible but in this case, it is more important to ensure that the large wall of metal behind the component takes the load of the cut. The number of components that can be machined together in a gang milling operation is determined by the width of machine table. An important factor is possible deflection of arbor due to its loading. The arbor must run truly to get a high quality finish and accuracy, which is done by amply supporting the arbor, otherwise the number of cutters employed will have to be reduced. Swarf disposal is also taken in account, as there is enough material removal during the process.

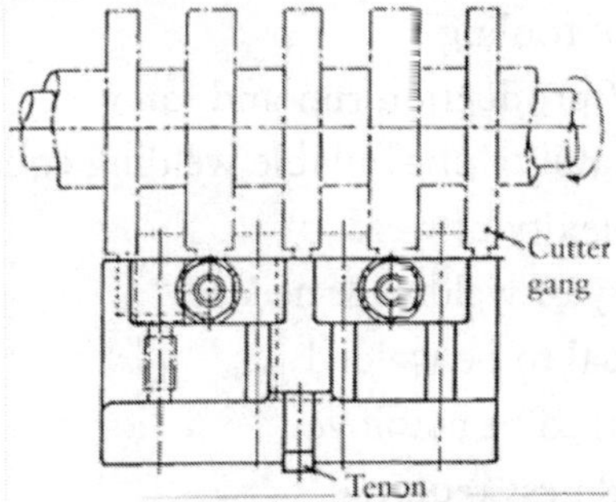

Fig. 41: Gang milling fixture for pin

12. Welding Fixtures

Joining processes may require some arrangement for tooling to hold the parts in correct relationship during joining. Another function of tooling is to assist and control the joining process. Generally, several parts are joined both mechanically and physically, e.g. two workpieces may be bolted together to assure alignment during welding. Before discussing various types of welding fixtures it is essential to know about objectives of welding fixture design and the factors affecting the welding fixture design.

Design of Welding Fixtures

The purpose of a welding fixture is to hold the parts to be welded in the proper relationship both before and after welding. Sometimes, a fixture will maintain proper part relationship during welding but the part will distort after removal from the fixture. Major fixture design objectives are:

1. To hold the part in the most convenient position for welding.
2. To provide proper heat control of the weld zone.
3. To provide suitable clamping to reduce distortion and warping.
4. To provide clearance for filler metal.
5. To provide for ease of operation and maximum accessibility to the point of weld.

The factors affecting the fixture design are:

1. Cost of tooling
2. Size of production run and rates
3. Adaptability of available welding equipment
4. Complexity of weld
5. Quality of weldment needed
6. Material to be welded
7. Process to be employed
8. Smoothness required
9. Conditions under which welding will be performed
10. Thermal conductivity of both workpiece and tool material.

Depending on the type of welding being done, various types of welding fixtures are discussed as follows:

i. Gas Welding Fixtures

The heating and cooling conditions are considered while designing gas welding fixture. A minimum of heat loss from the welding area is required, because rapid heat loss leads to cracks in weld. To do this, large fixture masses should not be placed close to the weld line. If the fixture supports are too far from the weld line, the part may distort. The contact area and clamps should be of minimum size, consistent with load transmitted through the contact point.

One gravity type fixture for gas welding is shown in Fig. 42. This design eliminates excess fixture material from the weld area to minimize heat loss, while providing sufficient support.

Figure 43 shows another gas welding fixture which holds two flat sheets for joining. C-clamps hold the workpieces to support bars. Alignment is done visually. Hold down plates are used to keep the workpices flat and to prevent distortion. C.I., carbon steel and stainless steel are used as fixture material.

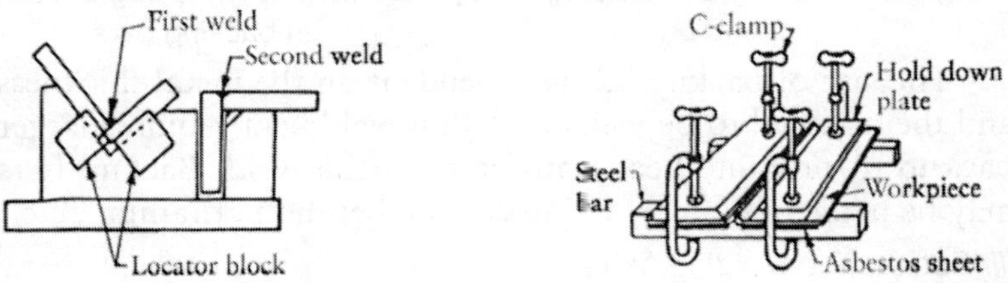

Fig. 42: Welding fixture using gravity to help locate parts

Fig. 43: Fixture for gas welding operation

ii. Arc Welding Fixtures

Arc welding concentrates more heat at the weld line than gas welding. The fixtures must provide support alignment and restraint on the parts and must permit heat dissipation. The arc welding fixture is shown in Fig. 44.

The important design consideration for arc welding fixtures are:

(i) The fixture must exert enough force to prevent the parts from moving out of alignment during welding process.

(ii) Backing bars should be parallel to the weld lines.

(iii) Backing bars should promote heat dissipation from the weld line.

(iv) Backing bars should support the molten weld and protect the weld root from the atmosphere.

Backing bars are made from solid metal. A simple backup could be a rectangular bar with a small groove directly under the weld. Generally the backup would be clamped against the part to make the weld root as airtight as possible. Some common groove shapes are shown in Fig. 45.

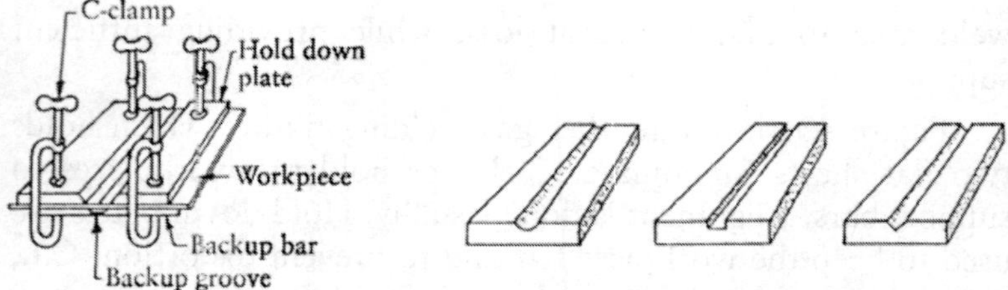

Fig. 44: Fixture for arc welding operation

Fig. 45: Different types of groove in backing bars

The size of backup bar is dependent on the metal thickness and the material to be welded. A thin weldment requires larger backup to promote heat transfer from the weld. Backup bars may be made of copper, stainless steel, berylium, titanium, etc.

iii. Resistance welding fixtures

The design consideration applied to the fixtures for resistance welding are:

(i) Keep all magnetic materials out of the throat of resistance welding machine.

(ii) Insulate all gauge pins, clamps, locators, index pins, etc.

(iii) Protect all moving slides, bearings, index pins, adjustable screws from flash.

(iv) Ease of operation and operator's safety.

(v) Provide enough water for circulation inside electrodes to avoid overheating.

(vi) Stationary parts of fixture and work are affected by the magnetic field of the machine. This aspect should be considered.

General Fixture Design Considerations

Simple fixtures may have the part located visually with scribed lines as a guide. This is quite similar to locating parts for gas welding. For higher production a quicker locating method is needed. A locating land may be used in the fixture to locate the edge position of the part to be welded (Fig. 46). Dowel pins, V-blocks can also be used as locators.

13. Materials for Jigs and Fixtures

The material of a component of jig or fixture depends upon the use and stresses it undergoes while in duty.

The essential components of jigs and fixtures and its materials are as follows:

1. Locating devices: Like locating pins, locating Vee, rest buttons have hardened surfaces.
2. Clamping devices are made of high carbon steel to withstand the clamping loads.
3. Tool guiding devices like drill bushes, liner bushes are generally made from hardened, silver steel, cast steel, case hardened mild steel.
4. Body is fabricated from cast iron or mild steel.

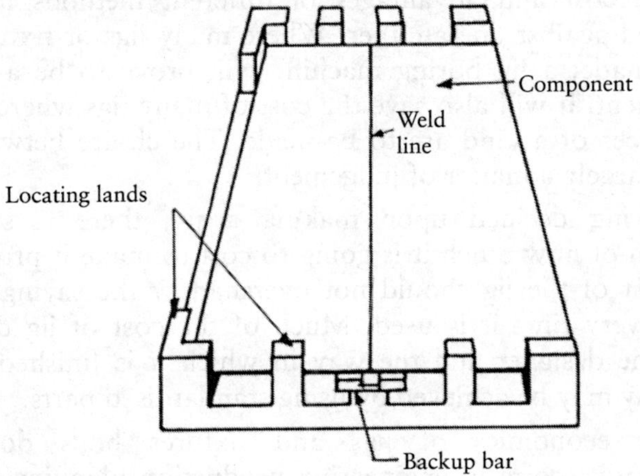

Fig. 46: Locating lands

14. Economics of Jigs and Fixtures

The uses of jigs and fixtures must be considered in determining whether jigs or fixtures should be made for a given job as well as the amount of money that can be spent on them. Even where the number of pieces to be made may not seem to warrant the cost of a jig, the saving in time in assembly, as well as the proper functioning of the machine itself, may make it necessary to make a jig to secure duplication of parts and of machine performance.

In case of a limited number of pieces to be made its importance may be considered in comparison to the growing application of the jig boring machine and of other boring machines which have been designed to permit the accurate location and boring of holes.

Designed for the purpose of securing accuracy in making of jigs at a reduced cost, the jig borer or its equivalent is now used to make jigs unnecessary in small lot work. These machines provide a very accurate method of locating holes in relation to each other and of boring them to size within very close tolerances. These machines can be used to an advantage as work can be duplicated at a future time if necessary.

The cost and advantages of different methods must be balanced against one another. Where many jigs or fixtures are to be made, a jig boring machine will prove to be a paying investment. It will also save the cost of many jigs where only a few pieces of a kind are to be made. The choice between the two is largely a matter of judgement.

Having decided upon making a jig, there is still the problem of how much it is going to cost to make it profitable. The cost of the jig should not overshadow the savings to be made every time it is used. Much of the cost of jig depends upon the designer and the way in which it is finished. Some economy may be achieved by using standardised parts.

The economics of jigs and fixtures boils down to comparative costs in engineering production planning. Before selecting the equipment and tools to be used it is necessary to have details of full production as follows:

(i) Rate or production—number of components to be produced in unit time.

(ii) Total period allocated for the product completion.

(iii) Alternative schemes involving different kinds of expenditure.

(iv) Life of equipment.

(v) Depreciation on equipment.

(vi) Obsolescence of equipment.

(vii) Renewable standard components.

15. Practical Case Studies

The following example of different methods of machining a component are intended to show how to analyse each scheme relative to a minimum number of machined components. The methods can then be compared and the most economic scheme selected. If a number of components require a milling operation, one of the following methods may be adopted:

(i) The components may be machined, using standard equipment available, as, for instance, a milling machine vice, holding one job at a time.

(ii) A special fixture may be designed and manufactured for machining one component at a time. The positioning of job is automatic.

(iii) A string milling fixture may be employed holding six components in line. The six components are positioned simultaneously and no adjustment is needed. There are six components and one cutter.

(iv) A special fixture may be used for gang milling six components together. This involves six components and six cutters.

Each of these methods uses a different technique, and it is probable that the cost of each method will also differ. Furthermore, it is possible that a heavier type of machine would be required for scheme (iv), but this factor is neglected to simplify the problem.

Comparison

Scheme 1: With this scheme no new equipment will be required, but each component is required to be set up individually. Let the time taken to set up and machine each component = 30 min.

Scheme 2: The advantages of a specially designed milling fixture will result in a reduction of operating time, but not necessarily of cutting time. The initial setting up time will also be reduced. Therefore the time to produce one piece will be reduced, compared with Scheme 1. The cost of the fixture will have to be considered. Assume the cost of fixture = ₹ 1000 and the time taken per component to be 9 min as against 30 min taken in Scheme 1.

Scheme 3: The provision of a string milling fixture will reduce the time taken per piece still further as compared with Scheme 2, but with increase in cost due to more elaborate equipment. Assume the cost of equipment to be ₹ 2200 and time taken is reduced to 7 min.

Scheme 4: The fixture required for gang milling 6 components at a time will require to be of a more complex type than for Scheme 3, specially if it is pneumatically operated. Five additional cutters, or sets of cutters will also be required. The equipment will therefore cost considerably more than needed for Scheme 3. Assume the cost to be ₹ 5000 and time taken per component to be correspondingly reduced to 1 min.

If ₹ 50 per hour be taken as running cost of factory covering all factory charges, the cost of each scheme can be compared as follows:

***Scheme* 1**

Cost of operation of one component @ 2 components per hour = ₹ 25.

***Scheme* 2**

Cost of operation of one component @ 7 components per hour = ₹ 7.15.

The saving is ₹ 25 – ₹ 7.15 = ₹ 17.85.

The cost of fixture is covered when $\frac{1000}{17.85} = 56$ components have been machined.

***Scheme* 3**

Cost of operation of one component @ 9 components per hour = $\frac{50}{9}$ = ₹ 5.56.

The saving is ₹ 25 – ₹ 5.56 = ₹ 19.44

The cost of the fixture is covered when $\frac{2200}{19.44}$ = 113 components have been machined.

***Scheme* 4**

Cost of operation of one component @ 60 components per hour = $\frac{50}{60}$ = ₹ 0.84

The saving = ₹ 25 – 0.84 = 24.16

The cost of the fixture is covered when $\frac{5000}{24.16}$ = 207 components have been machined.

REVIEW QUESTIONS

1. What are jigs and fixtures? What are the main differences between a jig and a fixture?
2. What are the uses of jigs and fixtures? How are they helpful in mass production?
3. What are elements of a (i) jig, (ii) fixture?
4. What is the six-point location principle? Explain.
5. Explain the principles involved in design of jigs.
6. Explain the principles involved in the design of a fixture.
7. What are drill bushes? What purpose is served by them? List various types of drill bushes.
8. Describe the various types of drilling jigs giving suitable sketches.
9. Write short notes on:
   (i) Milling Fixture (ii) Welding Fixture

OBJECTIVE TYPE QUESTIONS

Choose the correct alternative

1. The device in which a component is held and located for a specific operation in such a way that it will

guide one or more cutting tools during machining is known as

(a) Jig (b) Gauge (c) Fixture (d) Template

2. The production tooling equipment which locates and holds the component for a specific operation (but it does not guide the cutting tool) is knows as

   (a) Template (b) Gauge (c) Fixture (d) Jig

3. ______ are used to locate and guide the cutting tools such as twist drills, reamers and counter boring cutters.

   (a) Jig bushes (b) Drill jigs
   (c) V-blocks (d) Locating pins

4. A ______ can hold and locate the work and guide the drill at the desired position.

   (a) drill bush (b) metal pin locator
   (c) drill chuck (d) drill jig

5. Devices designed for locating and holding cutting tools are usually called

   (a) Gauges (b) Fixtures
   (c) Templates (d) Tool-holders

6. A workpiece in space free to move in any direction can have

   (a) 3 degree of freedom (b) 6 degree of freedom
   (c) 9 degree of freedom (d) 12 degree of freedom

7. The application of jigs and fixtures

   (a) Increases the production
   (b) Widens the technological capacity of machine tools
   (c) Either fully or partly automates the machine tools
   (d) All of the above

8. The work holding device commonly used while machining a key way in a cylindrical shaft on a shaper is

(a) A vice (b) A mandrel
(c) V-block (d) An angle plate

9. Most of the locating and clamping devices are made up of
(a) H.S.S. (b) Aluminium
(c) C.I. (d) Steel

10. Clamping should always be arranged
(a) Directly above the points supporting the work
(b) Directly above the points supporting the tool
(c) Directly below the points supporting the work
(d) None of the above

11. If the diameter of the hole is subject to considerable variation, then use
(a) Conical location (b) Vee-block
(c) Pin (d) Plain locating pin

12. For drilling holes in a long strip, use
(a) Channel jig (b) Plate jig
(c) Leaf jig (d) Box jig

13. The simplest type of jig out of the following is
(a) Leaf jig (b) Plate jig
(c) Template jig (d) Box jig

14. To drill a series of holes on a circular face of the workpiece, use
(a) Leaf jig (b) Box jig (c) Pot jig (d) Index jig

15. Cylindrical job should always be clamped in a ______ for drilling
(a) collet (b) jaw (c) vice (d) v-block

16. The important property to be considered in the selection of suitable material for manufacture of locating pins used in jigs and fixture construction is
(a) Shear strength (b) Tensile strength
(c) Elasticity (d) Wear resistance

17. In the selection of a suitable material for the manufacturing of a drill jig bush, the critical property to be considered is
    (a) Tensile strength (b) Elasticity
    (c) Wear resistance (d) Shear strength
18. Typical locating devices used in Jigs and Fixtures construction are
    (a) Drill Jigs (b) V-blocks
    (c) Mandrels (d) Angle plates

ANSWER KEY

1. (a)	2. (c)	3. (a)	4. (d)	5. (d)
6. (d)	7. (d)	8. (c)	9. (d)	10. (a)
11. (a)	12. (b)	13. (c)	14. (d)	15. (d)
16. (d)	17. (c)	18. (b)		

11

Micro-Finishing Processes

1. Introduction

To increase the dependability and service life of machinery, stricter requirements in specifying the permissible surface roughness of various machined parts are required. Surface finish has a great impact on the properties of machined parts like lubrication, wear resistance, fatigue strength and corrosion resistance. If the finish is not good then due to excessive wear of rubbing surfaces, the oil film on the peaks of micro irregularities will get ruptured and it leads to dry friction.

Just like peaks, the valleys of micro irregularities also act as stress raisers and lead to fracture of parts. So microfinishing proceses are used for machining the surface of critical parts to get better surface finish and high accuracy. These processes include honing, lapping, polishing, buffing and superfinishing. Table 1 gives the surface finish attainable in grinding and other micro-finishing processes.

Table 1: Range of Height Irregularity in Different Processes

Process	Average Height of micro-irregularities (Micron)
Precise Turning	1.25 to 12.5
Grinding	0.9 to 5
Honing	0.13 to 1.25
Lapping	0.08 to 0.25
Super finishing	0.01 to 0.25

The micro-finishing process has less surface speed, the contact pressure is very small and the contact area is very large, as compared to grinding process. Various microfinishing processes are described below in detail.

2. Honing

Honing is an abrading process used mainly for finishing round holes by means of bonded abrasive stones called hones. Honing is primarily used to correct hole out-of-roundness, taper, tool marks and axial distortion. Various hole geometries that can be corrected by honing are shown in Fig. 1. Honing stones are made from common abrasive and bonding materials,

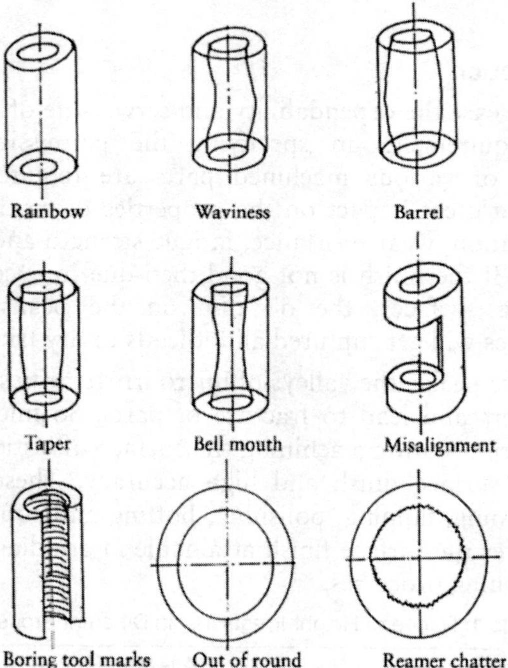

Fig. 1: Hole geometry that can be corrected by honing

often impregnated with ceramics, sulphur, rubber, resin wax to improve cutting action and lengthen tool life. The various abrasives used to make honing stones are silicon carbide, aluminium oxide, diamond or cubic boron nitride. Silicon carbide is used for honing cast iron and non-ferrous materials, whereas aluminium oxide is used to hone steel parts. Diamond is also used as an abrasive to hone parts made of ceramics or hard carbides. The abrasive grain size ranges from 80 to 600 grit. Almost every material can be efficiently honed. Steels of all varieties, cast iron, aluminium, magnesium, brass, bronze, glass, ceramics, hard rubber, graphite and silver are a few examples. Mostly honing is done on internal cylindrical surfaces, such as automobile cylinder walls. Holes and bores can be machined to high standards of dimensional accuracy and form accuracy.

When honing is done manually, the honing tool is rotated and workpiece is passed back and forth over the tool. Length of motion is such that the stones extend beyond the workpiece surface at the end of each stroke.

For precision honing, the work is usually held in a fixture and the tool is given a slow reciprocating motion as it rotates (Fig. 2). The stones are thus given a complex motion as rotation is combined with oscillatory axial motion. These two motions combine to give a resulting cross-hatch lay pattern. Honing stones may be held in the honing head by cementing them into metal shells, which are clamped into holder or they are cemented directly into holders. During honing operation, the honing head is not guided externally but floats in the hole being guided by the work surface. Coolants are essential to the operation of this process, to flush away small chips and to keep temperatures uniform. Kerosene oil, or sulphurized mineral oil is generally used for this purpose.

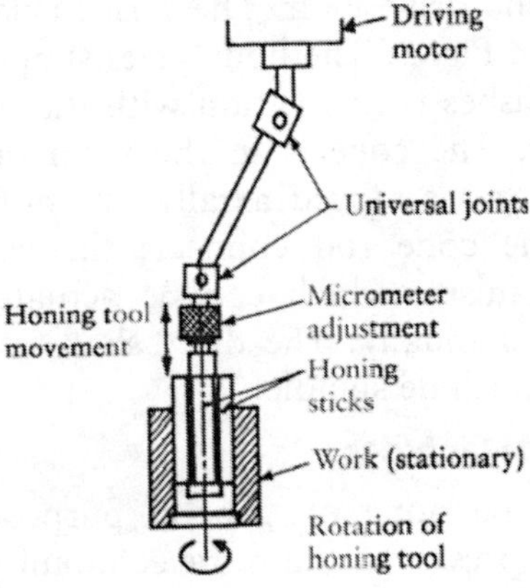

Fig. 2: Honing

Honing differs from grinding due to the reason that honing requires the use of low speed and low pressure, which keep the surface temperature relatively low, whereas grinding machines are run at high speed and high wheel pressure. Also abrasive stones used in the honing machine have a relatively large area of abrasive in contact with the job. As surface temperature is low, surface damage is kept to a minimum. The feature of honing, which allows it to develop round and straight bore is the relationship of cutting faces of the honing stones to the surface being honed. Either the tool or the fixture holding the work part floats. Due to this, the honing tool exerts equal pressure on all sides of bore. With honing process, it is possible to handle work parts having internal diameters from 1.5 mm to 15 mm. In honing, dimensional accuracy up to 2 microns is common. Surface finish up to 0.1 μm can be expected.

Cutting speeds employed for CI, structural steel and Copper-Tin alloys:

10 – 40 m/min,

for hardened steel:

5 – 25 m/min

Honing Tools

Various types of honing tools are used depending on the type of surface to be honed. Basically a honing tool used for honing an internal surface is made up of four elements—body, cone, cone rod and drive shaft. The various elements of honing tool are shown in Fig. 3. The body gives support to the honing sticks and establishes their relation with the work surface and with each other. The cone is in the form of a cone shaped wedge. As the cone is moved axially, the honing sticks move out radially. The cone rod connects the cone to the tool-expansion mechanism, which can be actuated mechanically, hydraulically or manually. The drive shaft is used to connect the body to the machine spindle.

Types of Honing Machines

Honing can be done on general purpose machine such as lathe, the drill press, etc. but more economical results can be

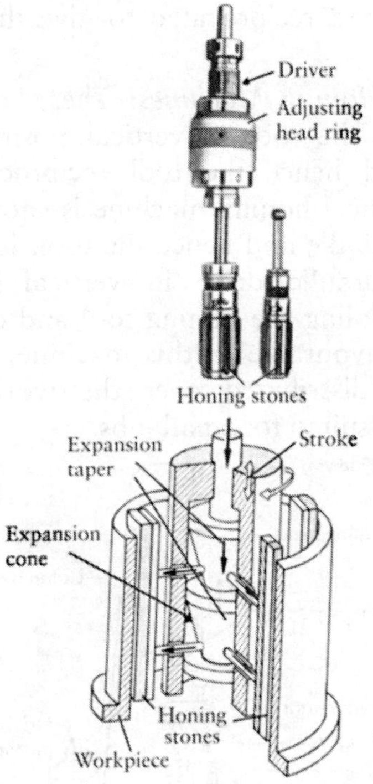

Fig. 3: Honing Tool

obtained by using honing machines for production work. The most common out of these are as follows:

(i) *Horizontal Honing Machines*: These machines are mostly used for honing comparatively longer jobs, such as gun barrells and long tubular parts. All such machines carry a horizontal spindle, on which honing tool in mounted. In some machines, the workpiece is mounted on a table which can move the work to and fro hydraulically. The hone reciprocates about its own axis and also simultaneously rotates. The motion of the honing tool may be controlled by an electric motor through the speed gear box, whereas for reciprocation electro-mechanical, rope or hydraulic drive is employed. In some machines, the work is held in a horizontal position and rotated about its own axis. No reciprocating motion is given to it. Against this the honing tool, which is mounted on a travelling head, is rotated and reciprocated to give the same results as above.

(ii) *Vertical Honing Machines*: These machines hold the work as well as the tool in vertical positions. Usually the spindle head and hence the tool reciprocate and not the workpiece. A vertical honing machine is shown in Fig. 4. The rotation of the spindle and hence the tool, is accomplished by means of a hydraulic drive in vertical honing machine. Conditions for cooling the honing tool and carrying away the chips are more favourable in this machine, as the coolant is more uniformly distributed over the work surface. These machines are best suited for small jobs.

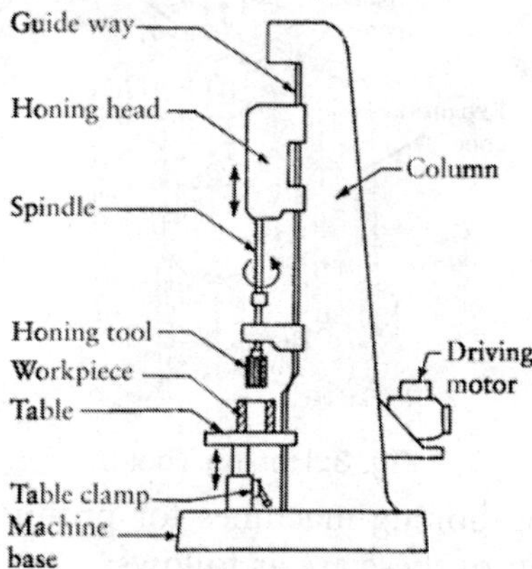

Fig. 4: Vertical spindle honing machine

In the latest honing machines, inprocess gauging equipment is incorporated to gauge the bore diameter automatically, throughout the honing cycle. When the desired diameter is obtained, a signal is generated which stops the expansion of the honing tool. After this, the pressure on the honing sticks gets reduced gradually and the tool is withdrawn from the bore. For controlling the bore size, two types of devices are used, depending on whether the tool or bore is gauged. In tool gauging, a tool having a gauging ring is used. The honing sticks of the tool have plastic tabs moulded on each end as shown in Fig. 5(a). These tabs wear down along with abrasives, therefore their outer diameter always remains equal to bore diameter. The gauge ring is positioned above the workpiece, so that the upper tabs enter the gauge ring at the top of each stroke. The inner diameter of ring is equal to the lower limit of the desired bore size. When the bore and the tabs acquire the same size as that of gauge ring the friction between the ring and the tabs causes the gauge ring, to turn, which subsequently generates a signal to stop the cycle.

In bore gauging [Fig. 5(b)] a floating sleeve is mounted around the honing tool whose outer diameter is equal to the required bore size. At the bottom of each stroke, the sleeve tries to enter the bore. Finally, when it enters the bore, a signal is generated that ends the honing cycle.

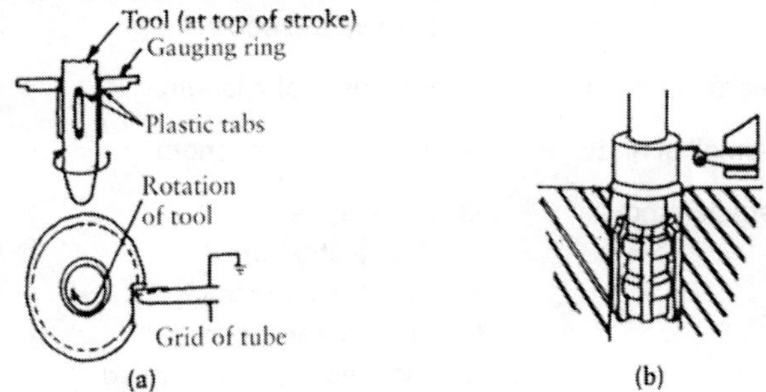

Fig. 5: Two methods for automatic size control of work (a) by gauging the tool, (b) by gauging the bore

Advantages of Honing

1. The honing process enables highly accurate holes, as the possibility of vibration is very less.
2. Many holes can be honed simultaneously on multiple spindle machines.
3. Hole of any dimension can be honed.
4. As compared to other hole finishing methods, high productivity at low cost is obtained.
5. Honing is a precision machining process used in automobile industries.

Disadvantages

1. It is impossible to improve straightness in holes.

2. It is difficult to hone tough non-ferrous metals due to glazing or clogging of the pores of the abrasive sticks.

Various defects that can occur during honing process and the possible causes of these defects are given in Table 2.

Table 2: Trouble Shooting Chart for Honing

Defects	Possible Causes
1. Ovality in bore operation	(a) Too much ovality from previous
	(b) Too much hone pressure
	(c) Hard stone
	(d) Inadequate coolant
2. Taper-large at ends	(a) Longer stroke length
3. Taper-small at ends	(a) Smaller stroke length
4. Slow stock removal	(a) Hard stone
	(b) Dense structure
	(c) Less hone pressure
	(d) High rotary speed
	(e) Less reciprocating speed
5. Poor stone life	(a) Rough grain size
	(b) Soft stone
	(c) Open structure
	(d) High hone pressure
	(e) High reciprocating speed
	(f) Less rotary speed
6. Too smooth surface finish	(a) Fine grain size
	(b) Hard stone
	(c) Close structure
	(d) Less hone pressure
	(e) Less reciprocating speed
	(f) High rotary speed
7. Too rough surface finish	(a) Coarse grains
	(b) Soft stone
	(c) Open structure
	(d) High hone pressure
	(e) Low rotary speed

8. Loading of abrasives or glazing of abrasives	(a) Fine grains (b) Hard stone (c) Open structure (d) Low hone pressure (e) Low reciprocating speed (f) High rotary speed
9. Stone breakages	(a) Too high hone pressure (b) Hone expansion too early (c) Hone hitting at the bottom (d) Hone stroke too long

3. Lapping

Lapping is basically an abrasive process in which loose abrasives non-bonded grinding media function as cutting points with momentary support of the lap (a soft plate).

The process has the following features:

(a) Use of loose abrasives between the lap and the work.

(b) The lap and workpiece are not positively driven, but are guided in contact with each other.

(c) Relative motion between the lap and work surface should be constantly changing. The most effective path is of cycloidal nature.

Vehicle of lapping is the lubricating liquid in which abrasive elements are suspended. Machine oil, soluble oil, grease, etc., are used as lapping vehicle. Figure 6 shows the principle of lapping process.

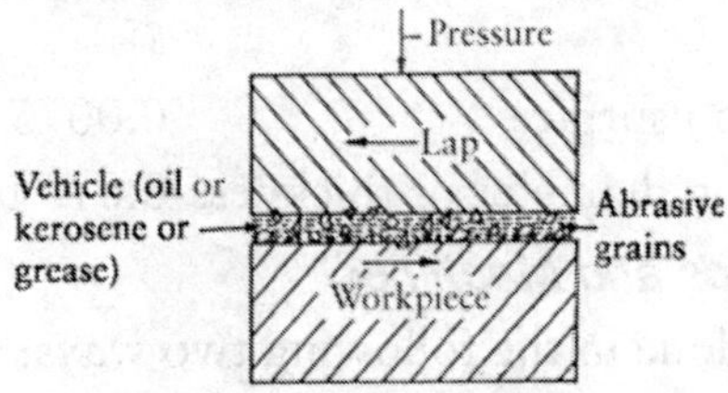

Fig. 6: Principle of lapping

Lap Material

Cast iron is the best lap material, but brass, bronze, lead, soft steel are also used. In any event, lap should be softer than the workpiece, so that the abrasive gets embedded in the lap.

Abrasives used in Lapping

Silicon carbide is used for rapid stock removal and aluminium oxide for improved surface finish. Corundum and Boron carbide can also be used with grain sizes ranging from 200-1200.

Lapping Speed

The speed of the lap relative to workpiece surface is chosen taking into account the accuracy and surface roughness required. The lapping speeds are shown in Table 3.

Table 3: Lapping Speed for Various Jobs

Type of Jobs	Roughness value (Ra in microns)	Lapping speed (metre/min)
For medium accuracy	< 10	200-400
For accurate jobs	10 ≤ Ra < 12	100-250
For very accurate jobs	12 ≤ Ra < 14	10-100

Lapping is a precision finishing process done on precision tools, gauges and valves where resistance to wear of moving parts, better sealing characteristics and longer life of cutting edges are prominent factors. A very thin layer of metal from 0.005 to 0.01 mm is usually removed by lapping. Lapping improves accuracy of form, surface quality and results in accurate dimensions.

As lapping is not primarily meant for removing metal, so it should be kept in mind that the material left on the work surface is minimum. Keeping in view the above discussions, the recommended range of lapping allowance to be left is as follows:

Allowance on surface 0.0075 to 0.0125 mm.

Allowance on diameter or thickness 0.015 to 0.05 mm.

Lapping Methods and Machines

Lapping is done in the following two ways:

1. Hand Lapping
2. Machine Lapping

1. *Hand Lapping*

Lapping can be done by hand held tools for both flat work and external cylindrical work explained as follows:

(a) *Flat work hand lapping*: In hand lapping, as shown in Fig. 7, either the lap or the workpiece is held by hand and motion of the other enables the rubbing of the two surfaces in contact. This method is used for lapping presswork dies, dies and metal moulds for castings, etc. Sometimes a lapping compound is placed between the two surfaces and then they are moved against each other. A few examples of this method are lapping of surface plate, engine valve and valve seat, lapping of laboratory specimen prepared for examination of the microstructures.

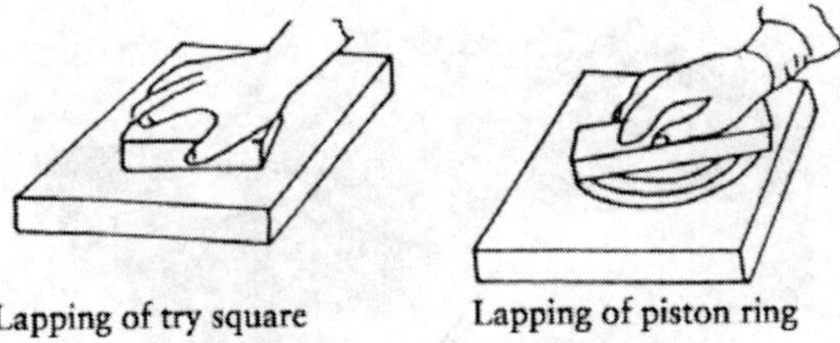

Fig. 7: Methods of hand lapping

(b) *External cylindrical hand lapping (Ring lapping)*: Ring lapping is done for finishing external cylindrical surfaces. Ring laps are made of soft close-grain cast iron. The ring lap has several cuts partially through and a complete slit as shown in Fig. 8. Screws are provided for precise adjustment. The size of the ring lap should be slightly shorter than the workpiece. The ring lap is reciprocated over the workpiece surface. The abrasive and vehicle are fed through the slot to maintain a straight round hole in the lap. This type of lapping is recommended for stepped plug gauges or gauges made in small quantities. An expanding mandrel is used for lapping holes.

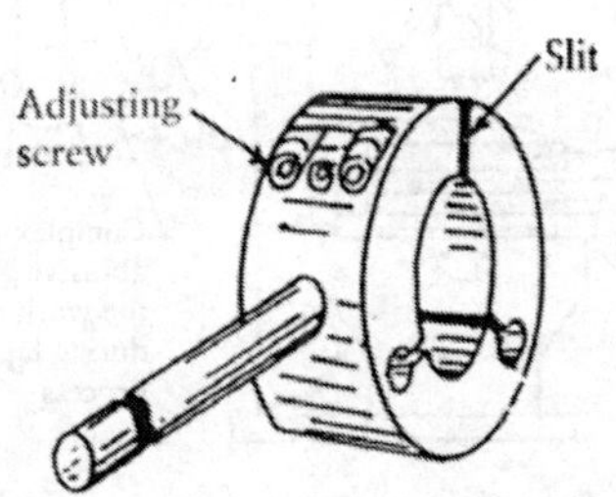

Fig. 8: Adjustable ring lap

2. Machine Lapping

This is performed for obtaining highly finished surface on many parts like races of ball and roller bearings, worm and worm gears, crankshafts, camshafts and various automobile engine parts like housings, injector pump parts, spray nozzle, etc. Fig. 9(a). Various types of machine lapping processes are explained below:

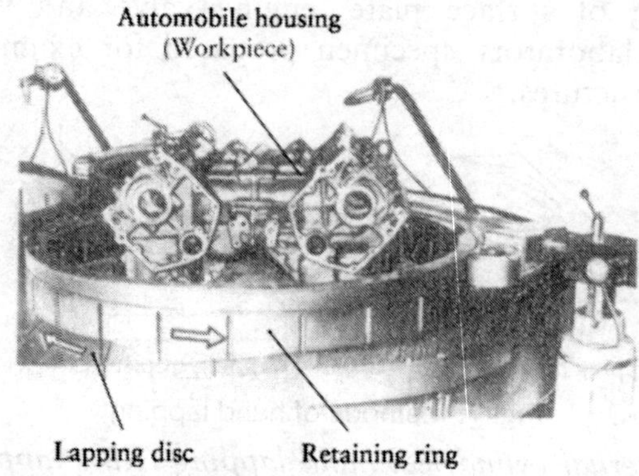

Fig. 9(a): Lapping machine

(a) *External cylindrical machine lapping*: It is a vertical spindle machine carrying one upper stationary lap and one rotating lower lap. The upper lap is free to float and rest on work which rides on the face of lower lap. Pressure is applied by gravity. The work is held loosely in the work guide, so it follows a random path between the lap faces. An external cylindrical lapping machine is shown in Fig. 9(b).

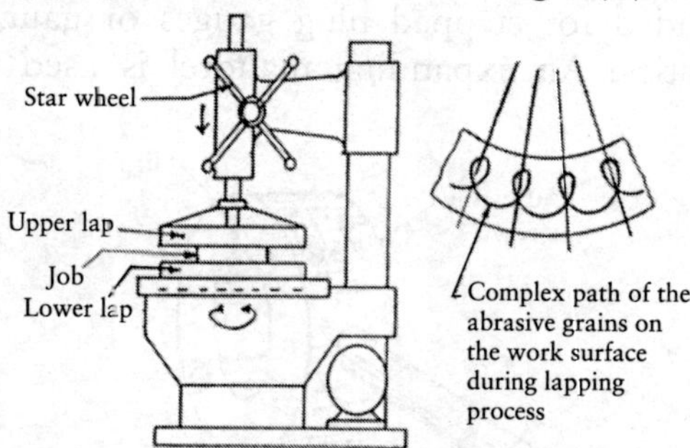

Fig. 9(b): Lapping machine for cylindrical work

(b) *Flat machine lapping*: The work holder propels the work in this case. The rotating driving spindle may give a friction drive to the work holder, or a positive drive may be given through gear teeth on the periphery of spindle and work holder. The driving spindle rotates at a different speed than the lower lap and the motion given to the work holder causes the work to cover the entire lap surface.

(c) *Centreless roll lapping machine*: The machine consists of two cast iron rollers—lapping roller and regulating roller. Lapping roller is twice in diameter as compared to that of regulating roller and both revolve in the same direction and at the same speed. The abrasive compound is applied to the rollers and the workpiece is laid between the two rollers are shown in Fig. 10. Lapping roller creates a rapid lapping action due to its increased surface speed. The workpiece is moved evenly over the entire surface of the roller by a fibre stick which is uniformly reciprocated.

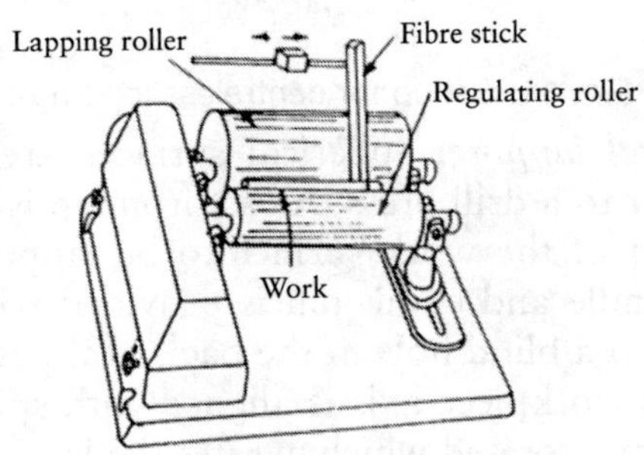

Fig. 10: Roller type machine for centreless cylindrical lapping

This machine is used for lapping one piece at a time and is designed for lapping plug gauges, measuring wires and cylindrical objects.

(d) *Centreless lapping machine*: This machine is similar to centreless grinding machine except that extra-long grinding wheel and regulating wheels are used to allow the workpiece to remain in abrading contact for a longer time. The lapping wheel speed falls in the range of 175-650 m/min, whereas regulating wheel has a speed of 70-175 m/min.

Figure 11 shows a centreless lapping machine. The spindles of lapping wheel and regulating wheel are swivelled in the vertical plane and are not parallel. Due to this the workpiece comes in contact with the wheel at an angle to its axis which leads to wraparound effect on the workpiece. This leads to better surface finish. This machine is a high production machine and is used to lap piston pins, shafts and bearing races.

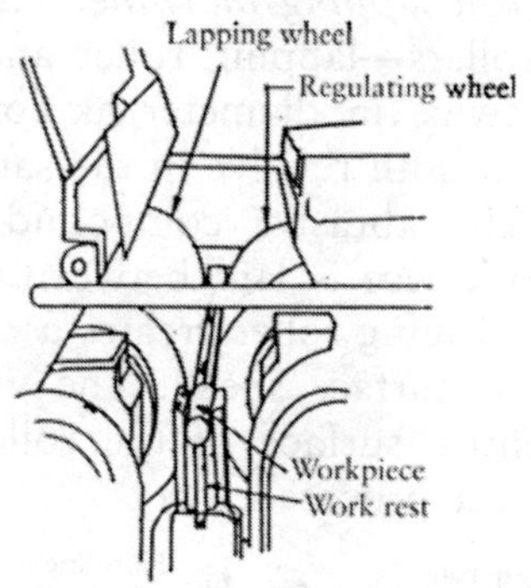

Fig. 11: Set up for centreless lapping

(e) *Spherical lapping*: Spherical surfaces are lapped on a machine similar to a drill press. A cast iron lap is used which is the counterpart of the work surface to be lapped. A crank is held in the spindle and crank pin is provided with a ball that enters freely into a blind hole in the back of the lap as depicted in Fig. 12. The workpiece axis is aligned with spindle axis and the spindle is then rotated which gyrates the lap.

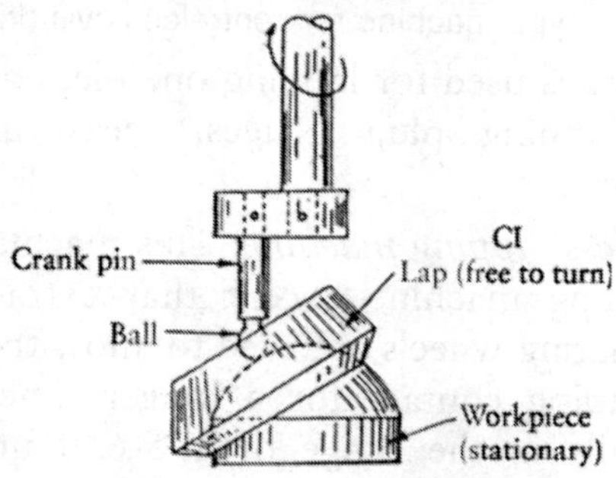

Fig. 12: Spherical lapping

There are two more processes which are connected with lapping. These processes are polishing and buffing. Both of them will now be discussed in detail.

4. Polishing

Polishing is an intermediate abrading operation which follows grinding and precedes buffing. Polishing is usually undertaken to make metal surfaces smoother and to produce a more uniform surface by removing deep scratches, nicks, discoloration and other surface imperfections occurring due to grinding.

The polishing process uses abrasive grains which are firmly attached to flexible belt or flexible wheel. Coated abrasive belts have become common for polishing of metals. An important benefit from using abrasive belt is the quick change from used belt to new one. The belts used for polishing are coated with resins and glues followed by abrasives. The polishing wheels are generally made of canvas, felt or leather.

The abrasives used for polishing are aluminium oxide or silicon carbide. Aluminium oxide is used for most carbon steel, alloy steel, high speed steel and non-ferrous metals. Silicon carbide is recommended for finishing low tensile strength materials such as brass, copper, cast iron and aluminium. Natural abrasives like emery and corundum are sometimes used in specialized operations for getting the finest quality finishes. The bonding agent used on polishing wheel is either hot glue or cold silicate based cement.

The new wheels need a thinned sizing coating of the bonding agent. For this, the hot glue or cold cement is coated on the wheel and when it becomes dry, the wheel is rotated on a shaft and the wheel face is sanded lightly with sandpaper. The wheel is again recoated with hot glue or cold cement and immediately the wheel is rolled in a trough containing abrasive grains.

Lubricants like beef tallow or beeswax are used to cool the polishing wheel. The lubricants also decrease the tendency of wheel to grab and hold particles of metal removed from the work.

5. Buffing

It is an operation in which the workpiece is brought in contact with a revolving, cloth buffing wheel that has been charged with very fine abrasives. The abrasives remove minute amounts of metal from the workpiece, eliminate fine scratch marks and produce a very smooth surface. Buffing is used to give high luster to the workpiece. Buffing is different from polishing due to the fact that polishing is an abrasive operation used for metal removal, whereas buffing involves gentle plastic deformation of a surface into a bright, smooth finish.

Buffing can be divided into two operations—Cutting—down and colouring. Cutting down is done to refine a surface by removing scratch lines from polishing, die marks or other imperfections. Colouring refines the cut down surface and produces a high finish or luster.

Buffing wheels are made of discs of linen, cotton, cloth or canvas that are made firm by stitching to fasten the layers of cloth together. For colouring on noble metals like gold, silver, etc., cotton flannel or wool cloth buffs are used. In addition, loose buffs made of sheep skin can be used for final colouring. Buffing wheels used for very soft polishing or for polishing into interior corners may have no stitching. In this case, the cloth layers are kept in position by centrifugal force resulting from rotation of wheel.

Many types of abrasives are used for buffing, like amorphous silica, aluminium oxide, rouge or chromium oxide. Amorphous silica is one abrasive that is highly suitable for buffing aluminium, copper, brass and die castings. Aluminium oxide powders can also be used as abrasive for buffing hard metals, whereas chromium oxide is used for buffing of stainless steel, chrome and nickel plate. Rouge is used for colouring on copper, brass, gold and silver.

The buffing compounds can either be greaseless or have a grease base. Both forms use the same abrasive. Only the vehicles differ. Greaseless compounds are mainly employed for satin finish and are used in cutting down before colouring. Grease compounds use oil and fat. These compounds are highly suitable for automatic buffing. The buffing compounds are sprayed on to the buffing wheel either under gravity or by air pressure. Automatic spray guns are used to throw the abrasive on to the wheel.

Buffing can be done manually or automatically. The manual buffing machine uses a double ended shaft having wheels on both ends. The shaft is driven by electric motor through belt and pulleys. If buffing is done manually, it becomes expensive due to labour cost. In the automatic machine, conveyors and work holders are incorporated.

6. Superfinishing

Superfinishing is the process used for improvement of machined surfaces to make them more wear resistant. It is, in a way, a fine finishing process. In this process, the main aim is to remove the burnt out layer of the surface and to correct the inequalities in geometry. This leads to even distribution of the load.

It is a slow speed abrasive machining process. Generally stock ranging from 2.5 to 5 microns is removed.

The abrasive is either aluminium oxide widely used for steel, or silicon carbide for finishing cast iron and non-ferrous metals.

The abrasive stick is made of very fine grains. The work is rotated between centres as in a lathe and abrasive stick holder is reciprocated back and forth with short strokes but rapidly. The rotational speed of job is very low; 2 to 20 metres/minute. The sticks stroke may be 2 to 5 mm long with a frequency of up to 1800 strokes/minute. Sometimes superfinishing is referred to as short stroke honing.

A lubricant made of kerosene oil is used to give fine finish.

Superfinishing can be accomplished on lathe. Special purpose machines are also available for finishing crank shafts, cam shafts, etc. The superfinishing process is shown in Fig. 13. A superfinishing attachment with abrasive stick fixed is mounted on a standard centre lathe.

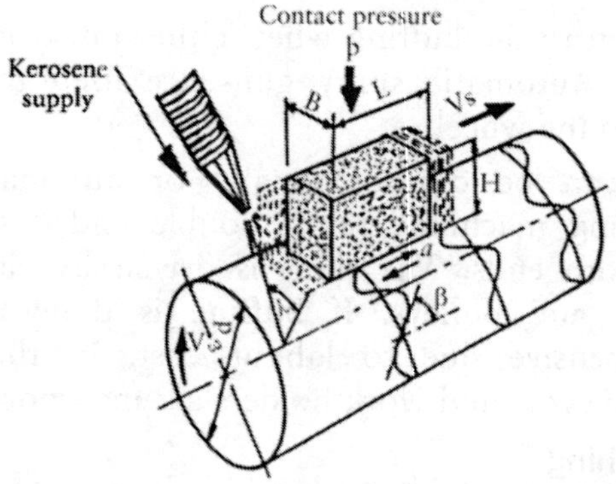

p = Contact pressure between stone and workpiece = 0.5 – 4.0 kg/cm^2

$$\tan \beta = \frac{\pi dn}{a.f.}$$

a = Amplitude of stone vibration in mm (Max 6 mm).

f = Number of strokes per min (about 500 cycls/min).

d = Diameter of workpiece in mm.

n = R.P.M. of workpiece.

Vs = 6000 sq mm/mt. (for long shafts, stone is traversed).

(As angle β increases, the finish improves)

Fig. 13: Superfinishing process

If the work is longer than the stone, then a traverse motion is also required parallel to the axis of work. It should be noted that the stone gradually "wears in" to the average radius of the part. The early stages of the operation consist of the abrasion of the peaks and ridges of the workpiece. The stone will have contact with the workpiece at isolated points. However, as the work approaches nearer to a true cylinder the area of contact increases thus reducing the pressure on unit area. The load on each individual grain becomes very small and abrasion practically ceases. Eventually the particles of metal become so small that they oxidize and are deposited on the surface of the stone. At this point, metal removal virtually ceases.

Principle of Flat Superfinishing

Flat superfinishing, as shown in Fig. 14, is performed with a machine containing an upper and lower vertical rotating spindle. The upper spindle has a spring-loaded quill on which cup-shaped stone is located; while the lower spindle carries the work. As the work revolves, the end face of the cup is brought into spring loaded contact with the work. As in cylindrical work, the stone automatically wears to the most desirable shape.

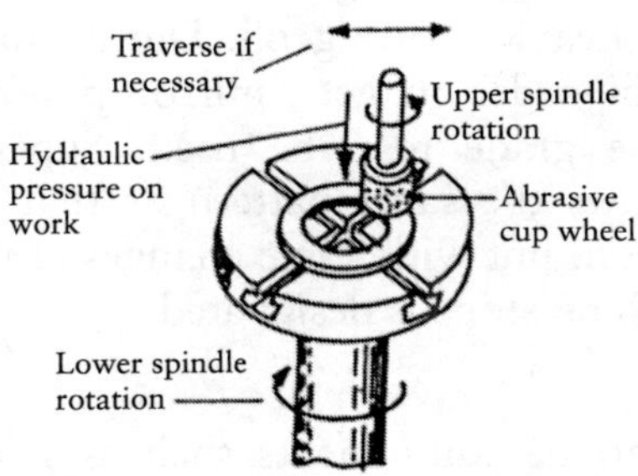

Fig. 14: Superfinishing a flat surface

Superfinishing Speeds, Production Rates and Surface Finish

The work surface speed is around 20 m/minute, under a flexible contact pressure of about 0.5-4 kg/cm^2 of stone contact area and in a fluid lubricant bath. The time consumed in a superfinishing process depends on the quality of the previous machining operation. It is better that the job is at least rough ground or fine turned before it is put on the superfinishing process. Surface finish produced will have a mirror like finish and the surface finish obtained is of the order of 0.05-0.2 micron.

Surface roughness profile of a superfinished surface as recorded on a surface profilometer is shown in Fig. 15.

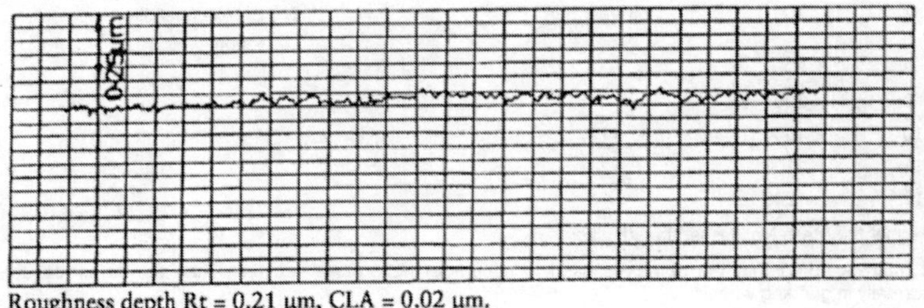

Roughness depth Rt = 0.21 µm, CLA = 0.02 µm.

Fig. 15: Surface roughness profile of a superfinished surface

Selection of Abrasive and Designation of Superfinishing Stone

Aluminium oxide grit is generally used on steels and silicon carbide is best on CI and non-ferrous metals. Vitrified bond is generally used. Resin shellac bonds will produce very smooth surface, however, their cutting action is slow. Cutting fluids used in superfinishing are light mineral oil or kerosene oil together with a heavier cutting oil. During the final stage of superfinishing, in order to get "mirror polishing", diamond paste of suitable grade may be used. Superfinishing stone designation follows the same pattern as that of the grinding wheel specification, but with some changes. Table 4 shows the way a superfinishing stone is designated.

Applications

1. Automobile components such as gudgeon pins and cylinder lines are superfinished.
2. Textile and paper mill calendars are superfinished.
3. Hydraulic cylinder and piston assemblies in aircraft and in heavy earth moving equipment are superfinished.

Table 4: Superfinishing Stone Designation

5 Number	– Corresponding Mesh No.	Q Hardness of bond	T Grain type	V Bond type	I Manufacturer's Code No.
0	60-80	A = Very soft	A = Al_2O_3	V = Vitrified	
1	100-120	↓	B = SiC grey	X = Vitrified + graphite*	XYZ
2	180-240		T = SiC green	S = Vitrified + sulphur**	
3	280-320		D = Diamond***		
4	400				
5	500		C = Chrome-oxide	K = Synthetic resin	
6	600		F = B_4C (Boron) Carbide)	L = Synthetic resin + graphite	
7	700	W = Very hard		B = Bronze (metallic bond)	
8	800				
9	900				
10	1000				
11	1100				
12	1200				

(Higher the Mesh No., lesser the grit size)

It may be noted that the abrasives used in superfinishing process are finer than the abrasives used normally in grinding wheels.

*Graphite content distinguished by its black colour.

**Sulphur content distinguished by its yellow colour.

*** Diamond wheels are usually metal bonded.

7. Superfinishing Attachment on a Lathe

Superfinishing attachment used on a standard centre lathe is shown in Fig. 16. Main parts and operating elements of superfinishing attachment are discussed below:

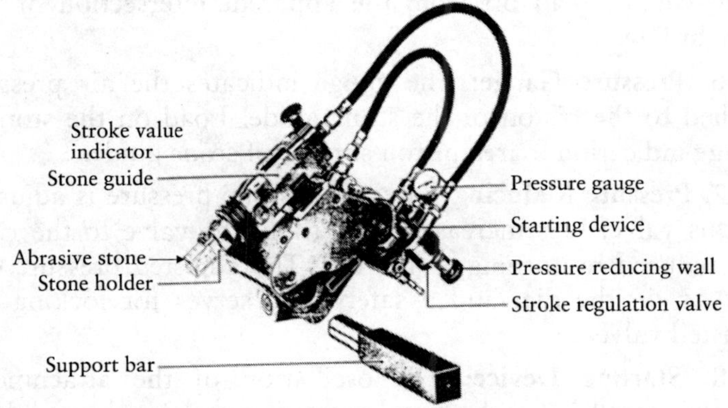

Fig. 16: Superfinishing attachment (pneumatically operated)

1. Support Bar: It is clamped into the tool holder of the lathe, and the superfinishing attachment is fastened to the round shaft. It can be turned into the right position to the workpiece and then fixed.

2. Stone Guide: The stone guide consists of an air cylinder with piston, to which the stone holder is fastened. It is operated by the control valve. By actuating this valve, the stone moves against the workpiece. The stone guide is connected with the attachment by means of a dovetail guide, allowing longitudinal adjustment and fastening in every position desired. The attachment can be provided with a second stone guide to attain a double efficiency when machining larger workpiece, or for finishing two bearing sections at the same time.

3. Stone Holder: The stone holders are fastened in the piston rod of the stone guide by means of their spherical head part. The universal movability allows the stone to be set precisely in the workpiece.

4. Stroke Regulation Valve: This valve is for regulating the oscillation stroke. The stroke is lengthened by turning the valve to the left and reduced by turning to the right up to its complete stop.

5. Stroke Value Indicator: The stone guide is provided with a scale showing two crossing straight lines. The stroke value (mm) can be read off from the apparent intersection of this straight line.

6. Pressure Gauge: The gauge indicates the air pressure applied to the piston of the stone guide. Load on the stone = Gauge indication × area piston surface of stone guide.

7. Pressure Reducing Valve: The stone pressure is adjusted by this valve. It is increased by turning the valve to the right and reduced by turning to the left. The adjusted pressure can be read on the dial and a safety ring serves for locking the adjusted valve.

8. Starting Device: The oscillation of the attachment, exactly parallel to the workpiece is attained by a short operation of the starting button. The unit should be maintained properly, so that when the starting button is pressed vibrations should commence. Photograph of superfinishing attachment being used on a lathe is shown in Fig. 17.

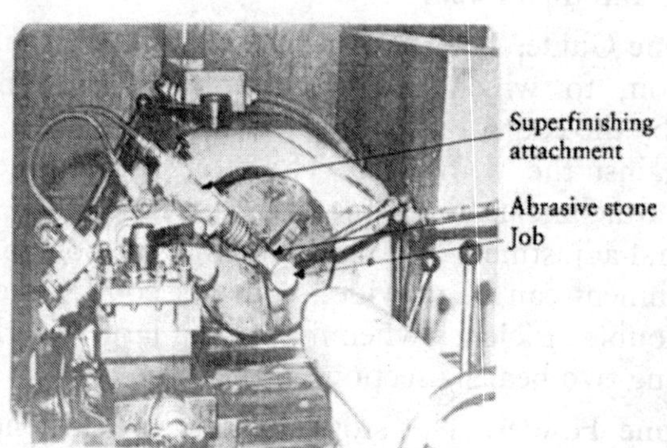

Fig. 17: Superfinishing attachment on centre lathe

8. How Micro-Finishing Processes Differ from Grinding?

The micro-finishing processes differ from the grinding process as explained:

Grinding Process	Micro-Finishing Processes
1. High surface speed of about 2000 m/min.	1. Less surface speed of about 100 m/min.
2. Contact pressure between grinding wheel and workpiece is usually very large	2. Contact pressure is usually very small, about 1 kg/cm^2
3. Contact area is very small. At the maximum only line contact is there	3. Contact area is very large. In some cases, the entire surface of the abrasive comes in contact with the workpiece
4. Less accuracy due to possible work deflection and distortion	4. High accuracy is obtained as no deflection and distortion of workpiece occurs
5. Less surface finish	5. Higher surface finish
6. Due to high temperature produced, burnt out layers will be present near the surface produced	6. Less temperature is produced. Hence no burnt out layers are left as in grinding
7. Running in period is required for ground surface before full load can be applied	7. No running in period is required
8. Grinding processes are generally less Costly	8. Micro-finishing processes are very costly

REVIEW QUESTIONS

1. What is the purpose of honing?
2. Briefly explain the process of honing.
3. Explain why honing cannot be used as a method to reduce the errors of hole location or alignment on a workpiece.
4. What is the chief purpose of lapping?
5. Explain the difference between the type of abrasive cutting tool used for honing and that used for lapping.
6. What is the main purpose of superfinishing?
7. Compare the relative surface finish obtained in honing, lapping and superfinishing.
8. What is the purpose of buffing?
9. What is the major difference between buffing and polishing?
10. Describe the two buffing steps—cutting down and colouring.

REVISION QUESTIONS

1. What is understood by lapping?
2. What materials are used to produce a lapping powder?
3. Which lapping tools are in use?
4. What is the function of the lapping cage?

   Ans: (Lapping cage is used to hold workpieces in position while they are being lapped.

   In machine lapping, the workpieces are held/ attached by means of suitable fixtures to a rotating disc (lapping cage) and pressed against the grinding medium by a pressure plate which moves in the opposite direction.)
5. How does honing differ from lapping?
6. What materials can be honed?
7. What is the correct infeed value for internal honing?
8. What is the special feature of short-stroke honing?

OBJECTIVE TYPE QUESTIONS

Choose the correct alternative

1. Ball bearing races are

   (a) lapped (b) honed (c) buffed (d) ground
2. In which of the following processes, least material is removed?

   (a) Grinding (b) Lapping

   (c) Honing (d) Superfinishing
3. The process of precision grinding of workpiece with loose dust type abrasive is known as

   (a) honing (b) lapping (c) buffing (d) polishing
4. Pressure applied on workpiece in case of lapping operation is

   (a) 0.05 kg/cm^2 (b) 0.1 kg/cm^2

   (c) 1.0 kg/cm^2 (d) 5 kg/cm^2

5. Which one of the following processes is used for finishing cylindrical holes and objects?
   (a) Lapping (b) Honing
   (c) Polishing (d) Buffing
6. Which abrasive particle is held in the form of sticks in honing process?
   (a) Al_2O_3 or SiC (b) Corrundum
   (c) Quartz (d) Diamond
7. Buffing is
   (a) A finishing operation after honing
   (b) A process of electroplating
   (c) The process of bringing out the luster of metal
   (d) Process of covering a metal with soft material
8. Honing is used to correct
   (a) Ovality
   (b) Waviness of axis
   (c) Non-parallelism of cylindrical features
   (d) All of the above
9. Honing operation produces normal quality of finish (CLA value) of the order of
   (a) 0.01-0.08 micron (b) 0.08-0.2 micron
   (c) 0.2-0.4 micron (d) 0.4-0.8 micron
10. Which of the following process would remove least material?
    (a) Honing (b) Lapping (c) Superfinishing
11. The process of precision grinding with loose dust type abrasive is
    (a) Honing (b) Superfinishing (c) Lapping
12. Honing operation
    (a) cannot be used to change the location of hole or shape of a hole.
    (b) cannot correct ovality.
    (c) cannot correct waviness of axis.

ANSWER KEY

1. (a)	2. (d)	3. (b)	4. (b)	5. (b)
6. (a)	7. (c)	8. (d)	9. (c)	10. (c)
11. (c)	12. (a)			

12

Machine Shop Measuring Devices (Metrology)

1. Introduction

Metrology is the 'science of measurement in engineering'. It is intimately concerned with the design, manufacturing and testing of gauges of all kinds. The practice of metrology involves precise measurement requiring the use of precise instruments and equipment to permit the degree of accuracy required to be obtained. Due to the industrial revolution and great advancements, industrial inspection is concerned with the checking of a product at various stages in its manufacture, right from the raw-material form to the finished product. Various measuring devices are used for measurement. The devices used are classified as follows:

1. Classification of Measuring Devices

i. Depending on Type of Measurement

(a) *Direct Measurement*

In which the measured value is determined directly, e.g., micrometer, vernier calliper, vernier height gauge, etc.

(b) *Indirect Measurement*

In which the dimension is determined by measuring other values functionally related to the required value, e.g. divider, sine-bar, etc.

ii. Depending Upon the Contact with Workpiece

(a) *Contact*

Here, the measuring tip of the instrument actually touches the surface to be measured, e.g. micrometer, callipers, dial indicators, etc.

(b) *Contactless*

Here, no contact is required for measurement, e.g. tool maker's microscope, profile projector, etc.

iii. According to their Functions

(a) *Linear or length measuring instruments*

For example steel rule, calliper, divider, micrometer, vernier calliper, etc.

(b) *Angular or Angle measuring instruments*

For example combination set, bevel protractor, sine bar, dividing head, etc.

On the basis of function to be performed by the instrument, the measuring devices are classified as follows:

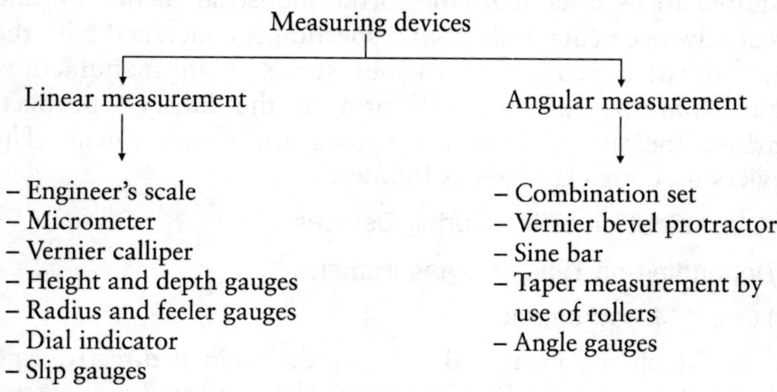

The various terms connected with the measuring instruments can be defined as given below:

1. **Range of measurement:** It indicates the size values between which measurement can be made on the given instrument, e.g., micrometers are usually available for the following ranges—0 to 25, 25 to 50, 50 to 75, 75 to 100, 100 to 125, etc. (in mm).

2. **Scale division value:** It is the measured value corresponding to one division of the instrument scale, e.g. scale division value of a micrometer is 0.01 mm.

3. **Precision:** Precision is associated with the measuring process. It is defined as the repeatability of a measuring process. The chief concern is with comparing the dimension measurement relative to each other, irrespective of the fact that unit has the exact value. Precision tells us the degree to which the various measurements performed by same instrument on the same component agree with each other.

4. **Accuracy:** Accuracy is also associated with the measuring process. Accuracy is the agreement of the result of a measurement with the true value of the measured quantity. The difference between accuracy and precision is shown by Fig. 1.

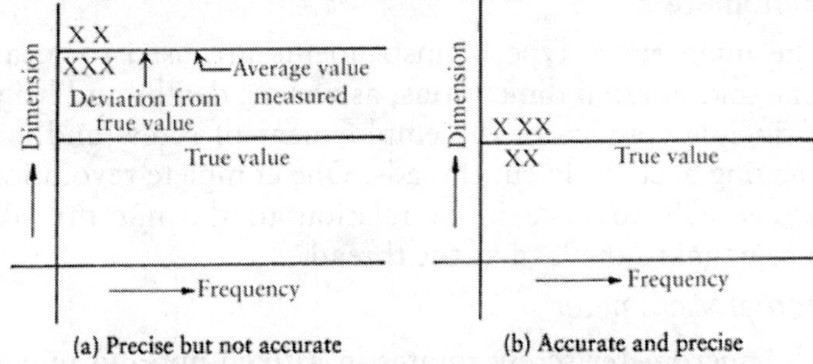

Fig. 1: Difference between accuracy and precision

5. **Sensitivity:** Sensitivity is associated with the measuring instrument. Sensitivity refers to the ability of a measuring device to detect small differences in a quantity being measured.

2. Linear Measurements

In our day to day life, one needs to measure linear dimensions, e.g. length, breadth, height, depth and thickness, etc. There are many instruments for measuring these linear dimensions. These instruments are classified as non-precision and precision instruments. The various linear measuring instruments are as follows:

(a) Engineers' Scale

Engineer's scale is the most common measuring instrument used in inspection. It works on the basic measuring technique of comparing an unknown length to the one previously calibrated. It consists of a strip of hardened steel having line graduations etched or engraved at intervals of a fraction of a standard unit of length. Depending upon the interval at which the graduations are made, the scales can be manufactured in different sizes and shapes. It may be graduated either on both sides and both edges or on both edges of only one side and various conversion factors from one system to another system are written on the other side for the ease of the user. The scale need not be graduated uniformly throughout its length. In some parts, e.g. it may have five divisions of a cm, in some portion ten divisions, so that it can be used for all types of work and the particular range chosen depending upon the accuracy required.

(b) Micrometers

The micrometer type of instruments are used to measure external and internal dimensions, as well as depths and heights. The principle consists of the employment of screw and a nut, both having accurately cut threads. One complete revolution of the screw will advance it, in relation to the nut threads, a distance equal to the lead of the thread.

(i) *External Micrometer*

The micrometer screw rotates in a fixed nut and moves in an axial direction relative to the nut. The screw is rigidly attached to the thimble while the fixed nut is fastened to the sleeve or barrel. Thus, the thimble rotates and moves axially along the barrel when the screw is rotated. Fig. 2 shows the principle of an external micrometer.

It has a U-shaped frame to the left of which an anvil is attached and to right end, a barrel or sleeve is attached. A spindle passes through the barrel. A thimble which surrounds the barrel or sleeve is attached to the right end of the spindle. The unattached faces of the anvil and spindle form the measuring faces of the micrometer. A linear scale with graduations in millimeter is engraved on the barrel. This is the main scale. A circular scale is engraved on the bevelled surface of the thimble. This is the vernier scale.

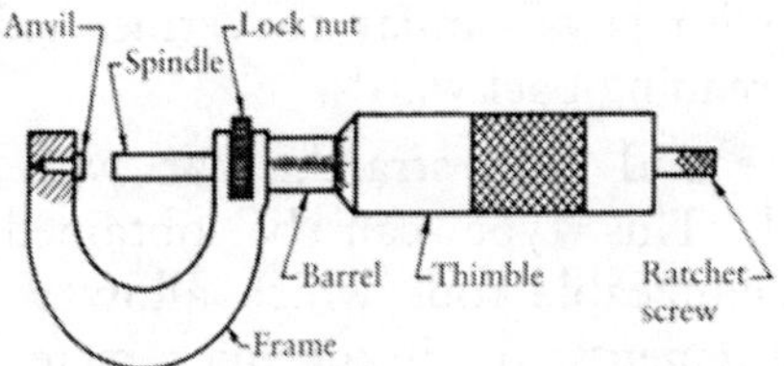

Fig. 2: External micrometer

Least count of micrometer is the smallest value of measurement which can be measured by it.

$$\text{Least count} = \frac{\text{Distance moved by screw by giving one revolution to the thimble}}{\text{number of division on circular scale}}$$

If the screw moves by 0.5 mm with one revolution of the thimble and there are 50 divisions on the circular scale, then

$$\text{Least Count} = \frac{0.5}{50}\text{ mm} = 0.01\text{ mm}$$

The micrometer is used as follows:

1. Check the zero reading.
2. Place the part to be measured in between the measuring faces.
3. Advance the spindle by rotating the ratchet until it begins to slip and click is heard. This indicates that there is no further movement of the spindle.
4. Note the reading both on the main scale and on the circular scale of the thimble. Reading on micrometer = Reading uncovered on the main scale + (Least count × number of divisions on the thimble scale which coincides with horizontal line on the main scale).

A micrometer is checked and adjusted to zero position by means of gauge blocks, standards or end bars. Micrometers are usually available in the ranges of 0 to 25, 25 to 50, 50 to 75, 75 to 100, 100 to 125 and 125 to 150 mm ranges.

(ii) *Inside Micrometer*

This micrometer is used for measuring the internal dimensions. This micrometer is available in two designs.

1. One with jaws similar to vernier caliper and with scale reading backwards.
2. The second is a straight bar with a micrometer barrel. This type can be obtained with several interchangeable rods which allow a wide range of measurements. An inside micrometer of type (1) is shown in Fig. 3.

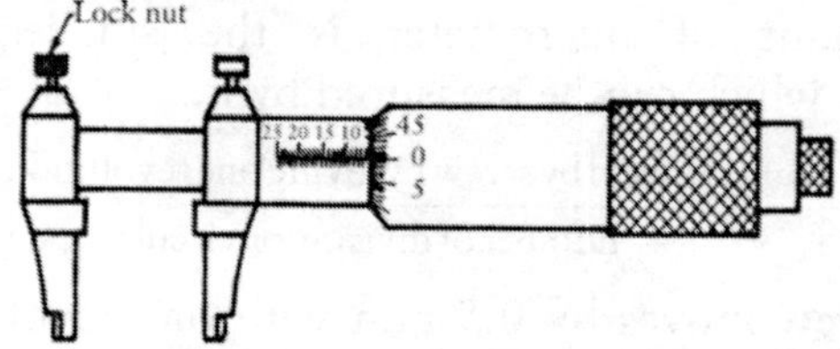

Fig. 3: Inside micrometer

(iii) *Depth Micrometer*

A depth micrometer has the same function as that of a vernier depth gauge, that is, to measure the depth of a slot or hole. They give the dimensions (depth) from some surface to the bottom of a slot or hole. It has one shoulder which acts as reference surface and is held firmly and perpendicular to the centre line of the hole. Here also, for large ranges of measurements, extension rods are used. A depth micrometer is shown in Fig. 4.

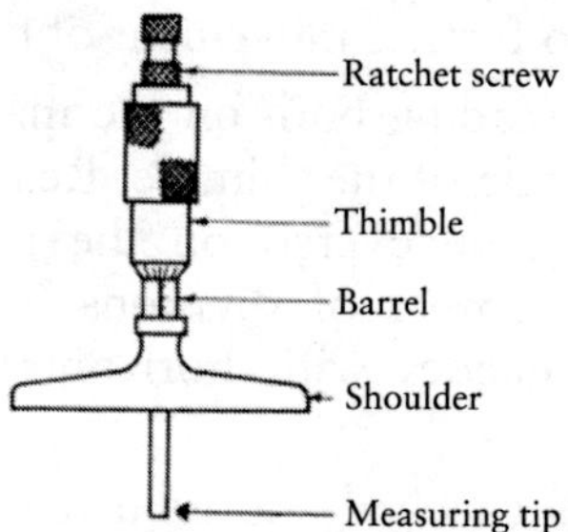

Fig. 4: Depth micrometer

(c) Vernier Calliper

The principle of vernier is that when two scales or divisions slightly different in size are used, the difference between them can be utilized to determine the accuracy of measurement. The vernier callipers are of two types depending on type of dimension to be measured:

(1) Out side calliper, (2) Inside calliper.

(i) *Outside Vernier Calliper*

It has a L-shaped frame. The main scale is engraved on this frame. One end of the frame contains a fixed jaw which is shaped into a contact tip at its extremity. A sliding jaw which moves along the guiding surfaces provided by the main scale is coupled to a vernier scale. The sliding jaw at its left extremity contains another measuring tip. When two measuring surfaces are in contact with each other, scale shows zero reading. The finer adjustment of the movable jaw can be made by the adjusting screw. An outside vernier calliper is shown in Fig. 5. Vernier calliper is a precision measuring instrument. The smallest dimension that can be measured with it is called least count.

Least count is determined by the difference between one main scale division and one vernier scale division.

For example, if 20 divisions of the vernier scale coincide with 19 divisions of the main scale

$$\text{then} \quad 20 \text{ VSD} = 19 \text{ MSD}$$

$$\text{and} \quad 1 \text{ VSD} = \frac{19}{20} \text{ MSD}$$

$$\text{Least count (LC)} = 1 \text{ MSD} - 1 \text{ VSD}$$

$$= \left(1 - \frac{19}{20}\right) \text{MSD} = \frac{1}{20} \text{MSD}$$

$$\text{If } 1 \text{ MSD} = 1 \text{ mm}$$

$$\text{Therefore, LC} = \frac{1}{20} \times 1 \text{ mm}$$

$$\text{or} \quad \text{LC} = 0.05 \text{ mm}$$

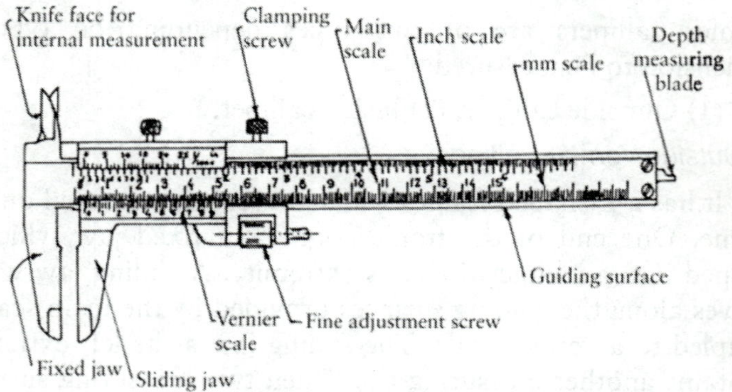

Fig. 5: Vernier Calliper

To read a measurement from a vernier calliper, note the reading of main scale to the left of the zero of vernier scale. Then note the vernier scale reading which exactly coincides with the main scale division.

So, the measurement is = (main scale reading to left of zero of vernier scale) + (LC × vernier reading exactly coinciding with any main scale division).

(ii) *Inside Vernier Calliper*

Inside vernier calliper is used for measuring the inside dimension, e.g. internal diameter, etc. The various parts of the calliper are same as that of the external vernier calliper, except that the measuring surfaces are on the outer side of the fixed and moveable jaws. The readings for the internal measurements are obtained by adding the combined width of the internal measuring jaws to the scale reading.

Precautions in the use of vernier calliper: No play looseness should be there between the sliding jaws on scale, otherwise the accuracy of the vernier calliper will be lost.

Usually the tips to measuring jaws are worn and that must be taken into account.

The accuracy in measurement primarily depends on two senses, viz. sense of sight and sense of touch. The shortcomings of imperfect vision can however be overcome by use of corrective eye lens and magnifying glass. But sense of touch is an important factor in measurement, which varies from person to person. The measuring instrument must always be properly balanced in hand and held lightly in such a way that only fingers handle the moving and adjusting screws.

(d) Height Gauge and Depth Gauge

(i) *Vernier Height Gauge*

The vernier height gauge is used for measuring the difference in height of two points or for marking purposes. It is practically similar to a vernier calliper with its fixed jaw set rigidly into a base. The base is a flat support. The main scale is carried on the vertical beam. A sliding arm which carries the vernier scale, has a clamp for holding either a jaw or marker. The jaw has two measuring faces for measuring internal and external height dimensions. The distance between the two faces must be taken into account when taking measurements. Vernier height gauge is made of good quality steel or stainless steel and is heat treated to remove the stresses. The measuring faces are adequately hardened. The various parts of a vernier height gauge are shown in Fig. 6.

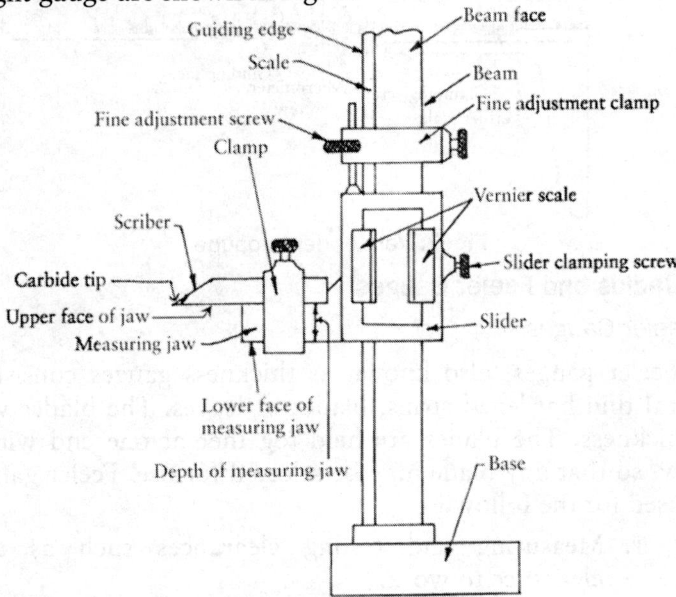

Fig. 6: Vernier Height Gauge

(ii) *Vernier Depth Gauge*

A vernier depth gauge is used to measure the depths of holes and the heights of shoulders in holes. It consists of a sliding head that slides on a beam or rule which carries the main scale. The vernier scale also slides on the rule, alongwith the head. The sliding head can be clamped at any desired position on the rule. In use, the flat end face of the rule or beam is made to contact the bottom of a blind hole whose depth is to be measured. Then, the base surface of the sliding head is brought into contact with end surface from which the hole depth is to be measured. The depth is then determined with the help of the main scale and the vernier scale. A vernier depth gauge is shown in Fig. 7.

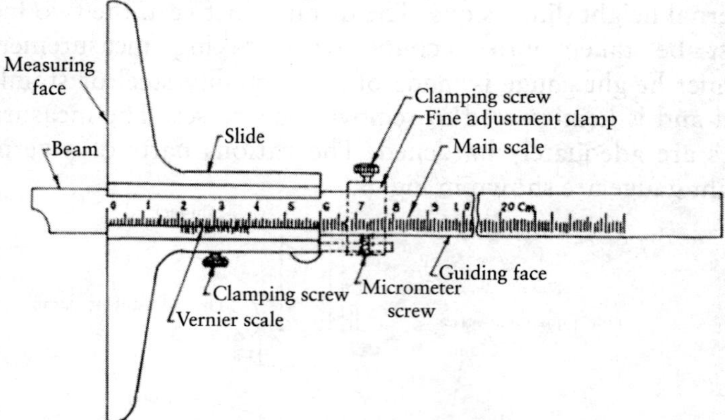

Fig. 7: Vernier depth gauge

(e) Radius and Feeler Gauges

(i) *Feeler Gauges*

Feeler gauges, also known as thickness gauges consist of several thin hardened strips, blades or leaves. The blades vary in thickness. The blades are held together at one end with a screw, so that any blade may be selected for use. Feeler gauges are used for the following:

1. Measuring and testing clearances such as tool clearance to work.

2. For measuring width of the gap between two parallel flat faces, e.g. in gauging of the clearance between the piston and cylinder.
3. For adjusting spark plugs and so on.

Their working depends on the sense of feel. The feeler blade should neither be forced between the surfaces nor should it slide freely, rather the correct blade will give a characteristic 'gauge fit' type of feel. Feeler gauges generally comprise of a series of gauging blades of different grades and thicknesses from 0.03 mm to 1 mm. The blades are made of heat treated bright polished tool steel having a tensile strength of about 180 kgf/cm^2 for thickness up to 0.5 mm and about 80 kgf/cm^2 for thickness upto 0.5 mm and above. Generally, the blades are available in the overall length of 100 mm. These are hinged in the sheath on a screw and nut of such a design that the blades are removable. The nut is in the form of a bush passing through both sides of the sheet and forms a hinge upon which the blades may be rotated. The various blades are assembled in such a way that each thin blade is given the maximum protection by being inter-leaved between two thicker blades. A set of feeler gauges is shown in Fig. 8.

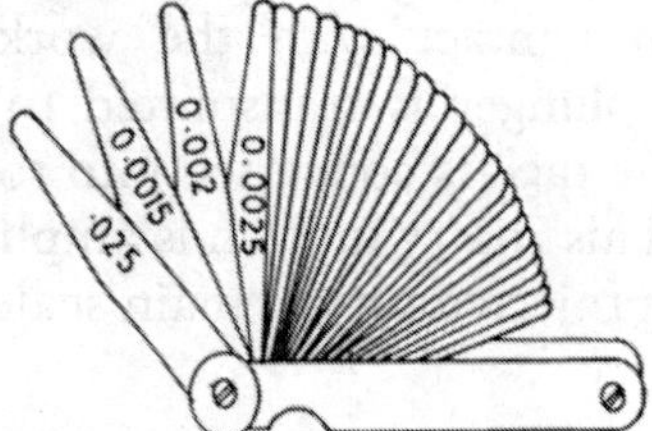

Fig. 8: Feeler of Thickness Gauges (dimensions in mm)

(ii) *Radius Gauges*

Whenever two surfaces meet, a proper curvature is generally given at the intersection to avoid stress concentration. In order to measure this curvature, radius gauges are employed. If curved surface is external, it is simply called 'radius'. If curved surface is internal, it is called a 'fillet'. Radius gauges are formed gauges which consist of sets of blades with curved endings on which is stamped the corresponding radius. On the one side, we have external radius and on the other internal radius so that it may be suitable for fillets as well as a radius. A set of radius gauges is shown in Fig. 9.

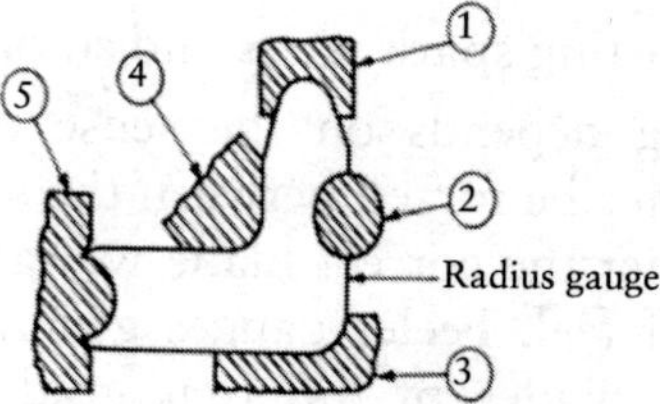

Fig. 9: Radius Gauges (1, 2, 3, 4, 5, part of work pieces)

(f) Dial Indicator

It is an instrument used for checking flatness of surfaces and parallelism of bars and rods. It can also be used to measure linear dimensions thickness of workpieces. Dial indicator can be of plunger type or lever type.

Working Principle

The mechanism of the dial indicator is simple, reliable and very sensitive. In the plunger type dial indicator, the plunger is attached to a rack. The rack meshes with a large gear wheel, which further meshes with a smaller gear wheel. The plunger is made to come in contact with the workpiece. The linear movement of the plunger is transmitted to the rack and the linear motion of the rack is converted into rotary motion of the large gear wheel. This rotary motion is amplified by the smaller gear wheel and is projected on the main scale by the movement of a pointer.

Construction

Plunger type dial indicator consists of a metal case that acts as a rigid frame and protective covering for the instrument mechanism. The case supports a gear train, gear rack, a graduated dial and a transparent dial cover held in a bezel. Various magnifications and dial markings are available. In a commonly used plunger type dial indicator, shown in Fig. 10, the main scale of the dial indicator is graduated into equal divisions corresponding to 0.01 mm movement of the plunger. A second but smaller dial is set in the main dial face to indicate the number of complete revolutions turned through, one revolution being equivalent to 1 mm of plunger movement. To enable the instrument to show zero reading for any position, the main scale can be rotated by rotating the bezel.

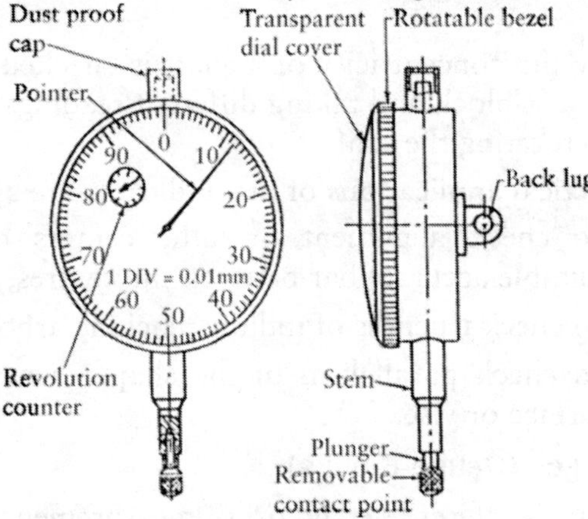

Fig. 10: Plunger type dial indicator

Fig. 11 shows a lever type dial indicator. This type relies on a lever and scroll system of magnification. It has only a limited range of stylus movement; little more than one revolution of the pointer. It is very popular for both inspection and machine setting.

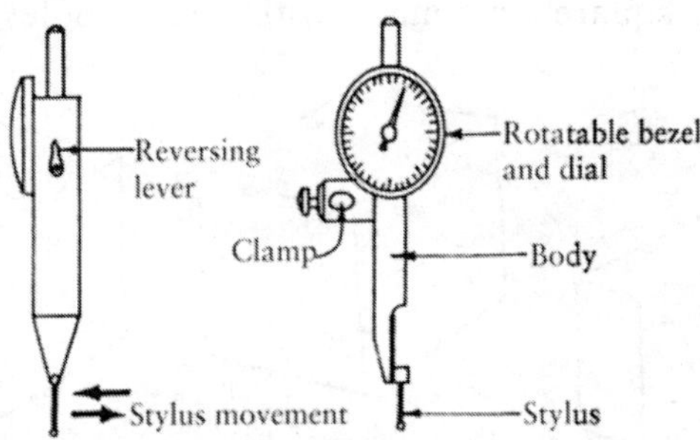

Fig. 11: Lever type dial indicator

The dial indicator is used mainly for finding errors in geometrical form for example ovality, out-of-roundness and taper, etc. It is also used to find positional errors of surfaces for example, parallelism, squareness, alignment, etc.

The flatness of the workpiece can be checked by placing the flat piece on the surface plate. The dial indicator is mounted on a stand and its plunger is adjusted to touch the flat piece surface at various points and readings are taken and compared.

Similarly the concentricity of a shaft is checked by placing the shaft on a V-block and taking different readings on the dial indicator by rotating the shaft.

The practical applications of dial indicator are as follows:

(i) To check alignment of lathe centers by using a suitable accurate bar between the centres,

(ii) To check trueness of milling machine arbor,

(iii) To check parallelism of the shaper ram with table surface or vise.

(g) Slip Gauges (Gauge Blocks)

Slip gauge is reference standard of measurement. Instead of lines for reference as on a ruler, the slip gauge is made with two reference surfaces.

Slip gauge is a small block of hardened steel carefully machined on the lapping surface that it has almost an optical finish. Depending upon its shape, the slip gauge can be rectangular, square or square with center hole as shown in Fig. 12.

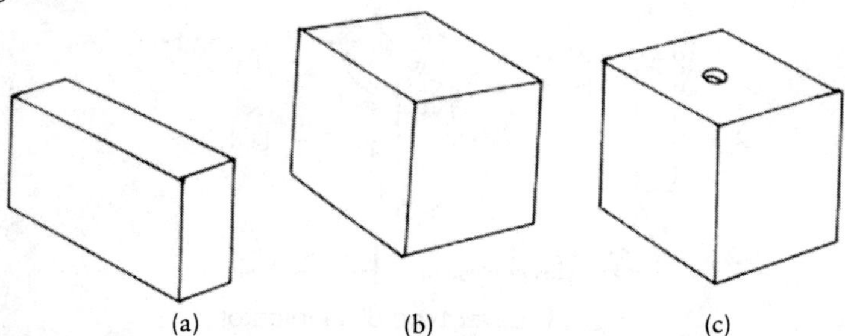

Fig. 12: Slip gauges (a) rectangular, (b) square, (c) square with centre hole

A standard set (M 112) having 112 slip gauges includes the following:

Range (in mm)	Steps (in mm)	No. of pieces
1.0005	–	1
1.001-1.009	0.001	9
1.01-1.49	0.01	49
0.5-24.5	0.5	49
25-100	25	4

Another set (M9\87) of slip gauges consisting of 87 pieces is also widely used. This set includes the following pieces:

Range (in mm)	Steps (in mm)	No. of pieces
1.001-1.009	0.001	9
1.01-1.49	0.01	49
0.5-9.5	0.5	19
10-100	100	10

Slip gauges are used in single or in combination to check and set micrometers, vernier calipers, snap gauges and indicators. They can also be used directly on workpieces to check centre distance, slot widths and shoulder clearances.

Gauge blocks are classified into following grades according to allowable tolerance for flatness, parallelism and length: Grade 00, Grade 0, Grade 1 and Grade 2. Grade 0, 1 and 2 are intended for general use. Grade 00 is normally used for calibration purposes. It should not be used in combinations.

In building up a definite size of gauge with gauge blocks, use as few blocks as possible. Start by selecting blocks which will eliminate the last figure to the right of the decimal point. For example, to build up a size of 3.582 mm, first eliminate 2 in the last place by selecting 1.002 mm block. The remainder is 2.58 mm. Again, the last digit 8 is eliminated with 1.08 block. The remainder is 1.5 mm which is eliminated with a single block.

Example:

3.582	
1.002	1.002 (Block 1)
2.580	
1.08	1.08 (Block 2)
1.50	
1.50	1.5 (Block 3)
0.00	3.582 mm requires three blocks

To join one block to another, the contacting surfaces of the two blocks are wiped and cleaned by a piece of chamois leather. The contacting surfaces are brought together by engaging a small part of the surfaces and sliding the blocks into full contact while pressing them firmly together. If the contacting surfaces are clean, they will cling together. This method of joining gauge blocks together is called 'wringing'.

3. Angular Measurement

Measurement of any angle precisely is one of the most important task in the field of measurement. Angular measurement is generally concerned with the measurement of individual angles on gauges, tools as well as small angular changes and deflections, etc. Various instruments used for measurement of angles on components are explained below:

(i) Combination Set

It is the most adaptable, common and versatile instrument used in layout and rough inspection work.

It consists of a rule, a square head, a center head and a protractor head as shown in Fig. 13. The rule is made of tempered steel with a groove cut along the length of one side. The other parts can slide along this groove. Each part is provided with a knurled nut for locking it into position.

The rule is available in the lengths from 225 mm to 600 mm. The combination set can be used simply as a rule. The combination set can be utilized for measuring angles by using protractor head. For this, the rule is placed on one side of the angle to be measured. The protractor is adjusted by rotating it till its working surface touches the adjacent side of the angle to be measured. The protractor is locked at this position and the reading is taken on the protractor scale.

The square head of the combination set is used as a try square for marking mitres and for checking sqaureness. The spirit level on the square head is used to see parallelism during squareness checking. The square head also helps the combination set to be used as a depth gauge.

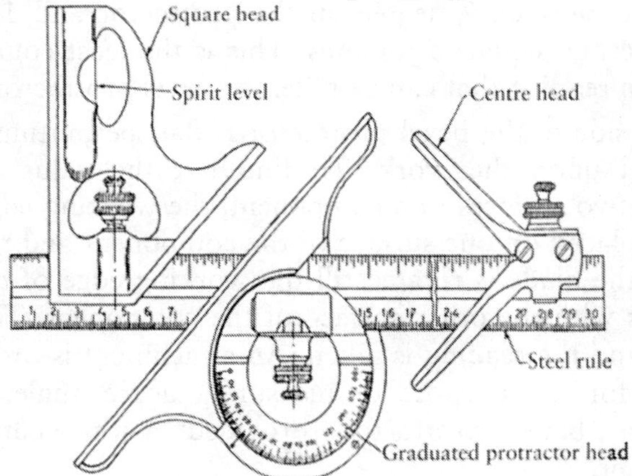

Fig. 13: Combination set

The centre head is used for locating the centre on the end of round bars. For this, the round bar is held beneath the rule in such a way that the sides of the centre head touch the round bar. A line is scribed on the end surface of the round bar along the side of the rule. The round bar is then rotated and again a line is scribed on the end surface of round bar. This is done for three positions. The centre of the triangle formed will be the centre of the round bar.

(ii) Bevel Protractor

It is probably the simplest instrument for measuring the angle between two faces of a component. A bevel protractor measures angles in degrees and minutes. The bevel protractor is an instrument having a main scale graduated in degrees and a sliding blade. The sliding blade is attached to a circular plate containing vernier scale. The vernier scale is also graduated to the right and left of zero upto 60 minutes. The sliding blade can be slid along its entire length around the centre of the main scale and can be clamped. The available size of the blade is 150 mm, 200 mm or 300 mm.

The vernier is graduated in such a way that 12 divisions on the vernier scale occupy the same space as 23 degrees on the main scale. Therefore, every division on the vernier is equal to one-twelfth of 23 degree which amounts to 115 minutes. Each two degree on the main scale is equal to 120 minutes. The difference between 2 degree on the protractor and 1 division on the vernier equals 5 minutes. This is the least count or the minimum reading that can be taken by bevel protractor.

One side of the bevel protractor is flat, permitting it to be laid level upon the work. To find out the value of angle between two surfaces of a component, the working edge of the stock is placed on one surface of the component and the turret holding the blade is rotated till the working edge of the blade coincides with the other surface of the component. The turret is fixed and the reading is taken. An attachment is provided at the top for the purpose of measuring acute angle. Fig. 14 shows a bevel protractor provided with acute angle attachment.

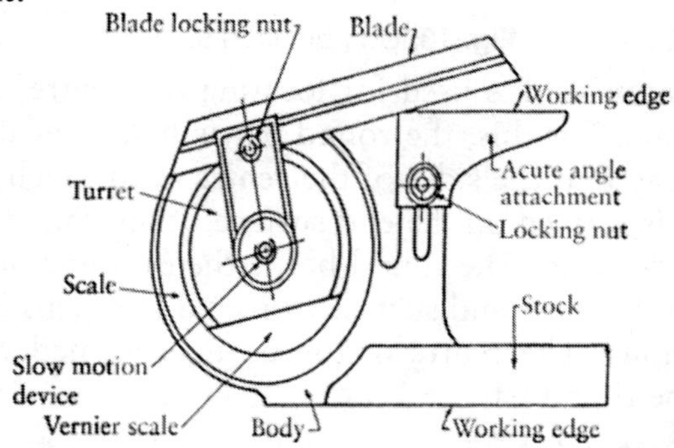

Fig. 14: Mechanical vernier bevel protractor with acute angle attachment

To facilitate designation, the bevel protractors are classified into four types as described below:

(a) **Type A**—Bevel protractor provided with a vernier graduated to read to 5 minutes of arc and also with a fine adjustment device and acute angle attachment.

(b) **Type B**—Bevel protractor provided with a vernier graduated to read to 5 minutes of arc and an acute angle attachment but without fine adjustment screw.

(c) **Type C**—Bevel protractor provided with vernier graduated to read to 5 minutes of arc but without fine adjustment device or acute angle attachment.

(d) **Type D**—Bevel protractor graduated in degrees and is not provided with either vernier or fine adjustment device or acute angle attachment.

Bevel protractors are designated by the type and size of the blade. A bevel protractor of type B with 150 mm blade shall be designated as: Bevel Protractor B 150.

(iii) Sine Bar

Sine bar is an instrument which is utilized for the measurement of angles with high degree of precision. It consists of a lapped steel bar at each end of which is attached an accurate roller as shown in Fig. 15. The axes of the rollers are mutually parallel and are also parallel to the upper surface of the bar. For ease in making calculations, the center distance between the roller is 100 mm or 250 mm. The sine bar is specified by the centre to centre distance between the two rollers.

A sine bar is always used in conjunction with a true surface such as surface plate from which measurements are taken. The operation of the sine bar is based on the trignometric relationship that the sine of an angle is equal to the opposite side divided by the hypotenuse. Measurement of the unknown side is done by slip gauges.

With the workpiece resting on a surface plate, H_1 and H_2 are built upto correct heights with slip gauge and their difference in elevation over the length L gives the sine of angle α.

In Fig. 16,
$$\sin \alpha = \frac{H_1 - H_2}{L}$$

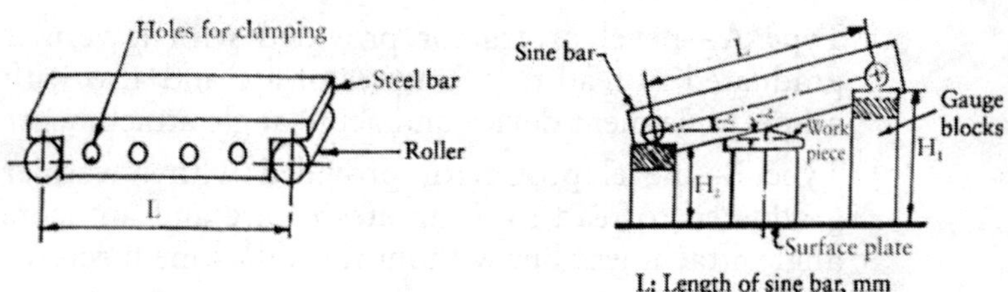

Fig. 15: Sine bar **Fig. 16:** Sine bar set up for measuring angle

where L is the known distance, usually 100 mm or 250 mm depending on sine bar used.

The angle α can be obtained from a trignometric table after sin α has been computed from the measurements.

A more expert way of handling a sine bar is illustrated in Fig. 17. The tapered workpiece whose, angle is to be measured is clamped or held firmly to the length of the sine bar by using angle plate. This can be accomplished by making use of holes in the sine bar or with the aid of a standard magnetic block. The sine bar is now tilted at an angle by building up slip gauges upto height H stack until the upper surface of the tapered pieces seems level. A dial indicator is run along the upper surface of the workpiece and the height of slip gauges stack is altered until there is no change of reading on the traversing dial indicator. Then the angle of taper of workpiece is equal to the angle between the sine bar and the surface plate. The angle can be calculated as

$$\text{Sin } \alpha = \frac{H}{L}$$

The sine bar may also be used to set a workpiece to a required angle. For this, the roller at one end is elevated above the other roller by a distance equal to the sine of the required angle, multiplied by the length of sine bar. The sine bar may be set in position by using slip gauges.

(iv) Measurement of Tapers

The most important element of a taper is its angle. For this, the diameter at the ends are derived by calculations from measurements near the ends. The measurements made in the determination of the end diameters can be used to obtain the angle of the cone. Referring to Fig. 18, if D and d are the diameters at the two ends and H is the distance between them, then

$$\tan \phi = \frac{D - d}{2H}$$

Direct measurement at the ends however will give only approximate results as the taper may "run off" slightly from near the ends. However, items such as plug and ring gauges require measurement of the length and of the diameter at the ends. The measurement of taper limit gauges thus involves determination of angle of taper and diameter at the ends.

The methods explained below for measurement of taper make use of trigonometric functions. The required dimensions are computed by taking several readings with instruments.

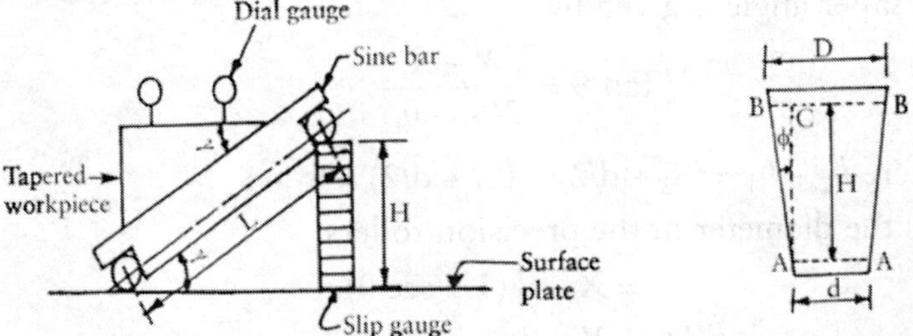

Fig. 17: Typical sine bar set up for measuring angle

Fig. 18: Taper measurement from measurement near the ends, D, and d

(i) *Measurement of Taper Plug Gauge by means of Rollers*

The taper angle of taper plug gauge is found out by using precision rollers and slip gauges. The method of measurement is to place the gauge with its small end on the surface plate. Two rollers standing on a pile of slip gauges (height S_1) are supported on each side of the taper plug, as shown in Fig. 19. The two piles of slip gauges are selected in such a way that the measuring instrument (micrometer or vernier calliper) is made clear of the surface plate while measuring over the rollers.

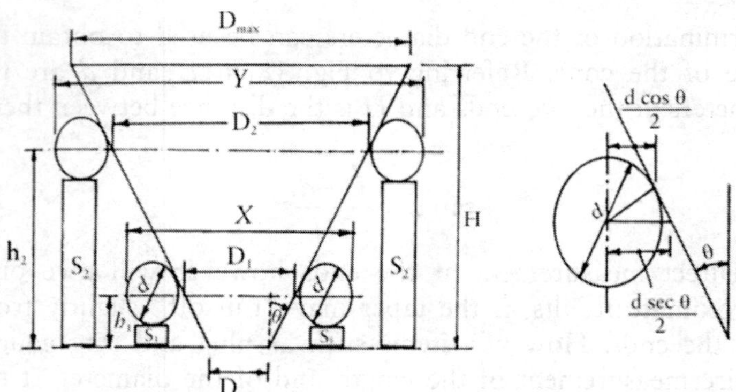

Fig. 19: Taper plug gauge measurement by rollers

Measure the size of the taper gauge over the rollers. Let it be X. Select another pile of slip gauges upto height S_2 to bring the roller close to the top of the gauge and again measure the taper gauge over the rollers and assume this measurement to be Y. The taper angle is given by

$$\tan\theta = \frac{Y - X}{2(h_2 - h_1)}$$

where $h_2 - h_1 = (S_2 + d/2) - (S_1 + d/2) = S_2 - S_1$

d is the diameter of the precision rollers.

$$D_1 = X - d\,(1 + \sec\theta)$$

$$D_2 = Y - d\,(1 + \sec\theta)$$

From D_1 and D_2, the diameters at the two ends can be found out.

$$D_{max} = D_1 + 2\,(H - h_1)\tan\theta$$

or $$D_{max} = D_2 + 2\,(H - h_2)\tan\theta$$

$$D_{min} = D_1 - 2\,h_1\tan\theta$$

or $$D_{min} = D_2 - 2\,h_2\tan\theta$$

(ii) *Taper Ring Gauge Measurement by Ball and Slip Gauges*

Internal taper of sufficiently large diameter is measured in the same manner to that used for taper plugs. However, instead of rollers, equal size balls are used. The dimensions are measured by slip gauges. As shown in Fig. 20(a), the ring gauge is stood with its large end on the surface plate. This is done to avoid the tendency of the balls to ride up the tapers when the slip gauges are inserted, as it happens if the large end is at the top. While measuring at the large end, if the balls are not placed on the slip gauges, the surface on which the balls and taper ring gauge stand should be ground or lapped.

The taper angle is found out as $\tan\theta = \dfrac{X - Y}{2(h_2 - h_1)}$

$D_1 = X + d\,(1 + \sec\theta)$

$D_2 = Y + d\,(1 + \sec\theta)$

The diameters at the large and small ends are found as follows:

$D_{min} = D_1 - 2H + 2h_1 \tan\theta$

$D_{max} = D_2 - 2H + 2h_2 \tan\theta$

(iii) *Taper Ring Gauge Measurement by Unequal Balls*

Many internal tapers are either too long or too small in diameter to permit use of slip gauges and equal size balls. In such cases, two balls are used of such sizes that one will rest in the cone near the large end and one near the small end making contact all round the circumference as shown in Fig. 20(b). Height measurements are made over the top of the balls and the diameters and taper are calculated in the following manner:

$$\sin\theta = \frac{r_1 - r_2}{(h_1 - r_1) - (h_2 - r_2)}$$

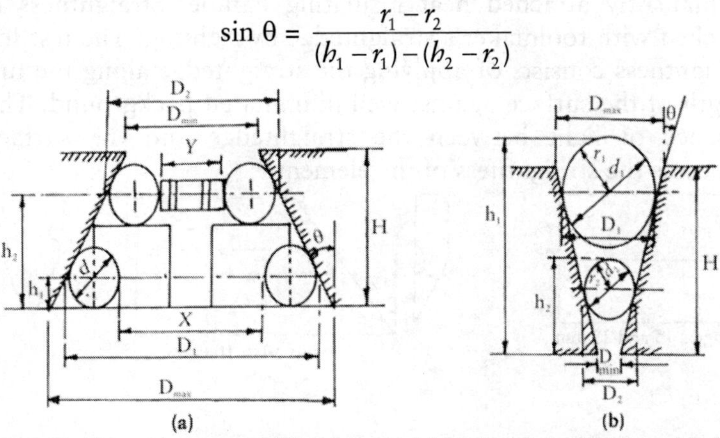

Fig. 20: Taper ring gauge Measurement (a) with slip gauges and balls, (b) by unequal balls

As $D_1 = d_1 \sec \theta$

and $D_2 = d_2 \sec \theta$

$$D_{max.} = D_1 + 2\,(H - h_1 + r_1) \tan \theta$$

$$D_{min.} = D_2 - (h_2 - r_2) \tan \theta$$

4. Checking of Surfaces

Checking of surfaces means checking of straightness, flatness and perpendicularity of surfaces. Surface checking is very important in metrology. The straightness, flatness and perpendicularity are checked with different surface checking devices as follows:

Straightedge

This is a measuring instrument having a rectangular (small size) alloy steel strip and an edge perfectly straight within the limits of accuracy. Straightedges are designed for checking straightness and flatness of parts. Straightedges are classified as:

(i) Toolmaker's straightedge

(ii) Straightedge with wide working plane

The toolmaker's straightedge has the highest accuracy and is available in various cross sections as shown in Fig. 21. Each straightedge of length from 150 mm to 300 mm shall have a permanently attached heat insulating handle. Straightness is checked with toolmaker's straightedge by sighting. The test for straightness consists of applying the straightedge along the full length of the surface against well-illuminated background. The absence of light between the straightedge and the surface indicates the straightness of the element.

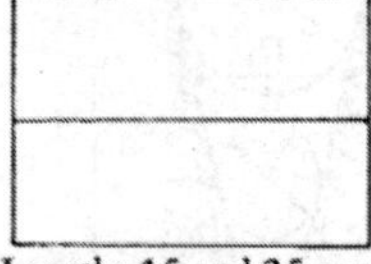

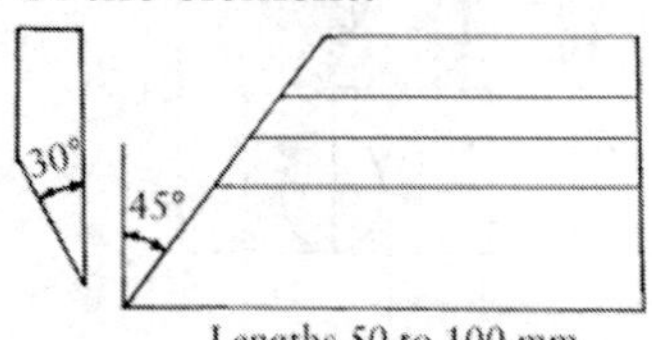

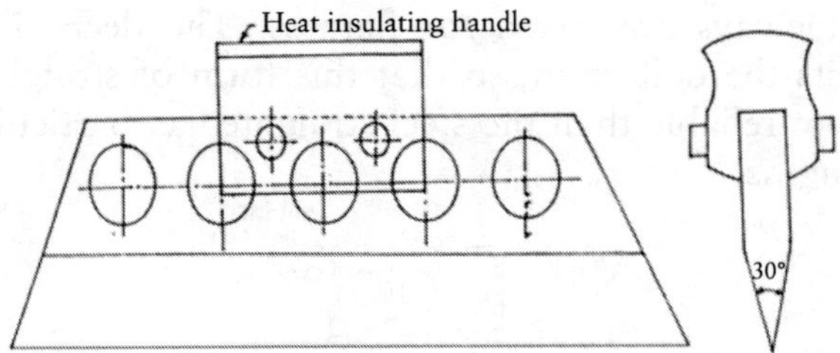

Fig. 21: Straightedges

Straightedges with wide working planes can be of steel or cast iron. Steel straightedges are made in the form of rectangular or I-section beams with cored holes to make them light as shown in Fig. 22. They are used to check surfaces for straightness by direct comparison method. The method involves placing the straightedge with its working surface on two identical slip gauges located on the surface being tested. The distance between the surface of the straightedge and the part being tested are then measured at different points with the help of slip gauges. To reduce measuring errors caused by straightedge gravity deflection, this measurement method should not be used with long straightedges (over 2000 mm).

Fig. 22: Steel straightedge with wide working plane (a) rectangular section, (b) I-section

Cast iron straightedges are made bow shaped as shown in Fig. 23. This straightedge is used to check flatness by means of pigment transfer technique. In this technique, a thin layer of blue paint is applied to the working surface of the straightedge. The straightedge is further placed in contact with the surface being checked and is slightly moved in longitudinal direction. This results in blue spots defining bearing points of the test surface which contact the straightedge working surface. For a flat surface, the blue spots are distributed uniformly over the whole length. In this technique, the presence of deviations from flatness is revealed but its numerical value cannot be determined. Its principal use is in machine tool manufacture where slideways are tested for flatness. The deep rib at the back limits the deflection, so that this form of straightedge is much more reliable than the steel straightedge, particularly for longer lengths.

Fig. 23: Bow shaped cast iron straightedge with wide working plane

Try Square

It is a device employed to see whether a certain angle between two surfaces or features is 90° or not. The name try square means that this instrument is not the precise type, though various grades of accuracy are available. The instrument consists of: (i) stock and (ii) blade, the two making an angle of 90° with each other externally as well as internally. A try square is shown in Fig. 24. Two types of try squares are commonly used, one with extra thick stock and the other without stock. The one without stock is less rigid and so requires thicker blade. There are three grades of accuracy (grade A, grade B and grade C) for try squares suitable for workshop and inspection purposes. The size of the square is designated by the length of the blade and the accuracy, that is grade.

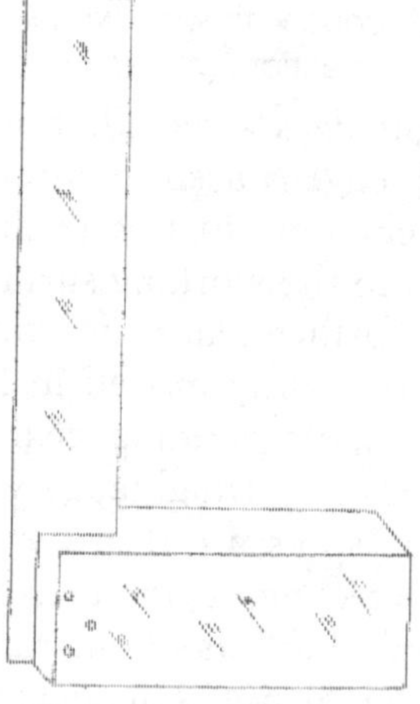

Fig. 24: Try square

The job to be checked is held in, the left hand and the try square is held in the right hand from the stock. The stock is brought in contact with one surface and blade with the other surface. The square is checked by seeing through the line of contact of the blade and the surface of the job. If no light is passing, the angle is correct. If the light passes from the blade near the stock, the angle is more than 90° and if from the tip of the blade, the angle is less than 90°.

The squares are made of good quality tool or alloy steel hardened and ground and should be free from cracks, flaws and pits. Sharp edges of the try square are removed and it is made smooth and bright all over. The stock or the blade is slightly relieved across the width at the junction of the inner faces of the stock and blade.

The tolerable departure from straightness of each edge of blade and working faces of stock for three grades is given in Table 1.

Table 1

Length of blade (mm)		Deviation in microns		
Above	**Up to and including**	**Grade A**	**Grade B**	**Grade C**
0	150	0.60	1.25	2.50
150	300	1.00	1.50	3.10
300	1000	—	1.80	3.75

5. Surface Plate

The surface plate forms the basis of measurement and establishes the reference plane from which all precision measurements start. It has got a flat surface and all other measuring instruments, having flat bases, can be placed on the surface plate. The big surface plate generally rests on four levelling screws with the help of which its surface can always be kept horizontal. The top surface is scraped to true flatness.

The surfaces plate serves two main purposes. First, it is used as a datum plane for taking accurate linear and angular measurement. Secondly, it serves for testing the degree of flatness of other surfaces with the aid of a suitable marking compound. For accurate work, it is essential that form and dimensions of surface plate must be such that distortion is reduced to a minimum under heavy loads. For this reason, it is customary to reinforce the plate properly by the provision of a frame and ribbing of considerable depth. A surface plate is shown in Fig. 25.

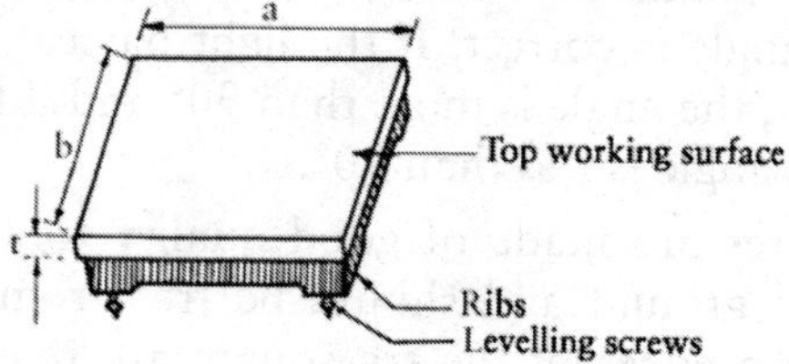

Fig. 25: Surface plate

The surface plate should be made from a material which will provide a high degree of rigidity and freedom from warping and capable of taking a high finish and the surface must be resistant to wear. Most commonly used material for this purpose is either the plain or alloyed close-grained cast iron of good quality. It is aged to set it free from all internal stresses. Granite plates are a modern alternative. As compared to cast iron surface plate, granite plate is cleaner to handle, not subject to rusting and offers a smoother surface.

Cast iron surface plate has three grades namely grade I, grade II and grade III. Grade I is applicable to surface plates of sizes upto and including 2000 mm × 1000 mm. Grade II is applicable for sizes upto and including 4000 mm × 1250 mm and grade III is for sizes from 2000 mm × 1250 mm to 5000 mm × 3150 mm. These plates are designated by their length, breadth and grade. For example, a cast iron surface plate of length 2000 mm, breadth, 1000 mm and of grade I shall be designated as:

Cast Iron surface plate 2000 × 1000 Grade I

Granite plate has also three grades and like cast iron surface plate, it is designated by its length, breadth and grade.

The recommended sizes of cast iron surface plates of grade I and II and the tolerance on their flatness are given in Table 2.

When not in use, a wooden cover is always placed over it. When in use, the whole surface of the plate must be used in order to have uniform wear and not only the centre or the corners. Also grease must be applied to the top surface of the plate when not in use for a long time. A check for flatness must be made after regular intervals.

Table 2: Size and Accuracy of Surface Plates

Nominal size a x b in mm (Length x breadth)	Minimum Thickness (t) of top of plate in mm	Tolerance on flatness in microns	
		Grade I	Grade II
100 x 100	9	2.5	5
160 x 100	9	3.5	7
160 x 160	9	3.5	7
250 x 160	12	3.5	7
250 x 250	15	5	10
400 x 250	15	7.5	15
400 x 400	18	7.5	15
630 x 400	18	7.5	15
630 x 630	21	7.5	15
1000 x 630	24	10.0	20
1000 x 1000	30	12.5	25
1600 x 1000	30	17.5	35
2000 x 1000	36	20.0	40

6. Limit Gauges

The method used for ascertaining the dimensional accuracy of a component may be either by direct measurement or by means of limit gauges. A limit gauge is an instrument by means of which certain dimensions or form can be checked for which it is designed or adjusted. For small quantities, the direct measurement method can be employed, but for high rates of production, gauges essential. These gauges are simple to use and either accept or reject the workpiece being checked by them. These gauges can be used by skilled operators.

Classification of Limit Gauges

The limit gauges can be widely classified as: (i) Plug Gauge, (ii) Ring Gauge, and (iii) Snap Gauge.

(i) Plug Gauge

A plug gauge is an accurate cylinder used for size control of holes. Commonly used plug gauges are of 3 types:

(a) Plain cylindrical, (b) Cylindrical taper

(c) Thread plug gauges

The plain cylindrical plug gauges are of two types—reversible and progressive. Reversible plug gauges have a 'Go' plug in one end of the holder and 'Not Go' plug in the other end. The 'Go' end should enter the hole with little or no interference. The 'Not Go' will not enter the hole. Progressive gauges have both 'Go' and 'Not Go' plugs on one end.

The 'Not Go' end is not subjected to wear as it seldom enters the hole of the work under inspection. Therefore, its parallel portion is made relatively short in comparison to that of 'Go' end. This feature distinguishes the 'Not Go' end from 'Go' end. The 'Go' end is made 1.5 to 2 times longer than the 'Not Go' end as shown in Fig. 26.

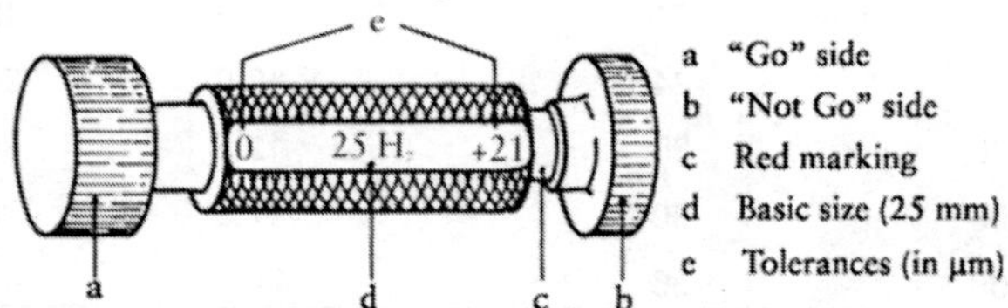

Fig. 26: Plain cylindrical plug gauge

In order to increase the wearing properties of plug gauges, they are chrome plated. An additional advantage of chrome plating is that when finally the 'Go' end wears it can be readily renewed by plating and brought to original dimensions by grinding and lapping.

While using a plug gauge, it is generally the tendency of the inspecting person to insert the gauge obliquely into a hole. To avoid this, a groove is machined behind the end. This type of gauge is called pilot gauge. As shown in Fig. 27, the gauge has a small chamfer, then a narrow ring or pilot having the same diameter as that of gauge followed by a groove and the main body of the gauge.

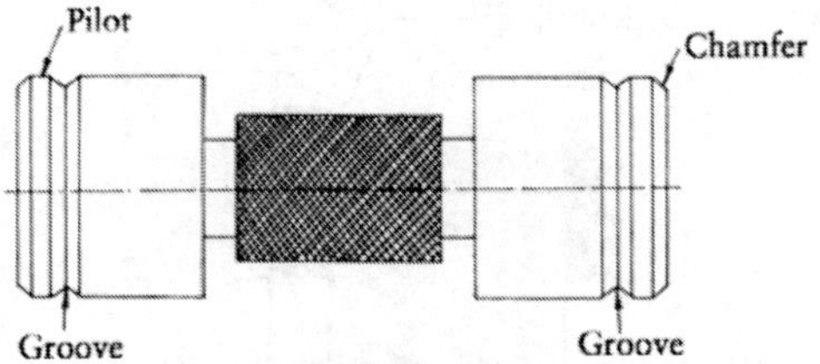

Fig. 27: Pilot plug gauge

Cylindrical taper plug gauges are used for checking the amount of taper and the size of tapered cylindrical holes such as in drill sleeve and machine tool spindles. Fig. 28 shows the testing of a taper by taper plug gauge.

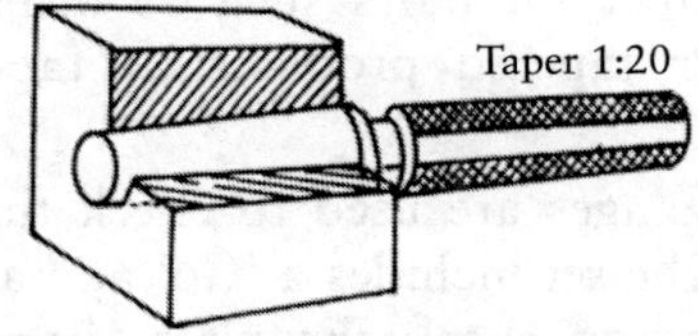

Fig. 28: Testing of taper hole by taper plug gauge

Tapped holes can be checked with a thread plug gauge. Fig. 29 shows a double ended thread plug gauge with one 'Go' end and the other 'Not Go' end.

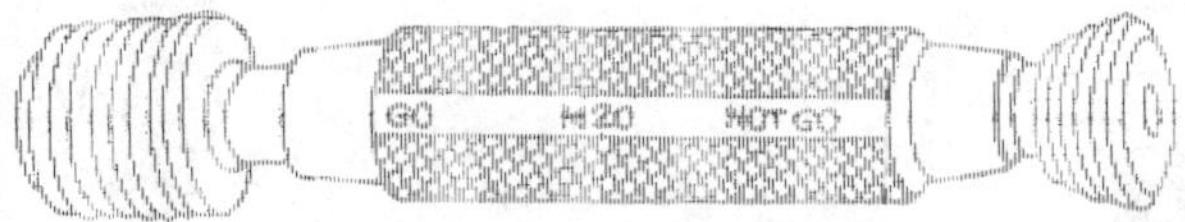

Fig. 29: Thread plug gauge

(ii) Ring Gauge

This gauge is used for checking external diameters of parts. Ring gauges are classified as plain ring gauge, taper ring gauge and thread ring gauge.

Plain ring gauges are in the form of cylindrical rings and are used to check external diameters of straight round parts. They are made in pairs as shown in Fig. 30. The 'Not Go' ring is used to check the minimum size limit while the 'Go' ring checks the maximum size limit. If both the rings pass over the workpiece, it is undersize. If neither does, the workpiece is oversize.

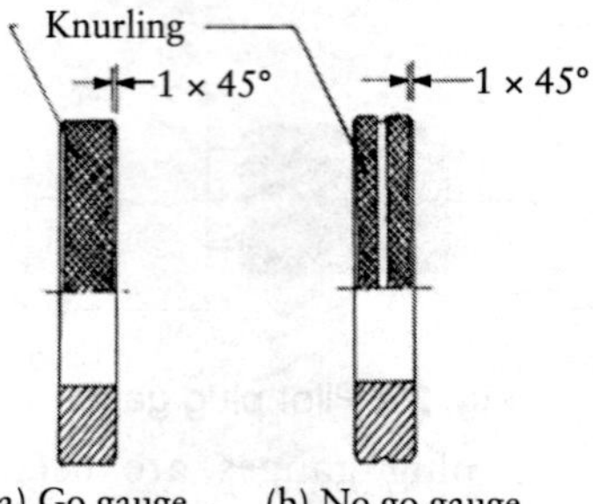

Fig. 30: Plain ring gauge

Taper ring gauges are used for checking the diameter and the amount of taper on parts such as drills, reamers, lathe centers and other tapered products. A taper ring gauge is shown in Fig. 31.

Thread ring gauges are used to check the accuracy of an external thread. The set includes a 'Go' and a 'Not Go' gauge. The gauge is used to check the pitch diameter, lead, flank angle, maximum pitch diameter, and the clearance at the minor diameter simultaneously. Fig. 32 shows a thread ring gauge.

Fig. 31: Tapered ring gauge

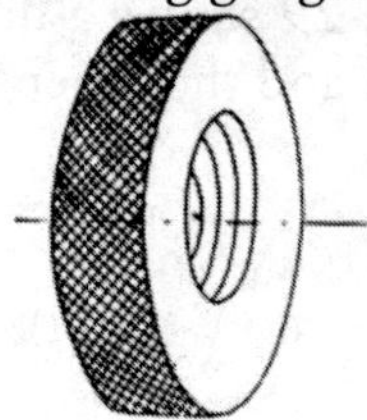

Fig. 32: Threaded ring gauge

(iii) Snap Gauge

This gauge is used for checking the external diameters. It consists of two jaws of calliper forms at its two ends as shown in Fig. 33. One jaw is marked 'Go' and corresponds to maximum allowable diameter. The other jaw is marked 'Not Go' and corresponds to minimum allowable diameter.

An improvement over the above type is the one shown in Fig. 34, which consists of a C-shaped frame having two parallel jaws which are made to standard sizes. These sizes cannot be adjusted. This type of gauge is largely being replaced by using two button anvils on the jaws which are adjustable. In operation, the outer button is set to 'Go' size and the inner button to 'Not Go' size as depicted in Fig. 35. This gauge is widely used compared to other snap gauge as it is lighter in weight, easy to adjust and is designed to permit interchangability of as many parts as possible.

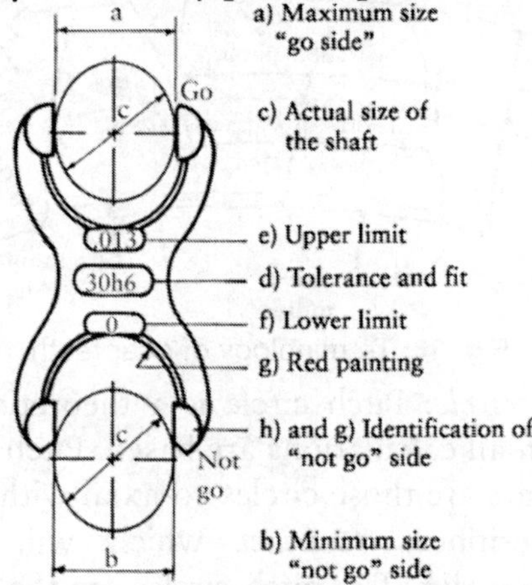

Fig. 33: Designation on snap gauge

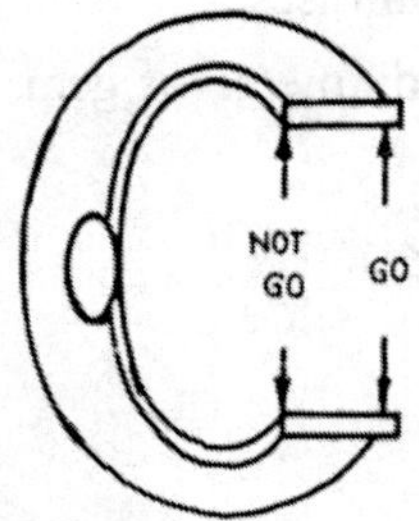

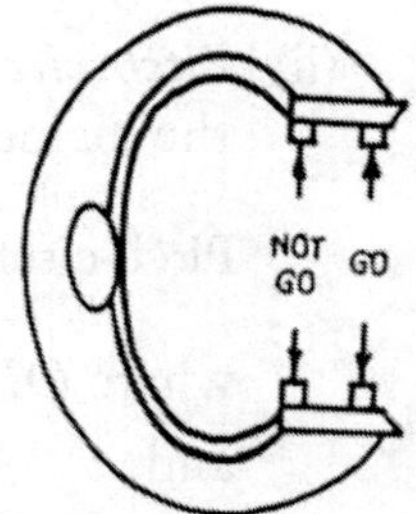

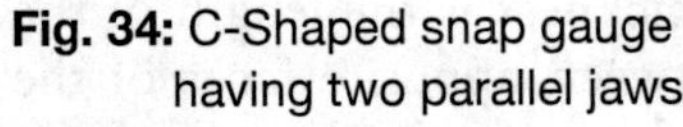

Fig. 34: C-Shaped snap gauge having two parallel jaws

Fig. 35: C-Shaped snap gauge having two button anvils

7. Gear Measurements

The subject of gearing and gear measurement forms a highly specialized field. Therefore, fundamental commonly accepted terms and simple techniques of gear measurement are described below:

Terminology of Gear Teeth

The terminology of gear teeth is shown in Fig. 36. The definitions of the terms are given below:

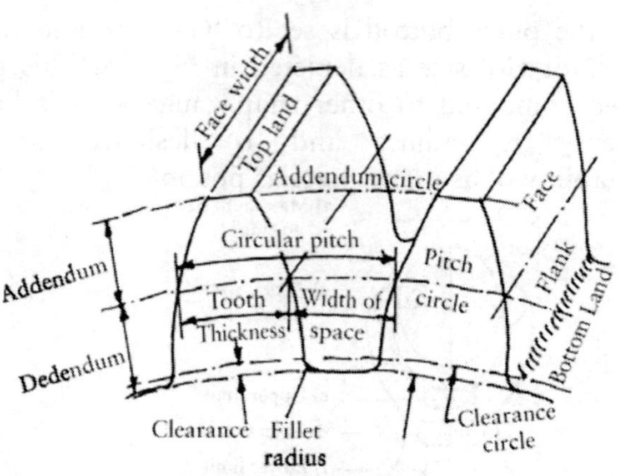

Fig. 36: Terminology of gear teeth

(i) *Pitch circle*: Pitch circle is a theoretical circle upon which all calculations are based. Pitch circle of a pair of gears are those circles co-axial with the gears, and in peripheral contact which will roll together without slip. The pitch circles are the outlines of the imaginary smooth rollers or friction discs.

(ii) *Pitch circle diameter*: Pitch circle diameter of gear is the diameter of its pitch circle.

$$\text{Pitch circle diameter } (D) = \frac{T \times OD}{T+2}$$

where OD = outside diameter

and T = number of teeth

(iii) *Tooth thickness*: Tooth thickness is the length of the arc of the pitch circle between opposite faces of the same tooth. Chordal thickness is the length of the chord subtended by the tooth thickness arc.

(iv) *Circular pitch*: It is the distance measured on the pitch circle from a point on one tooth to a corresponding point on an adjacent tooth. Thus, the circular pitch is equal to the sum of the tooth thickness and the width of the space.

$$\text{Circular pitch} = \frac{\pi D}{T}$$

where D is pitch circle diameter.

(v) *Module*: It is the ratio of the pitch circle diameter to the number of teeth. Module $m = D/T$ and is expressed in mm (i.e. diametral pitch per tooth of gear wheel).

(vi) *Diameter pitch* (D_p): It is the ratio of number of teeth on the gear to the pitch circle diameter. Thus, it is the reciprocal of the module.

(vii) *Addendum*: It is the radial distance between the top land and the pitch circle. It is equal to one module.

(viii) *Clearance*: It is the amount by which the width of a tooth space exceeds the thickness of the engaging tooth measured on the pitch circle. Its standard value is 0.157 m (m is the module).

(ix) *Dedendum*: It is the radial distance from the bottom land to the pitch circle.

Dedendum = Addendum + Clearance

= m + 0.157 m

= 1.157 m (m is the module of the gear tooth)

(x) *Clearance circle*: The clearance circle is the circle tangent to the addendum circle of the mating gear.

(xi) *Backlash*: The amount by which the width of a tooth space exceeds the thickness of the engaging tooth measured on the pitch circle is called backlash. (Backlash in gear system means some looseness in the fitting)

Measurement and Testing of Gear

The important elements of gear are tooth thickness and involute form of tooth profile. The tooth thickness is measured by gear tooth vernier and the involute form of tooth profile is tested by involute tester.

Gear Tooth Vernier

It is a special type of vernier calliper which consists of a horizontal and vertical vernier slide. Fig. 37 shows a gear tooth vernier. The vertical slide controls the height on the gear tooth at which the measurement is taken. As the tooth thickness is measured at the pitch circle, the vertical vernier slide is set to the addendum of the tooth. The jaws of the instrument are inserted on the tooth to be measured. The horizontal vernier slide is adjusted so that the jaws just touch the tooth. The reading on the horizontal vernier slide gives the chordal thickness of gear tooth.

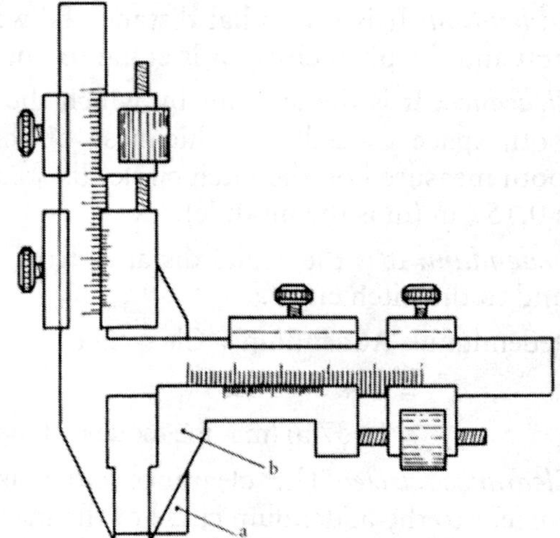

Fig. 37: Gear tooth vernier (a) vertical vernier slide (b) horizontal vernier slide

Involute Tester

The accuracy of involute curve of the tooth flanks of a gear is very important for the accuracy of a gear. The instrument used for testing the involute profile of gear tooth is based on the principle that if a straight edge is rolled without slip on a base circle disc, a point (A) on the straight edge will trace out an involute curve as shown in Fig. 38.

An involute testing equipment is shown in Fig. 39. The gear to be tested is mounted integral with a base circle disc. The disc is then rolled without slip along a straight edge. The involute of a tooth flank will slide past a stylus on the straight edge. This stylus is capable of registering any movement as the involute slides past it and these movements are amplified and the errors are recorded in the form of a graph. If the resulting graph is a straight line there is no error in the involute form of the gear. Otherwise, the tooth flank has departure from a true involute curve.

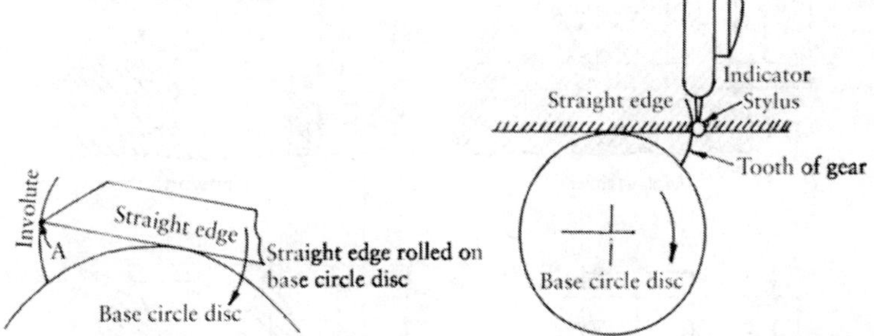

Fig. 38: Generation of an involute

Fig. 39: Gear involute testing equipment

8. Screw Thread Measurement

Screw threads are helical grooves of various shapes wound around a cylindrical or a conical component. An important application of screw thread is as fastening device. Thread fasteners are used to join parts and they find universal application in engineering. Also screws of length more than normal length like lead screws are used to translate accurate motion.

Screw threads are used for four general purposes.

(i) For fastening devices such as screws, studs, bolts and nuts.

(ii) To transmit motion, when thread cutting on a lathe.

(iii) To provide accurate measurement as in case of micrometer.

(iv) To increase torque. (Screw jacks may be used to raise heavy loads.)

The form of thread used for different applications depends very much on the purpose of application. More than sixty forms of screw threads are used in the industry all over the world. But the most commonly used are Metric, Whitworth, British Association (BA), Unified, Acme, Butteress and Square, etc. Metric, Whitworth, BA and Unified threads are widely used in fasteners. Various types of screw threads are shown in Fig. 40

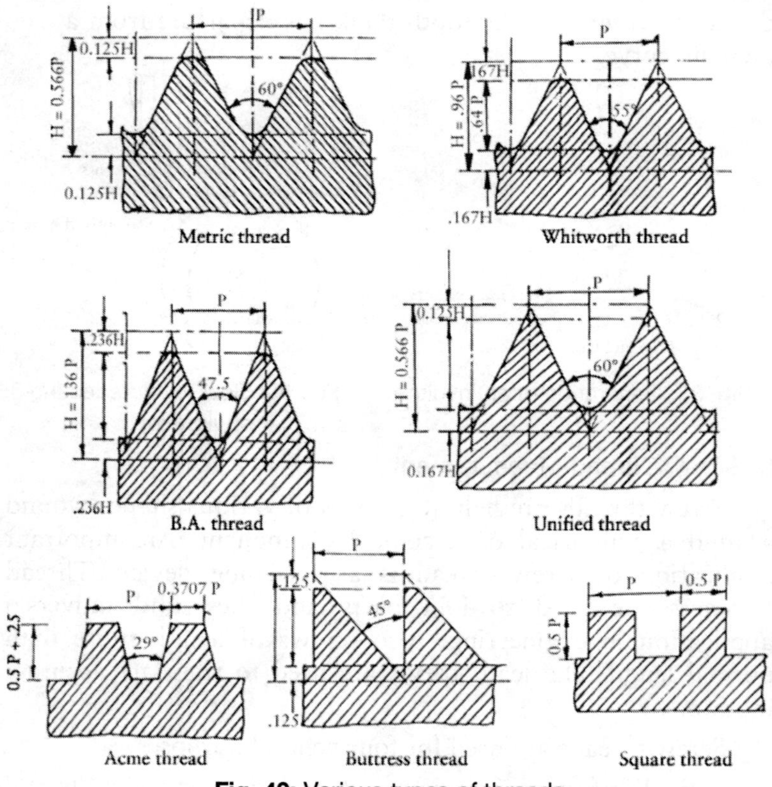

Fig. 40: Various types of threads

In addition to general purpose fasteners applications, certain threads called transmission threads are used to repeatedly move machine parts against heavy loads. Square, Acme and the Butteress threads come under this category. Square thread is the stringent and the most efficient but it is also the most difficult to cut due to its parallel sides and it cannot be adjusted to compensate for wear. The Acme thread is slightly weaker and less efficient. The Buttress form is used for translation of loads in one direction only because of its non-symmetrical form and combines the high efficiency and strength of square thread with the ease of cutting and adjustment of Acme threads.

Geometry of Screw Threads

Fig. 41 indicates terms commonly used with regard to threads.

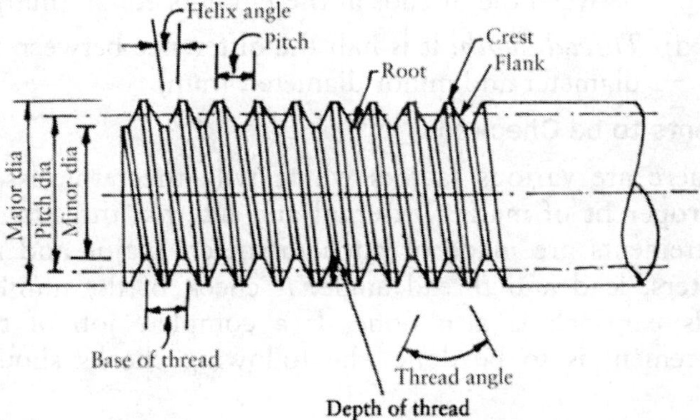

Fig. 41: Terminology of screw thread

(i) *Crest:* It is the top surface joining the two sides of a thread. Crest may be rounded or flat.

(ii) *Root:* It is the bottom surface joining the sides of adjacent threads. Root may be rounded or flat.

(iii) *Flank:* The surface of the thread which connects the crest with the root.

(iv) *Pitch:* Pitch of a thread is the distance measured parallel to its axis, between corresponding points on adjacent threads. (expressed in mm)

(v) *Lead:* Lead of a screw thread is the distance it advances axially in one revolution. (mm)

(vi) *Thread angle:* It is the angle between the sides, measured on an axial plane. (degrees)

(vii) Flank angle: It is the angle between flank of thread and a plane perpendicular to the axis, measured in axial plane. (degrees)

(viii) *Major diameter:* Major diameter is the largest diameter of a screw thread on a screw or nut. (mm)

(ix) *Minor diameter:* It is the smallest diameter on the screw or nut. (mm)

(x) *Effective or pitch diameter:* Effective diameter is the diameter of an imaginary cylinder which would pass through the threads at such points as to make the width of the threads and width of the spaces between the threads at these points equal. (mm)

(xi) *Thread depth:* It is half the difference between major diameter and minor diameter. (mm)

Elements to be Checked

There are various factors which will affect the assembly and proper fit of mating external and internal threads. Thus, measurements are made of pitch diameter, major and minor diameters, lead and thread angle. A check of the number of threads per inch is also done. If a complete job of thread measurement is to be done, the following checks should be made:

(a) Threads per inch Count or Pitch Measurement

The test for thread count is simple. It is done by laying a steel rule against an outside thread and counting the number of peaks in one inch. A faster and better method is the use of screw pitch gauge. For accurate results, profile projector is used.

(b) Measurement of Effective Diameter

For measuring or checking effective diameter of an external screw thread, three wire method is used. The three wire method of measuring effective diameter is explained as follows:

Three Wire Method

Fig. 42 shows the principle of measurement over wires. Three wires of identical size are placed in the vees of the thread. Two wires are placed in contact with the sides of the thread on one side and the third, coming immediately between them, on the other side. The wires are made of hardened steel and are lapped to sizes suitable for various pitches. For each pitch of thread, there is a 'best size' wire. The 'best size' wire is of such diameter that it touches the thread flanks at the pitch line. Wires of other sizes can also be used for measuring effective diameter, provided there is no error of flank angle.

A micrometer reading is taken over the wires. The effective diameter can be found from the micrometer reading taken over the wires by calculations. Consider the general case of a V thread: (as shown in Fig. 43)

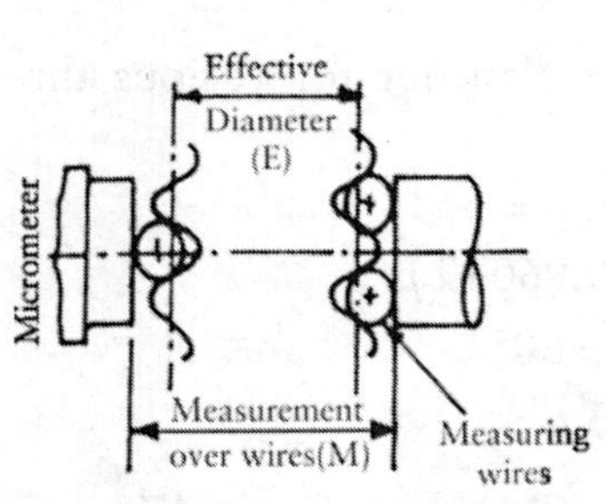

Fig. 42: Measurement over wires

Fig. 43: Effective diameter measurement by 3-wire method

Let M = Measurement taken over the wires

r = radius of wire

d = diameter of the wire

α = thread angle

p = pitch

E = Effective diameter

The best size wire for a Vee thread is given by

$$d = \frac{1}{2} p \sec\frac{\alpha}{2}$$

$$AD = r \operatorname{cosec}\frac{\alpha}{2}$$

$$CD = AD - AC$$

As $\quad AC = \frac{p}{4} \cot\frac{\alpha}{2}$

therefore $\quad CD = r \operatorname{cosec}\frac{\alpha}{2} - \frac{p}{4} \cot\frac{\alpha}{2}$

Effective diameter $E = M - 2\,CD - 2r$

$$= M - 2\left(r \operatorname{cosec}\frac{\alpha}{2} - \frac{p}{4}\cot\frac{\alpha}{2}\right) - d$$

$$= M - d \operatorname{cosec}\frac{\alpha}{2} + \frac{p}{2}\cot\frac{\alpha}{2} - d$$

$$= M - d\left(\operatorname{cosec}\frac{\alpha}{2} + 1\right) + \frac{1}{2}p\cot\frac{\alpha}{2}$$

From this formula, effective diameter for various threads are deduced:

(i) For Whitworth thread, $\alpha = 55°$
$E = M - 3.16568\ d + 0.96059\ p$

(ii) For Unified thread, $\alpha = 60°$
$E = M - 3\ d + 0.86603\ p$

(iii) For Birtish Association (BA) thread , $\alpha = 47\frac{1°}{2}$
$E = M - 3.48295\ d + 1.1363\ p$

For small diameter thread a hand micrometer is used to take measurement over the wires. It is always awkward to hold the screw thread, with three wires, in one hand. Therefore screw thread gauges are measured by using a floating carriage diameter measuring machine. This machine helps in taking measurements perpendicular to the screw axis.

Thread micrometer is also used for measuring effective or pitch diameter. A brief description of the thread micrometer is given as follows:

Thread Micrometer

The thread micrometer resembles an ordinary micrometer as shown in Fig. 44, but it has special contacts to fit in the screw thread form that is to be measured. In this micrometer, the end of the spindle is pointed to the Vee-thread form with a corresponding Vee-recess in the fixed anvil. When measuring threads only, the angle of the point and the sides of the Vee anvil that is the flanks of the threads should come into contact with the screw thread. If correctly adjusted, this micrometer gives the pitch diameter.

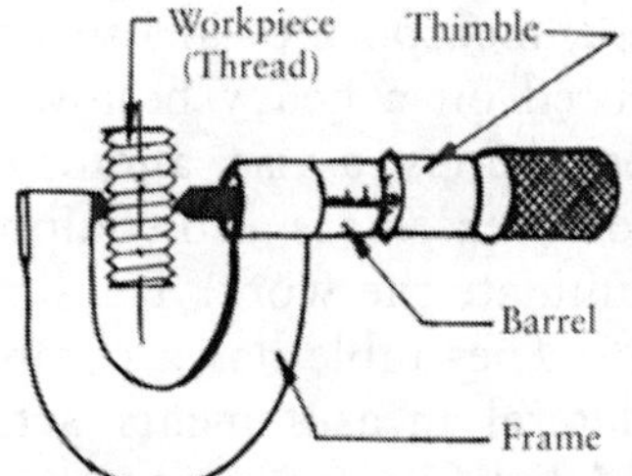

Fig. 44: Screw thread micrometer

Profile projector and tool maker's microscope are also used for the measurement of pitch diameter.

(c) Measurement of Major and Minor Diameter

The major diameter of screws can be checked with caliper type gauges (vernier, micrometer) in the same manner in which the outside diameter of a cylinder is measured. The minor diameter measurement, however, brings some special considerations. For this purpose, 'chisel' shaped anvils must be used on the gauge. The flanks of these anvils are cleared to such an extent that they will not make contact at the pitch diameter or some other diameter along the flank of the thread. This method of measuring the minor diameter is applicable only to threads having a definite radius at the root of the thread. Other methods of measuring minor diameter is by checking visually on tool maker's microscope or profile projector.

(d) Measurement of Thread Angle and Thread Form

Thread angle and form of a screw thread is checked by tool maker's microscope. Profile projector is also used for this purpose.

9. Tool Maker's Microscope

The sketch and equipment picture is shown in Fig. 45(a) and (b). The tool maker's microscope derives its name from the fact that initially it was being used to measure different tool angles. It consists of an optical head which can slide along the guideways of the vertical column and can be clamped with a screw at any position. Focussing is performed by adjusting on an optical head. The stage or table on which the workpiece to be measured is kept, incorporates in itself a transparent glass plate which is secured on a heavy hollow base. The table is equipped with clips, clamps, a vise, and/or centers for holding various types of workpieces. It is also equipped with adjustable lighting unit to illuminate the work. The illumination diagram is shown in Fig. 46. The table has a compound slide to give longitudinal and lateral measurements actuated by accurate micrometer screw. A light source is incorporated at the back of the base, from which a horizontal beam of light comes. This beam of light is reflected at right angle

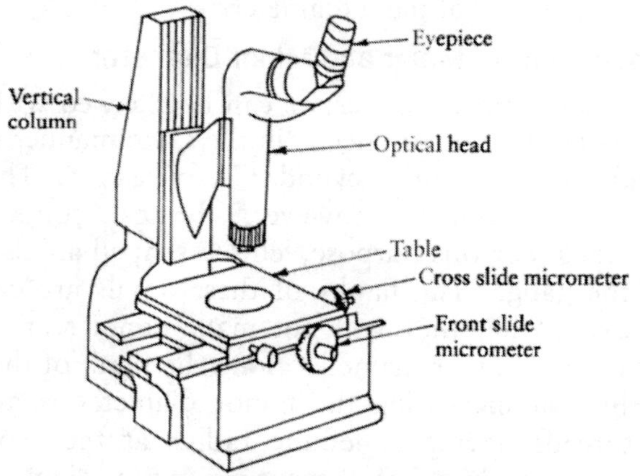

Fig. 45 (a): Toolmaker's microscope

by a mirror upwards towards the table. The light beam passes through the transparent glass plate of the table on which the workpiece is placed. A shadow image of the workpiece contour passes the objective of the optical head and gets projected on a screen after reflections from three prisms. The observations are made through the eyepiece to facilitate measurements. Cross lines are engraved on the screen which can be rotated through 360°. The angle of rotation of screen can be read on the optical head. The eyepiece field of view contains an illuminated circular scale with minimum reading of 1 minute. The procedure adopted for measuring the distance between two points on a workpiece is as follows:

The image of the workpiece is adjusted and focussed on the screen and the first point on the workpiece is made to coincide with the cross-line on the screen. The reading from the longitudinal micrometer is noted. The workpiece is then traversed with the help of micrometer screw till the second point on the workpiece coincides with the cross-line. The micrometer reading is noted again. The distance between the points is found out by finding the difference in the two readings. Similarly angles are measured by successively setting one of the cross-lines on the screen with the sides of the image of the angle and noting the angle through which the cross-line has been turned.

The tool maker's microscope has its greatest utility in measuring odd profiles, hole locations and locations of odd profiles, angles, etc., especially on thin flat stock when other measurement techniques are difficult. Some of the typical applications of tool maker's microscope include measurements of various parameters of external threads and measurements on tools (tool, geometry and tool wear), templates and gauges.

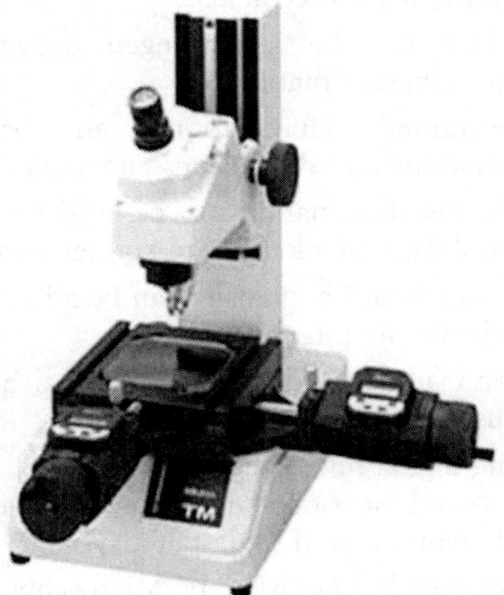

Fig. 45(b): Tool Makers' Microscope (Max. Magnification: 200X) (Courtesy: Mitutoyo, Japan)

Dimensions and angles can be measured; also shapes of screws and gears can be checked using reticles of various standards.

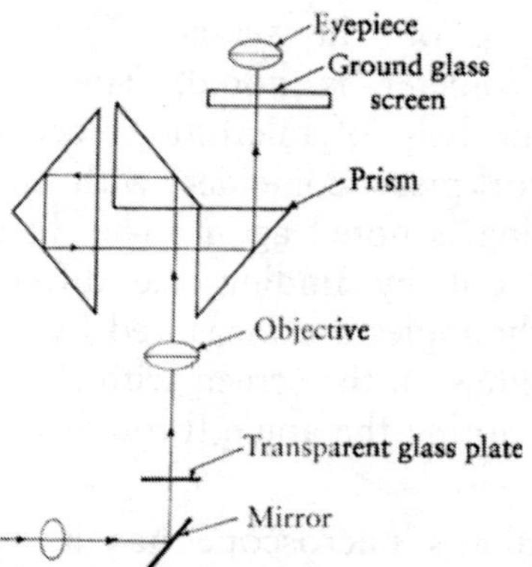

Fig. 46: Illumination in toolroom microscope

Features of Tool Makers' Microscope:

1. The eyepiece section of the optical tube incorporates a vernier protractor. Angle measurement is done by turning the protractor to align the cross-hair reticle with the workpiece image.
2. Reticles can be interchanged depending on the measurement situation.
3. Transmitted illumination can be used for measurement of dimensions and angles.
4. Reflected illumination can be used for measurement of tool wear (flank wear and crater wear).
5. The illumination intensity can be adjusted depending on the requirement.
6. X and Y stages enable longitudinal and transverse adjustment/location of workpieces.
7. Using digital micrometer heads, X and Y coordinates for two-dimensional data processing can be stored and analyzed on the computer system.

Accessories used in Tool Makers' Microscope (Fig. 47):

1. Eyepieces (with Magnifications of 10 X, 15 X and 20 X)

2. Objectives (with Magnifications of 2 X, 5 X and 10 X). Different objectives help to view the images at different working distances.

   (Total Magnification is the product of Objective Magnification and Eyepiece Magnification).

   Magnifications are stamped on the eyepieces and objectives.
3. Rotary Tables
4. Twin Bulb Reflected Illumination Unit.
5. Different Types of Reticles

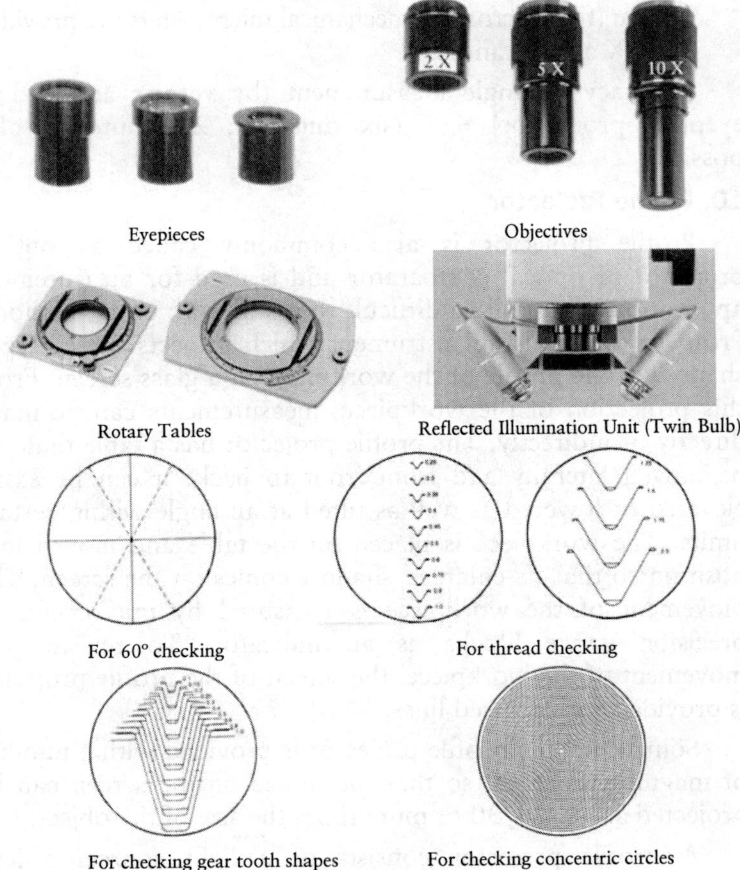

Fig. 47: Accessories used in Tool Makers' Microscope

Eyepiece – Objective details

	Eyepiece 10 X			Eyepiece 15 X (standard)			Eyepiece 20 X		
	Total Magnification	Viewing Field	Depth of Focus	Total Magnification	Viewing Field	Depth of Focus	Total Magnification	Viewing Field	Depth of Focus
Objective 2 X (Standard)	20 X	ϕ 6.5 mm	158 μm	30 X	ϕ 6.5 mm	124 μm	40 X	ϕ 5 mm	107 μm
Objective 5 X	50 X	ϕ 2.6 mm	56 μm	75 X	ϕ 2.6 mm	47 μm	100 X	ϕ 2 mm	42 μm
Objective 10 X	100 X	ϕ 1.3 mm	24 μm	150 X	ϕ 1.3 mm	21 μm	200 X	ϕ 1 mm	19 μm

(Courtesy: Mitutoyo, Japan)

Accuracy of Tool Makers' Microscopes

= 1 μm (one micron) if LCD digital read outs, are used.

(or)

= 5 μm (five microns) if mechanical micrometers are provided for X and Y ranges.

Accuracy of angle measurement (by vernier attached to eyepiece protractor) = 6 (six minutes); 2 minutes is also possible.

10. Profile Projector

Profile projector is also commonly called as optical projector or optical comparator and is used for measurement applications that will be difficult to assess with other methods. Profile projector is an instrument which projects the enlarged shadow of the profile of the workpiece on a glass screen. From this projection of the workpiece, measurements can be made directly or indirectly. The profile projector has a table that can be moved laterally and from front to back. It can be easily elevated or lowered as well as tilted at an angle within certain limits. The workpiece is placed on the table and moved into position so that its enlarged shadow comes on the screen. The movement of the workpiece is measured by micrometer or precision gauge blocks as an indicator. To register the movement of the workpiece, the screen of the profile projector is provided with crossed lines.

Sometimes, the profile projector is provided with a number of magnifying lenses, so that the image on the screen can be projected as 10, 20, 30 or more times the size of the object.

A profile projector consists of a light source, a lens system to direct the light past the workpiece, a work table, projection optics (including both lenses and mirrors) and the screen where the workpiece image is projected. The screen would hold a chart gauge or a scaled contour of the workpiece depending on whether a workpiece is measured or gauged. The principle of a profile projector is shown in Fig. 48(a).

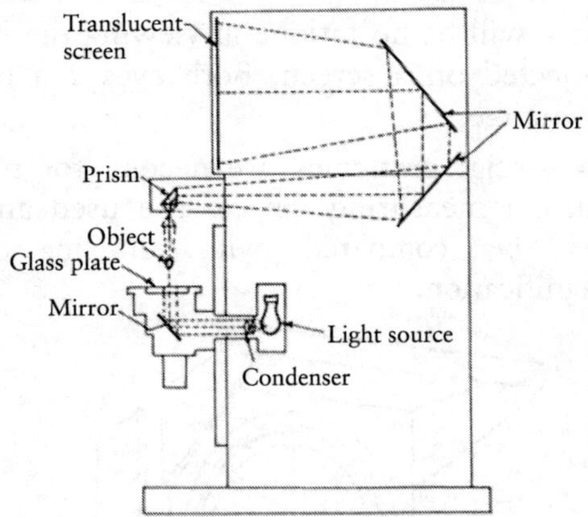

Fig. 48(a): Principle of profile projector

The light source used in the instrument is a tungsten lamp, filament lamp or a mercury arc lamp. Two light sources are used in profile projector. One light source is used to produce shadow image of the workpiece, whereas the second source produces a reflected image of the face of the workpiece. High intensity of light is needed to get the image of the face of the workpiece. Three types of measurement techniques are used in profile projector. These are measurement by comparison, by movement and by translation. Measurement by comparison is done by producing magnified two-dimensional image of the workpiece on the screen. A master chart gauge is put on the screen representing high and low limits of the absolutely perfect workpiece at the magnified size and then they are compared. Measurement by movement is done by relating the projected image to a reference point (cross-line) of the projector. This technique is best for single piece inspection.

Measurement by translation is done to trace contour of the parts that cannot be projected directly on the screen, for example parts having internal grooves. A pantograph mechanism is used for tracing purposes, where one arm of pantograph traces over the contour, the other traces a projected image on the screen. The photograph of profile projector is shown in Fig. 48(b).

The main advantages in using profile projector are as follows:

(i) There will be no fatigue in viewing the image as it is projected on a screen. Both eyes can be used with advantage.

(ii) No special graticules are needed for measurement. Normal measuring devices are used and the image may be compared with drawing of relevant magnification.

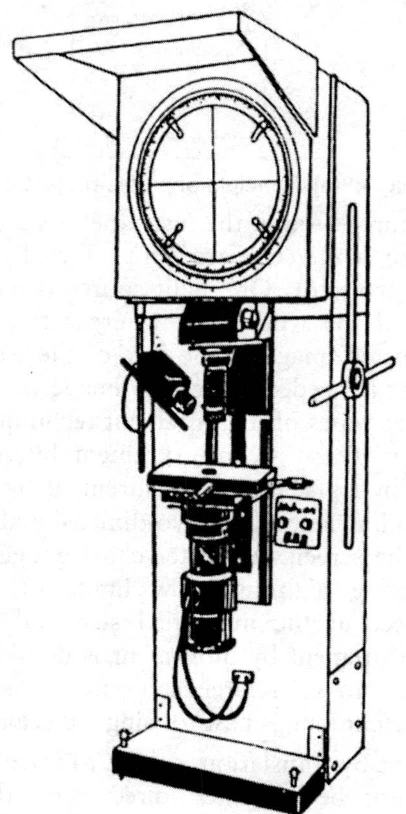

Fig. 48(b): Profile projector

(iii) The image can be seen by many persons at a time and thus it is helpful for discussing the cause of discrepancies, etc.

(iv) By replacing the screen with a photographic plate, permanent records can be made.

Profile projector can be used for tool wear measurements also. The main drawback of the profile projector is that it is bulky, as the entire magnification is achieved by one lens.

The specification of a typical profile projector are given as follows:

Screen, type and size	– Vertical, 450 mm diameter having angular graduation with vernier reading to an accuracy of one minute.
Worktable, type and size—	– Horizontal
200 mm × 200 mm	– Longitudinal travel (X-axis)-25 mm
	– Transverse Travel (Y-axis) -25 mm
	– Least count for both axis-0.01 mm
	– Loading capacity-10 kgs
Electrical supply	– Main single phase, 220/240 volts
Magnification	– 10X, 20X, 50X and 100X
Surface accuracy of measurement	– ± 0.15%
Contour accuracy of measurement	– ± 0.10%
Cooling system for lighting	– Air cooling
Illumination	– Condensor and halogen lamp system for adjustable illumination.

11. Autocollimator

The popularity of autocollimator is due to its ability to sense remotely at a distance, with high accuracy, the angular rotation of a flat mirror (reflector) around axes that are in the plane of the mirror. The sensitivity of an autocollimator may be 1/10 arc second or even less than that. With this sensitivity, the autocollimator not only monitors angular tilts as such but also converts linear displacement into angular ones so that they can be monitored.

For small angular measurements, autocollimator provides a very accurate and sensitive measurements. Optically, an autocollimator is a special form of a telescope which consists of an infinity telescope and a collimator combined into one instrument.

In the autocollimator shown in Fig. 49, the illuminating reticle is in the instrument itself and only a reflector is required as a target. An optical flat is used for this purpose. As the light rays come from the light source, they illuminate the reticle. After passing through the reticle, they are reflected by the beam splitter which directs the rays to an objective lens. The rays emerge as parallel stream from the objective lens. A reflector placed in front of the objective lens and exactly normal to the optical axis, reflects the parallel stream of light back through the objective lens to be seen in the eyepiece. If the reflector is not normal to the optical axis, and is tilted through a small angle α, the reflected rays will be deflected by twice the angle of tilt, i.e. 2α and the reflected image will be displaced by an amount $2\alpha f$, where f = focal length of autocollimator objective and α = angular tilt of the reflector.

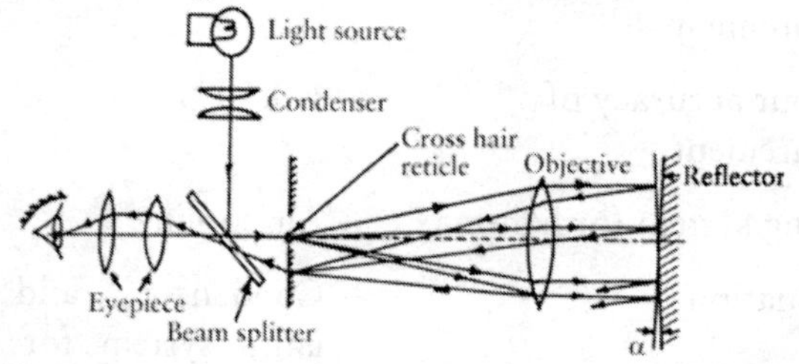

Fig. 49: Working principle of autocollimator

Autocollimators are used for measuring straightness and flatness of machine tool slide ways/guideways. Autocollimator can provide precise angular indexing in conjunction with polygons. Angular comparison can be done by simultaneous viewing of a component and a master angle. The squareness and parallelism of component can also be assessed by autocollimator.

The flatness of surface tables can be checked/measured by using autocollimators.

12. Coordinate Measuring Machine (CMM)

With the advent of numerically controlled machine tools, CMMs have also been developed. There has been growing need to have CMMs which can do faster first piece trial production inspection and many times, 100% dimensional inspection. The CMM plays a vital role in the mechanization of the inspection process. Some of the CMMs can even be used as layout machines before machining and for checking feature locations after machining.

A coordinate measuring machine consists of four main elements: (i) the machine structure, which is infact an X-Y-Z positioning device, (ii) probing system, (iii) machine control and computer hardware, and (iv) software for three-dimensional geometry analysis. The machine incorporates the basic concept of three-coordinate axes that is precise movement in x, y and z directions is possible and is measurable and controllable. Each axis is fitted with a linear measurement transducer which senses the direction of movement and gives digital display.

The worktable of the machine contains tapped holes to facilitate clamping and locating of parts. It is made of granite to make it stable in various types of environment. A probe is inserted into the probe arm, which is supported on the machine structure. The configuration of the CMM may be cantilever type, bridge type, column horizontal arm type and gantry type as shown in Fig. 50. The workpiece is placed on the table of CMM and may be clamped to the table depending on workpiece size and type of probe used. For making multiple

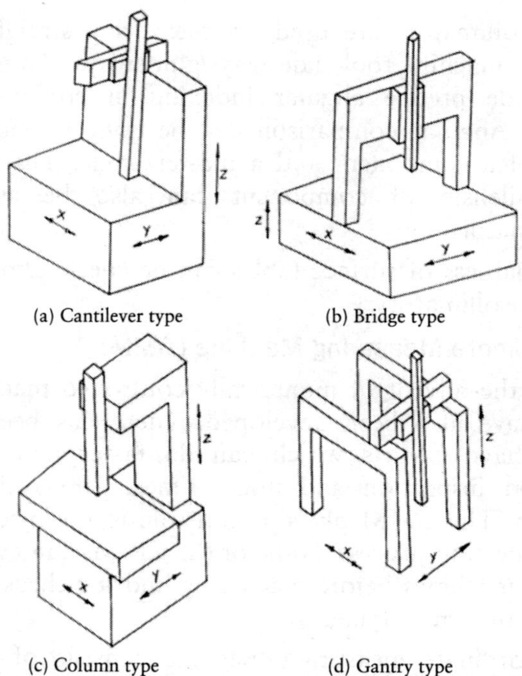

(a) Cantilever type
(b) Bridge type
(c) Column type
(d) Gantry type

Fig. 50: Various configurations of coordinate measuring machine

inspection of similar parts, a reference location may be set with a reference precision cube or sphere. The probe is then moved till it touches the desired features of the workpiece. The instantaneous machine location is shown on the digital display. The dimensions may then be calculated or compared, as desired. A photograph of bridge type CMM is shown in Fig. 51. The type of probe used plays an important part in CMM utility. Basically three types of probes are used—hard, electronic and non-contact type. A variety of hard probes are available in which a shaft and a probe tip is attached to the probe arm. Hard probes are used only in small, manually operated CMMs when inspecting simple parts. The electronic probe consists of a probe body and stylus. As the stylus comes in contact with the workpiece, a signal is sent to the computer indicating 3 D location of stylus. New CMMs offer the distinction of using non-contact probe that rely on optical interface to sense dimensional classification.

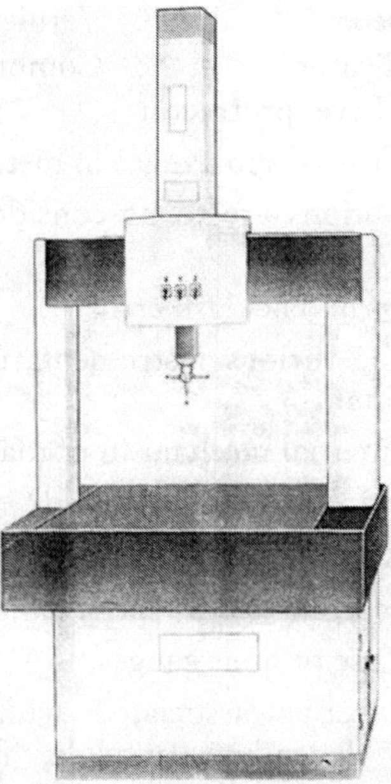

Fig. 51: Bridge type coordinate measuring machine

Provision of a rotary worktable increases the versatility of CMM, as all the areas of a workpiece can be approached with a single set-up. Rotary tables are especially useful when inspecting complex, multifaced parts or workpieces with a rotation axis, such as cams, gears and rotors.

The main advantage of coordinate measuring machine is quicker inspection alongwith accurate measurements. Automotive, defense and aerospace industries are the primary sectors where CMMs are used.

REVIEW QUESTIONS

1. Explain the difference between accuracy and precision.
2. Define the term 'least count' applied in various measuring devices.
3. Explain the constructional features of the following instruments:

(a) Micrometer (b) Vernier calliper
(c) Dial indicator (d) Combination set
(e) Vernier bevel protractor

4. What is the use of slip gauges in metrology?
5. What precautions are to be considered while using slip gauges?
6. Define the term 'backlash error'.
7. What are the various instruments used for angular measurements?
8. Sketch the internal mechanism of dial indicator.
9. On what principle is sine bar used?
10. What is the material of surface plate and how it is specified according to ISI specifications?
11. Explain the use of limit gauges.
12. Define the terms associated with gears that is addendum, dedendum, module, flank and face. Show by means of a diagram.
13. Sketch a gear tooth vernier.

OBJECTIVE TYPE QUESTIONS

Choose the correct alternative

1. Layout work for drilling operations must be performed on a ______ to ensure an accurate layout.
   (a) work bench (b) milling machine table
   (c) surface plate (d) wooden block
2. A precision instrument used for measuring angles to an accuracy of five minutes is called a
   (a) taper gauge
   (b) combination set
   (c) vernier bevel protractor
   (d) vernier caliper
3. To check accurately whether the workpiece is properly centered in a four-jaw chuck, a _____ is used.

(a) combination set (b) dial indicator
(c) micrometer (d) vernier caliper

4. The accuracy of a vernier bevel protractor used in machine shop practice is usually
   (a) one second (b) five seconds
   (c) one minute (d) five minutes
5. The diameter of a finish turned shaft must be checked for size with
   (a) a combination set (b) slip gauges
   (c) a taper gauge (d) a micrometer
6. Extremely accurate centering of work mounted in an independent chuck can be determined only by using a
   (a) centre gauge (b) height gauge
   (c) dial indicator (d) surface gauge
7. The protractor head on a combination set is designed to measure or check ______ in the machine shop.
   (a) length (b) diameter
   (c) thickness (d) angles
8. A precision measuring instrument designed especially for depth measurement of holes, slots and recesses is called a
   (a) vernier caliper (b) micrometer
   (c) depth gauge (d) radius gauge
9. Internal tapers are checked for accuracy using
   (a) slip gauges (b) a micrometer
   (c) a telescopic gauge (d) a taper plug gauge
10. To check accurately the run-out of a lathe spindle, a _______ is used.
    (a) slip gauge (b) micrometer
    (c) dial indicator (d) spirit level
11. Dial indicators can be used only to determine
    (a) the actual size of a part
    (b) the maximum size of a part

(c) the minimum size of a part
(d) the variation in size of a part

12. A micrometer is usually used to
    (a) measure an outside diameter or inside diameter
    (b) measure the distance between two points on a plane surface
    (c) check a class of thread
    (d) determine a centre of a circle

13. The least count of a metric vernier calliper having 25 divisions on vernier scale, matching with 24 divisions of main scale (1 m.s. division = 0.5 mm) is
    (a) 0.05 mm (b) 0.01 mm
    (c) 0.02 mm (d) 0.001 mm

14. A vernier having a least count of 0.01 mm has a zero error of +0.03 mm, while measuring shows a reading of 24.08 mm. The actual value of measurement will be
    (a) 24.08 mm (b) 24.11 mm
    (c) 24.05 mm (d) 24.24 mm

15. The least count of an external micrometer (range 0-25 mm) having 50 divisions on circular scale and pitch 0.5 mm is
    (a) 0.01 mm (b) 0.02 mm
    (c) 0.05 mm (d) 0.001 mm

16. Dial gauge used in metrology laboratories is
    (a) a direct measuring instrument
    (b) a comparing instrument
    (c) both a direct and an indirect measuring instrument
    (d) same as a micrometer and a vernier calliper

17. Sine bar can be used for measuring taper
    (a) with the help of height gauge
    (b) with the help of V-calliper
    (c) with the help of bevel protractor
    (d) without any accessory

18. The thread micrometer measures
    (a) the major diameter of the thread
    (b) the minor diameter of the thread
    (c) the effective diameter of the thread
    (d) the root diameter of the thread
19. In the case of vernier callipers, which one of the following statements is true?
    (a) The vernier has 10 equal divisions with a total length equal to 10 divisions on the main scale.
    (b) The vernier has 9 equal divisions with a total length equal to 10 divisions on the main scale.
    (c) The vernier has 10 equal divisions with a total length equal to 9 divisions on the main scale.
    (d) There is no relationship between the divisions on the two scales.
20. Which one of the following instrument is suitable for measuring the diameter of a steel wire?
    (a) Vernier gauge (b) Vernier calliper
    (c) Micrometer gauge (d) Metre rule/scale
21. Good quality scales for use in fitting shop measurements are usually made of
    (a) low-carbon steel (b) plastic sheets
    (c) stainless steel (d) good quality aluminium
22. Vernier calliper is commonly used in a
    (a) smithy shop for measurement of finished jobs
    (b) foundry shop for measurement of dimensions of mould cavity
    (c) machine shop for measurement of finished metallic jobs
    (d) welding shop for marking and measuring dimensions on a sheet metal.
23. Vernier callipers are made from
    (a) mild steel (b) stainless steel
    (c) cast iron (d) cast steel

24. Vernier callipers are used for
   (a) measuring angles
   (b) taking linear measurements
   (c) checking surface finish
   (d) checking surface irregularities
25. Height gauges are used for measuring
   (a) depth and height (b) external taper
   (c) internal taper (d) surface finish
26. Height gauges and depth gauges are based on the principle of
   (a) micrometer (b) vernier calliper
   (c) nut and bolt (d) sliding friction
27. Micrometers are manufactured from
   (a) mild steel (b) stainless steel
   (c) CI (d) cast steel
28. Micrometers are used for
   (a) checking angles
   (b) taking linear measurements
   (c) checking surface finish
   (d) marking jobs on a surface plate
29. Combination set is basically a
   (a) precision measuring instrument
   (b) marking device
   (c) checking and marking device
   (d) vernier bevel protractor
30. The least count of a vernier protractor having 12 divisions on vernier scale and matching with 23° on the main scale (1 main scale division corresponds to 2°) is
   (a) 5 minutes (b)10 minutes
   (c) 12 minutes (d) 5 degrees
31. The joining of two slip gauges for building up a size combination is called

(a) fixing (b) assembling
(c) wringing (d) pasting
(e) lapping

32. The taper of internal dovetail can be measured with the help of
   (a) sine bar
   (b) combination set
   (c) balls of standard dimensions and slip gauges
   (d) vernier callipers
33. External taper can be accurately measured with the help of
   (a) sine bar and slip gauges
   (b) vernier calliper
   (c) micrometer
   (d) combination set
34. A sine bar is specified by
   (a) its total length
   (b) the centre distance between the two rollers of the sine bar
   (c) the minimum angle that can be measured
   (d) the maximum angle that can be measured
35. Gear tooth calliper is used to find the chordal thickness of gear tooth of
   (a) spur gears (b) helical gears
   (c) worm gears (d) bevel gears
36. In manufacturing industries, plug gauges are used to
   (a) measure the diameter of the workpieces
   (b) measure the diameter of the holes in the workpieces
   (c) check the diameter of the holes in the workpieces
   (d) check the length of the holes in the workpieces
37. In production shops, snap gauges are used to
   (a) check the length of workpieces
   (b) check the diameter of the workpieces

(c) check the diameter of the holes in the workpieces
(d) measure the diameter of the holes in the workpeices

38. The distance between the corresponding points of any two adjacent screw threads is known as
(a) pitch of the screw thread
(b) accuracy of the screw thread
(c) start of the screw thread
(d) diameter of the screw thread at the pitch line

39. ______ can be used to scribe lines parallel to the edges of a part.
(a) Vernier callipers
(b) Screw gauge
(c) Divider
(d) Hermaphrodite calliper
(e) Combination set

40. A surface gauge is used
(a) in levelling the surface plate
(b) in finding the surface finish
(c) in laying out the work accurately
(d) in finding the depth of the surface

41. Work which must be held in a vertical position for laying out should be clamped to
(a) surface plate (b) an angle plate
(c) a V-block (d) a machine bed

42. The metric system of engineering measurements used in machine shop practice is based on a length unit called the
(a) millimetre (b) centimetre
(c) metre (d) micron
(e) decimal system

43. To determine the size of a bored hole by using a telescopic gauge, the gauge setting must be checked with

(a) a Vernier Calliper (b) a micrometer
(c) slip gauges (d) an alignment telescope

44. To check the diameter of a twist drill with a micrometer, the measurement must be taken across ______.
(a) the margins of the drill
(b) the flutes of the drill
(c) the cutting edges of the drill
(d) the lips of the drill
(e) the web of the drill

45. To find the straightness of the lathe spindle bore, a ______ is required.
(a) slip gauge (b) feeler gauge
(c) spirit level (d) test mandrel

46. Circumferential error of spindle location in centre lathes can be checked with the help of
(a) a spirit level (b) a block level
(c) a dial gauge (d) a micrometer

47. To obtain an accurate measurement using ______, the machinist must develop a keen sense of touch.
(a) Vernier Calippers (b) Micrometer
(c) Slip gauges (d) Vernier Bevel Protractor

48. The finish depth of an external V-thread form is most accurately measured by
(a) the 3-wire system (b) the 2-wire system
(c) a thread ring gauge (d) a thread plug gauge
(e) a thread micrometer

49. The thickness of light gauge sheet steel can be best checked with a
(a) finely divided steel scale
(b) depth gauge
(c) hermaphrodite calliper
(d) Micrometer

50. The minimum reading on a commonly used dial gauge is

(a) 0.025 mm (b) 0.001 mm
(c) 0.01 mm (d) 0.05 mm

51. While using a dial gauge for checking the run-out on a shaft, the dial gauge should be
    (a) held in the hand
    (b) fixed on a stand so that the dial gauge plunger is resting on the job
    (c) fixed on a stand and the plunger is slightly loaded i.e. pressed on the job, and the needle reading zero on the dial.
    (d) fitted securely to any fixed part and plunger slightly preloaded and dial reading zero

52. The overall magnification of a microscope that has an objective magnification of 30 X and an eyepiece magnification of 10 X is
    (a) 3 X (b) 30 X
    (c) 300 X (d) 3000X

53. For taking measurements of the height of a job with the help of an height gauge the following item is normally required in addition to the height gauge and the job
    (a) V-block (b) Angle plate
    (c) Surface plate (d) Sine bar

54. Circular scale of the micrometer is marked on the part known as the
    (a) shaft (b) barrel
    (c) ratchet (d) thimble
    (e) spindle

55. In an accurate micrometer a part called ______ is attached to the thimble so that the measuring pressure during clamping of different jobs is kept constant.
    (a) locknut (b) magnifying glass
    (c) spindle (d) spanner
    (e) ratchet

56. Millimeter scale in a micrometer is marked on the part called
    (a) barrel (b) thimble
    (c) spindle (d) anvil
57. Combination set is basically a
    (a) Precision measuring instrument
    (b) Marking device
    (c) Checking and marking device
    (d) Vernier Bevel Protractor
58. The centre head in a combination set is used for
    (a) locating the centre of a bar stock
    (b) measuring the diameter of a bar stock
    (c) measuring the centre-to-centre distance in bar stocks
    (d) measuring the angles
59. While using a combination set, the squareness of the given job is checked by the
    (a) centre head (b) graduated protractor
    (c) square head (d) scale
60. For correct and precise measurements of thread characteristics we use
    (a) screw pitch gauge
    (b) micrometer with accessories
    (c) tool room microscope
    (d) thread gauge
61. Dial gauge measures
    (a) circular dimensions as needle rotates on the dial
    (b) linear dimensions as contact pointer moves up and down
    (c) angular divisions
    (d) surface roughness of finished jobs
62. Which part of the combination set can be used to lay out the centrelines on the end of a round bar stock?
    (a) The centre head (b) The square head
    (c) The Bevel Protractor (d) The bubble

63. In Metrology Laboratory, the joining of two slip gauges for purposes of precision measurement is known as
    (a) assembly (b) welding
    (c) adhesion (d) wringing
    (e) slipping
64. In Limits, Fits and Tolerances, the term "allowance" is usually referred to as
    (a) Minimum clearance between shaft and hole
    (b) Maximum clearance between shaft and hole
    (c) Tolerance of hole or shaft
    (d) Difference between maximum size and minimum size of the hole
    (e) Difference between maximum size and minimum size of the shaft
65. In the case of a clearance fit between a shaft and a hole, the minimum clearance is obtained when
    (a) the shaft is of minimum diameter and the hole is of maximum diameter
    (b) the shaft is of maximum diameter and the hole is of minimum diameter
    (c) both the shaft and hole are of minimum diameter
    (d) both the shaft and hole are of maximum diameter
66. The condition of the assembly between a shaft and a hole is such that there is always a clearance fit. The maximum clearance between shaft and hole is obtained when
    (a) both the shaft and hole are of minimum diameter
    (b) both the shaft and hole are of maximum diameter
    (c) the shaft is of minimum diameter and the hole is of maximum diameter
    (d) the shaft is of maximum diameter and the hole is of minimum diameter

67. In the case of a spur gear, if N is number of teeth and D is the diameter of the pitch circle, then the module is given by
(a) D/N (b) πD/N
(c) N/D (d) ND/π

68. The working depth or the depth of engagement of two gears is equal to
(a) sum of their addendums
(b) sum of their dedendums
(c) addendum-dedendum
(d) dedendum clearance

69. The height by which a gear tooth projects beyond the pitch circle or pitch line is known as
(a) circular pitch (b) addendum
(c) dedendum (d) clearance

70. Bevel gears are used to transmit rotary motion from one shaft to the other whose axes intersect at
(a) right angles (b) any angle
(c) 45 degrees only (d) 180 degrees only

71. Theoretical minimum number of teeth on a pinion for avoiding interference in any gearing device could be
(a) 10-12 (b) 18-20
(c) 25-30 (d) 32-40
(e) around 50

72. Pitch circle diameter (in mm) of a gear wheel is given by the expression ______.
(a) $\frac{m}{z}$ (b) $\frac{z}{m}$
(c) mz (d) $mz + 1$
(e) $\frac{m+1}{z}$

where m = module of gear and z = number of teeth in the gear wheel

73. The term "module" in gear technology refers to
    (a) diametral pitch
    (b) no. of teeth/pitch circle diameter
    (c) pitch circle diameter/no. of teeth
    (d) pitch of gear teeth
    (e) no. of teeth
74. When measuring gear teeth for size with a gear tooth vernier caliper, the reading for chordal tooth thickness is obtained from the _______ of the caliper.
    (a) Vertical scale (b) Horizontal scale
    (c) Circular scale (d) Counter
75. The module of a gear, having 36 teeth and a pitch diameter of 225 mm, is
    (a) 0.16 mm (b) 25.00 mm
    (c) 12.50 mm (d) 6.25 mm
76. To find the module of a gear, divide the pitch circle diameter by the
    (a) addendum
    (b) dedendum
    (c) 1/2 the number of teeth
    (d) number of teeth
    (e) (number of teeth + 1)
77. The measurement from centre-to-centre of adjacent gear teeth taken on the arc of the pitch circle, is called the _______.
    (a) Circular pitch (b) Chordal pitch
    (c) Diametral pitch (d) Tooth thickness
78. A gear-tooth vernier caliper is used to measure the _______ and chordal thickness of gear teeth.
    (a) addendum (b) dedendum
    (c) circular pitch (d) module
    (e) diametral pitch
79. The common pressure angles in involute gears are _______.

(a) $14\frac{1}{2}, 21\frac{1}{2}$ degrees (b) $14\frac{1}{2}$, 20 degrees

(c) $14\frac{1}{2}, 20\frac{1}{2}$ degrees (d) 14, 20 degrees

(e) 14, 21 degrees

ANSWER KEY

1. (c)	2. (c)	3. (b)	4. (d)	5. (d)
6. (c)	7. (d)	8. (c)	9. (d)	10. (c)
11. (d)	12. (a)	13. (c)	14. (c)	15. (a)
16. (c)	17. (a)	18. (c)	19. (c)	20. (c)
21. (c)	22. (c)	23. (b)	24. (b)	25. (a)
26. (b)	27. (b)	28. (b)	29. (c)	30. (a)
31. (c)	32. (c)	33. (a)	34. (b)	35. (a)
36. (c)	37. (b)	38. (a)	39. (d)	40. (c)
41. (b)	42. (c)	43. (b)	44. (a)	45. (d)
46. (c)	47. (a)	48. (e)	49. (d)	50. (c)
51. (d)	52. (c)	53. (c)	54. (d)	55. (e)
56. (a)	57. (c)	58. (a)	59. (c)	60. (c)
61. (b)	62. (a)	63. (d)	64. (a)	65. (b)
66. (c)	67. (b)	68. (a)	69. (c)	70. (c)
71. (b)	72. (c)	73. (c)	74. (d)	75. (d)
76. (d)	77. (a)	78. (a)	79. (b)	

13

Surface Roughness and Its Measurement

1. Introduction

Every machining process leaves its own impact on the machined surface in the form of finely spaced micro-irregularities. Each cutting tool leaves its own individual pattern on the surface. This pattern is known as surface finish or surface roughness. The machining process used can be known by close investigation of the surface.

Surface irregularities are linked with the efficiency of the machining operation. Surface irregularities are also caused as the chip is forced from the work by the penetration action of the cutting tool. The shape of the job can also contribute to vibrations in the job and cutting tools leave their mark on the finished surface. Even the wear on improperly machined machine tool guide-ways can influence the finishing of the job.

2. Factors Contributing to Quality of Surface Finish

The quality of surface finish mainly depends on the machining operation, tool angles and cutting parameters. The most important factors are as follows:

1. Uniformity and keenness of the cutting edge of the tool.
2. Smoothness of the tool surfaces that come in contact with the workpiece.
3. The feed given to the cutting tool as it traverses the workpiece.

4. Condition of the machine tool guide-ways that guide the cutting tool along its path.
5. The type of cutting fluid used also influences the surface finish.

3. Some Important Surface Characteristics

Figure 1 shows some of the important surface characteristics as explained below:

1. *Profile*: It is the contour of a surface in a plane perpendicular to the surface.

2. *Surface Texture*: Repetitive or random deviations from the nominal surface, which form the pattern of the surface. It includes roughness, waviness, lay and flaws.

3. *Roughness*: Roughness consists of the micro-irregularities in the surface texture usually including irregularities that result from the inherent action of the production process. These are considered to include traverse feed marks.

4. *Waviness*: Waviness is usually widely spaced component of surface texture. Waviness may result from such factors as machine or work deflections, vibrations, chatter, heat treatment or warping strains.

5. *Lay*: The direction of the predominant surface pattern ordinarily determined by the production method used. Various directions of lay depending upon production process are shown in Fig. 2.

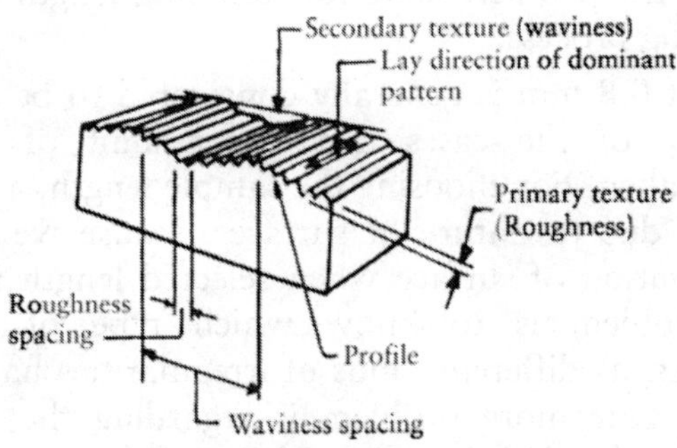

Fig. 1: Characteristics of surface finish

The symbols used for representing direction of lay are as follows:

= – Lay parallel to the line representing the surface

⊥ – Lay perpendicular to line representing the surface

X – Lay angular in both directions to line representing the surface

M – Lay multi-directional

C – Lay approximately circular relative to centre of surface

R – Lay radial relative to centre of surface

Type	Symbol	Name	Process
	=	Parallel	Peripheral Grinding, Turning, Boring, Shaping, Peripheral, Milling
	⊥	Perpendicular	Shaping, Milling, Grinding
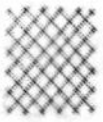	×	Straight criss cross	Honing Process
		Arcuate	End Milling, Face Turning.
		Arcuate criss cross	End Milling, Cup Wheel Grinding, Side Wheel Grinding
	C	Circular	Spot Facing, Face Turing

Fig. 2: Different types of lay (directions of Lay)

6. *Sampling Length*: It is also called 'cutoff length'. It is the length of profile required for evaluating the irregularities of a surface. It is measured in a direction parallel to the general direction of the profile. Value for sampling length depends on the machining process.

Value of 0.8 mm is generally considered to be satisfactory for majority of the cases. An upper limit of 25 mm is commonly taken. For choosing the sample length, a difficulty is experienced due to nature of surface because we do not get proper indication of surface when selected length is different. Another problem is to know which type of irregularity predominates, as different kinds of irregularities have different frequencies. One more problem is regarding the position of sampling length, that is where the sample length must be chosen. The last problem is dealt with by taking readings at several places and taking the mean value. The sampling length is standardised for various machining operations.

7. *Roughness Spacing or Roughness Width*: It is the distance parallel to the nominal surface between successive peaks and ridges, which constitute the predominant pattern of the roughness.

Waviness Spacing or Waviness Width: It is defined as the spacing of successive wave peaks or wave valleys.

The symbol shown in Fig. 3(i) represents a surface as per specifications, as follows:

a – Roughness Value (microns), Ra
b – Production Method (Treatment or coating)
c – Sampling Length (mm); cut off length
d – Direction of Lay
e – Machining Allowance (mm)
f – Other Roughness Value (such as Rt), microns

Symbol in Fig. 3(ii) represents a honed surface having R_a value 0.25 micron with cut off length 0.8 mm and direction of lay being straight criss-cross. The machining allowance is 0.15 mm

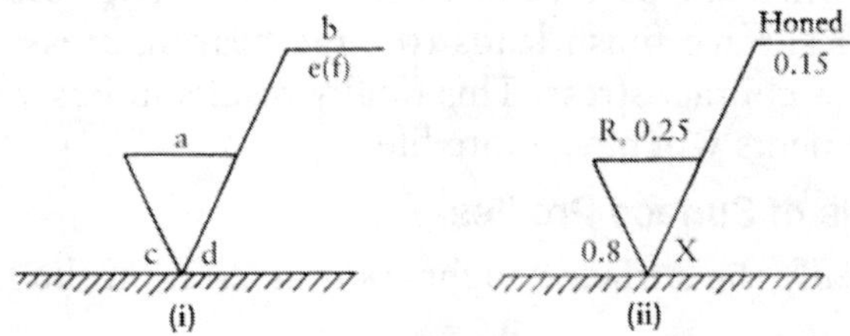

Fig. 3: (i) Symbol for representing surface roughness (ii) An example showing surface roughness

The machinist is always interested in the quality of the finish that his cutting tool has given to the surface of the machined job. The smoothness of the finished surface will indicate the following:

(i) Efficiency of cutting tool and whether the tool is correctly ground
(ii) Approaching breakdown of the cutting edge (tool wear)

(iii) Presence of vibrations of machine tool

(iv) Need for additional machining

4. Purpose of Finishing Surfaces with Respect to Their Function

Constant improvement in the surface texture has led to great advancements in the manufacturing technology. As the need for greater strength and higher bearing loads was felt the production of smoother and harder surfaces was stressed. It was also seen that fatigue life, bearing properties, lubrication and wear qualities of any component of a machine have direct relationship with its surface texture. A surface with high degree of finish reduces the wear and increases the life of the component. So these effects make the control of surface texture very important.

Failure due to fatigue always starts to occur at sharp corners where stress concentration occurs. The root of any such irregularity acts as a sharp corner and such a part fails earlier. So to increase the life of any part which is subjected to repeated reversal of stresses, the working surfaces must be given very good finish. Good bearing properties in any part are obtained when the surface has better finish. The reason being that better surface finish leads to large bearing areas and thus leads to less contact stress. This finally results in less wear, thus the components will have more life.

5. Analysis of Surface Profiles

Generally the surface roughness is assessed in the following ways:

Peak to Valley Height of Roughness

Usually represented as R_t in microns.

From the name, it appears that it is a very simple analysis, but taking it in a wider sense it will mean that peak and valley chosen is exceptional and result will not give actual surface assessment. To overcome this problem, peak to valley height is defined as the distance between a pair of lines running parallel to general lay of the trace, so that length lying within the peaks is 5% of trace length and length lying within the valley is 10% of the trace length.

Root Mean Square (RMS) Value (represented in microns)

RMS value is defined as the square root of the mean of the squares of the ordinates of the surface measured from a mean line.

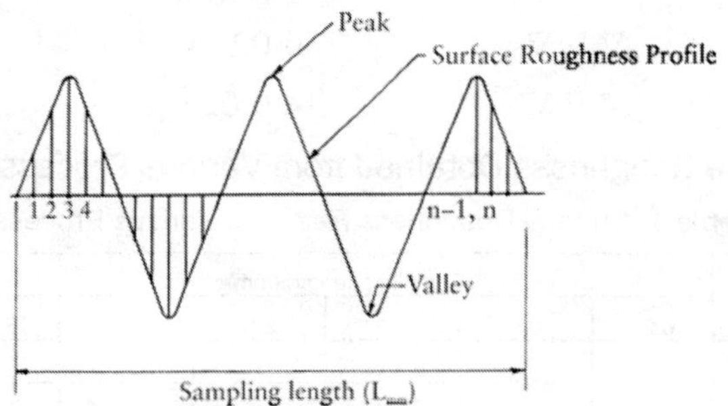

Fig. 4: Graphical representation of h_{rms}

Length L is divided into n equal parts and heights of ordinates 1, 2, 3, ..., n is h_1, h_2, h_3, ..., h_n respectively as shown in Fig. 4.

$$\text{Then } h_{rms} = \sqrt{\frac{h_1^2 + h_2^2 + h_3^2 + \ldots + h_n^2}{n}}$$

Centre Line Average Method (CLA Value) (Expressed in microns)

Usually represented as Ra in microns.

It is defined as "average height of all ordinates of the surface from a mean line, regardless of the sign". Thus from Fig. 4.

$$\text{CLA} = \frac{h_1 + h_2 + h_3 + \ldots + h_n}{n}$$

Also CLA value is given by the expression

$$\text{CLA} = \frac{A_1 + A_2 + A_3 + \ldots}{L} = \frac{\Sigma A}{L}$$

Root mean square value is about 11% higher than CLA value. CLA is now taken as standard method all over the world for representing surface finish.

6. Symbols for Designating Surface Roughness

Surface roughness is indicated by triangles. The following symbols are used.

Symbols	*R_a value in μm*
∇	8 to 25
∇ ∇	1.6 to 8
∇ ∇ ∇	0.025 to 1.6
∇ ∇ ∇ ∇	< 0.025

7. Surface Roughness Obtained from Various Processes

Table 1: Surface Roughness Range of Various Processes.

Process	Sampling Length (mm)						R_a (mm)
	0.25	0.8	2.5	8.0	10	25	
Planing							1.6-50
Shaping							1.6-25
Milling							0.32-25
Boring							0.4-6.3
Turning							0.4-6.8
Reaming							0.4-3.2
Buffing							0.16-0.32
Polishing							0.04-16
Grinding							.063-5
Broaching							0.4-3.2
Diamond Turning							0.16
Diamond Boring							0.16
Spark Machining							0.4-1.6
Super Finishing							0.16-3.2
Honing							0.025-0.8
Lapping							.012-0.16
Burnishing							0.04-0.8
Drawing							0.16
Extrusion							0.16-5
Moulding							0.32-6.3
Electro-polishing							0.1-0.8

Various parameters affecting surface roughness is shown in the form of a Fish-bone diagram (Fig. 5).

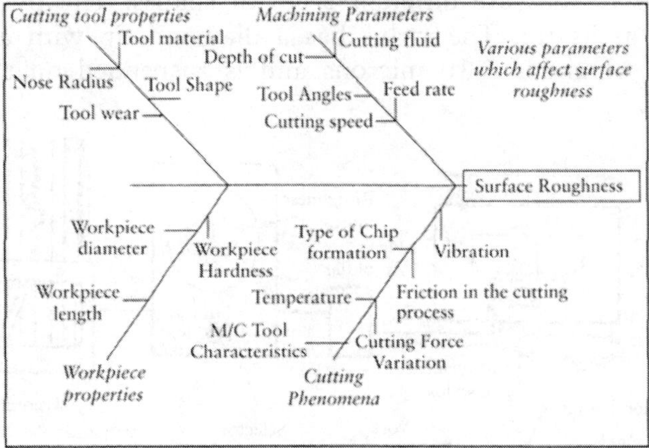

Fig. 5: Various parameters affecting surface roughness

8. Methods of Measuring Surface Roughness

At times, visual comparison with the naked eye will disclose that one surface is rougher than another. It is possible only in instances of wide difference, which could not be misinterpreted by individual reaction. The methods of evaluating surface roughness used most often are as follows:

i. *Inspection by Touch Comparison*

A finger nail is moved along the surface of the job and the amount of resistance and the depth of irregularities is noted mentally. The finger nail is then moved across a series of master roughness scales which have numbers in microns, corresponding to their surface roughness. The machining finish must compare satisfactorily with the correct master. This form of testing is used mainly by machinists, and is very subjective.

ii. *Profilometer*

It is a mechanical-electronic instrument, which is most widely used to measure surface roughness. It is one of the instruments that utilizes the tracer method and actually measures the difference in the depth of the surface irregularity. It can be used for both shop and laboratory purposes.

The two main units of the profilometer are the tracer and amplifier as shown in Fig. 6. Tracers are made in several designs to measure different types of shapes. The tracer has stylus at its tip. The stylus has a diamond tip with a point radius of about 5-10 microns and is suspended on the flat spring.

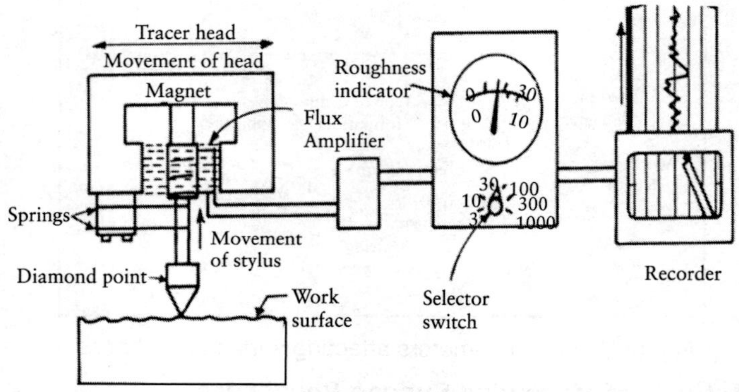

Fig. 6: Profilometer

The upper end of stylus is linked to an induction coil which is located in the field of a permanent magnet, so that movement of coil induces a voltage. As the tracer moves across the surface, whose roughness is to be measured, the stylus will follow the contour of the irregularities. These up and down movements of the stylus are converted into a small fluctuating voltage. The voltage signal is fed into the amplifier where it is amplified. The roughness indicator shows the variations in average roughness height. There is a motor driven unit, which permits the mechanical movement of the tracer and its stylus.

iii. *Hilger and Watts Surface Finish Microscope*

Optical instruments are also used to measure surface finish. Hilger and Watts surface finish microscope is one such type of instrument. It makes use of the wavelength of light. The microscope magnifies the surface and its irregularities nearly 200 times. The illumination is done by high intensity mercury lamp placed in a small tube. The lamp is connected with a capacitor switch, which is connected to the main supply as shown in Fig. 7. The light passes through a green filter and illuminates an adjustable iris diaphragm, which is focussed by a lens system after reflection at a semi-reflecting mirror on to the objective, which is specially designed to work with parallel light. Below the objective is a comparator plate, which is an optical flat coated with hard semi-reflecting oxide film, contacting the work surface and inclined to it at a very small angle, so as to enclose a wedge-shaped air space between the flat and the work, which must itself be the reflector. Multiple reflection of light takes place between the two surfaces enclosing this wedge of air. The work together with superimposed fringes so produced, is seen through the eyepiece.

One view of fringes is shown as Fig. 8.

The V-shaped notches imply a scratch on the surface of the work.

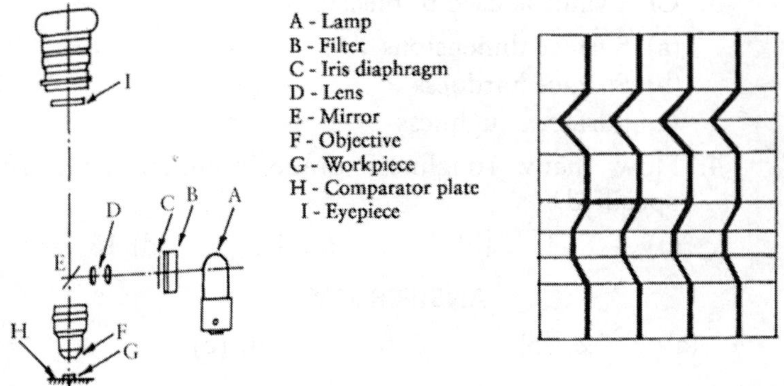

Fig. 7: Optical system of a surface finish microscope

Fig. 8: Diagramatic view of the fringes as seen in the microscope

REVIEW QUESTIONS

1. What is surface roughness? How is it designated?
2. What is waviness? What are the main reasons for waviness?
3. What do you mean by R.M.S. value?
4. What are the different methods for measuring surface roughness? Explain.

5. How is surface roughness designated?
6. What are the reasons for the formation of surface roughness?

OBJECTIVE TYPE QUESTIONS

Choose the correct alternative

1. Surface roughness on a drawing is represented by
   (a) Triangle (s) (b) Rectangle
   (c) Square (d) Parallelogram
2. RMS value stands for
   (a) Root minimum square value
   (b) Root mean square value
   (c) Root maximum square value
   (d) Roughness mean square value
3. CLA value is used to measure
   (a) Surface dimensions
   (b) Surface hardness
   (c) Surface roughness
4. How many roughness grade numbers have BIS specified?
   (a) 4 (b) 8 (c) 12 (d) 16

ANSWER KEY

1. (a) 2. (b) 3. (c) 4. (c)

14

CNC Machines

1. Introduction to Computer Numerical Control

Computer Numerical Control (CNC) is the latest machine tool control technique which enables high quality parts to be produced in small batches at low cost. Although special purpose machines, cam-operated single spindle or multi-spindle automats, copying machines and transfer lines also can be used to produce high quality parts at low cost, their usage cannot be economically justified unless the quantity required is very high. This is because these are hard-automation equipment making use of cams, limit switches, relays and special wiring which are costly and time consuming and they have very poor reusable value when the company has to change from one product to the other or even when the same product is slightly modified.

On the other hand, manufacturing systems built using CNC machines, robots, AGVs (Automated Guided Vehicle) and other computer-controlled equipment are called flexible automation systems since switching over from one part or product to the other is achieved simply by changing the programs (software) that control these equipment.

2. Definition of NC

Numerical Control (NC), as defined by Electronic Industries Association (EIA), U.S.A. is a machine tool control technique in which actions are controlled by the direct insertion of numerical data at appropriate point. The system must automatically interpret at least some portion of this data. Although NC was first used for the control of machine tools of the metal cutting industry, today it is widely used for all kinds of conventional machining (milling machines, lathes, machining centers, boring machines, grinding centers, etc.), sheet metal manufacture (turret punch press, pipe bending machines, flow forming machines, etc.), non-traditional machining (wire-cut EDM, water-jet cutting machines, plasma and laser cutting machines, electron beam welding machines, etc.), inspection [Coordinate Measuring Machines (CMM)] and drafting machines (plotters).

In earlier NC machines, the control was by means of a microprocessor in which no RAM memory was available and the data was stored in the memory registers. Therefore, at any time only one NC block could be stored in the machine and many functions like interpolation, interlocking, etc. were hard-wired. The NC programme was stored in paper tapes and the machine read only one block at a time from the paper tape, executed the machining instructions of that block, read the next block from the tape for execution and so on. Therefore, a paper tape reader had to be an integral part of the NC machine and it was continually read during machining. Due to the drastic and sustained fall in the prices of computer hardware, the NC machines manufactured during the last decade are each controlled by a dedicated micro-computer which has a large memory (RAM, EPROM and disk space in some cases) for storing NC programmes of several parts, it has a graphic monitor and also it is very fast. Because of this, the entire set of programmes to machine any part is read once and stored in its memory in a CNC machine and hence during the machining of the part, the tape reader is not used. Furthermore, many basic hard-wired functions of earlier NC machines are implemented as software for better flexibility and upgradability. The CNC machines are also able to perform additional functions to enhance the productivity of the machine tool. Some of them are graphic display of cutter path, conversational programming, machine tool diagnostics, storage of large number of cutter compensation values, axis calibration, adaptive control, etc. Latest CNC controllers make use of multi-processors (transporters) or time-sharing techniques so that while the machining is going on, the operator can load or unload other programmes, edit the programmes etc. Since the earlier NC machines are obsolete, the terms 'NC' and 'CNC' are interchangeably used today which actually refers to CNC machines. A typical CNC machining center is shown in Fig. 1.

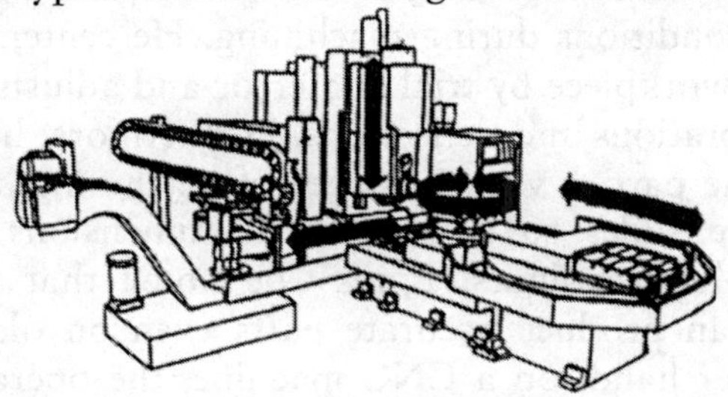

Fig. 1: A typical 5 axis CNC machining center

3. Comparison of NC and Conventional Machines

As depicted in Fig. 2, there are three major differences between a CNC machine and its conventional counterpart. They are

i. Presence of operator (in a conventional machine),
ii. Metal removal rate (high in a CNC machine),
iii. Independent drives (in CNC machines).

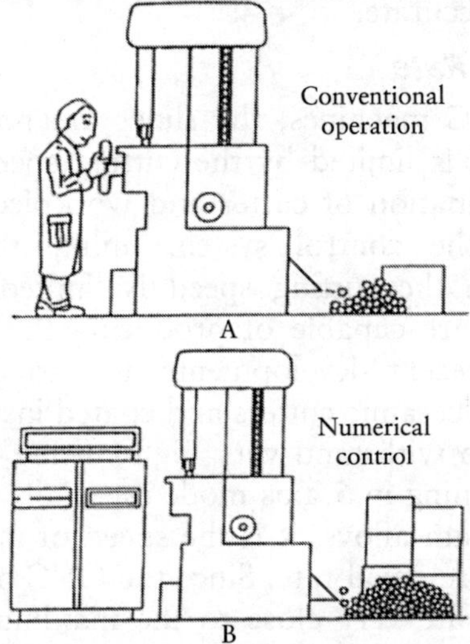

Fig. 2: Comparison of CNC and conventional machines

Presence of Operator

On a conventional machine, every movement required to machine the desired part is performed by the operator. He selects the spindle speed, feed rate and monitors them to compensate for slide friction, lack of power, and other changes in cutting conditions during machining. He centers the bar or locates the workpiece by trial and error and adjusts the clamps to avoid vibrations and deflections. Furthermore, he frequently measures the part at various stages of machining and controls the machine slides so as to get the dimensions within the specified tolerance limits. It may be noted that very skilled operators can produce accurate parts even on old machines. On the other hand, on a CNC machine, the operator's job is limited to loading and unloading of the part and setting of the tools and all other minute machining steps are carried out automatically with the help of the NC programmme. Since the operator cannot compensate for deficiencies like backlash, slide friction, lack of power, etc., the machine tool design shall make sure that these deficiencies are as minimum as possible. Therefore, the CNC machine tool is built to be more accurate and rigid than its conventional counterpart for the same application. There are many features that make a CNC machine more accurate.

Metal Removal Rate

On the CNC machines, the slides are power driven and hence the speed is limited by the cutting speed allowable for the given combination of cutter and workpiece materials and the speed of the control system unlike the conventional machines where the cutting speed is limited. The feed and spindle motors are capable of producing the required power and owing to recent developments in cutting tool materials (carbide cutters, ceramic cutters and coated inserts), the cutting tools are able to withstand very high cutting speeds. In some free form machining in 5 axis mode especially of aircraft parts made of aluminum alloys, it is the speed of interpolation that limits the metal removal rate. Since the CNC machines in such application operate very close to the maximum possible, the metal removal rate is far higher than the conventional machines. This necessitates the use of fast chip disposal systems to remove the chips from the cutting zone. Screw conveyors are used for chip conveyor mechanism on both milling and turning machines and on very heavy duty profilers, powerful vacuum extractors are used. Vacuum extractors are also used when the chip is in the form of fine powder as in the case of grinding. Since continuous coil chips are generated in turning and drilling operations, special chip breaker design are required to be incorporated on the cutting tools. It has been found that radical design changes improve the ease of chip disposal and machine tool rigidity. Two examples that can be cited in support of this are the popular slant bed lathe and horizontal machining centers (HMC).

Independent Drives

On an engine lathe, there is only one motor which basically drives the spindle and the feed motions of the saddle is derived from the same motor through gear train and lead screw. In the same manner, the feed motions of all three slides, viz., vertical, longitudinal and cross slides of a conventional milling machine are obtained through gear trains and lead screws. Due to restrictions on the gear combinations that can be obtained, only very discrete steps of feed motions will be available in this method of drives. Furthermore, there will be limitation on the number of slides that can be simultaneously moved. Contrary to this, on a CNC machine, each axis has its own drive, drive control, measuring device and feedback loop. Therefore, the spindle rotation and the axes movement are all independent of each other and they can be controlled in infinite steps through the instructions in the NC programme. This avoids the requirement of special tools like copying devices, taper turning attachments and special form cutters. On a CNC machine, one can move the axes in linear or circular or any other higher order path with the help of interpolators built in the system. Therefore, even very complicated free form surfaces can be machined with greater ease without any special template or form cutter very accurately.

4. Features of a CNC Machine Tool

A CNC machine tool is more rigid and accurate than the conventional counterpart. CNC machines which have an accuracy of 0.06 mm and repeatability of 0.004 mm are very common. This high level of accuracy is achieved by improved mechanical and electrical features or CNC machines. Some of the features which help improve the accuracy of a CNC machine are:

- Rigid construction
- Smooth/Frictionless Guideway elements
- Preloaded double-nut ball lead screws
- Advanced drives (stepper, dc servo and ac servo)
- Accurate position and velocity transducers
- Sophisticated control loops
- Axis calibration and pitch error compensation
- Compensation of deflections due to cutting forces and self-weight
- Control of room temperature (CNC machines are mostly kept in A.C. rooms).
- Temperature compensation of the slide commands to take care of differences in the temperature at various points due to coolant flow, etc.
- Accurate manufacture of the machine tool parts,
- Diameter and length compensation

The above features are built into the CNC machine tool by an appropriate design of its mechanical, electrical and electronic hardware and computer hardware and software. The mechanical hardware consists of the machine tool structure, its guide ways and moving parts such as table and spindle head, power transmission elements such as gear trains, belts and clutches, power conversion elements such as lead screws or rack and pinion arrangements. The motors, limit switches relay, etc. constitute the electrical hardware. The position and the velocity transducers are the major electronic hardware elements that may be digital, pulsed or analog devices.

There are two types of computers present in any CNC machine. One is the CNC controller which takes care of input and output activities, decoding the NC blocks and filling appropriate word address registers, interpolation to calculate of command values for each axis etc. This is also known as Data Processing Unit (DPU). The other computer is called Programmable Logic Controller (PLC). This is responsible for interlocking of various switching functions by constantly monitoring the status of relays, switches and other switching devices. This also converts the command values into lower level voltage signal that will be processed by the drive unit. The PLC along with the drive control circuitry is sometimes known as Control Loop Unit (CLU). DPU and CLU together are called Machine Control Unit (MCU).

Mechanical Hardware

The machine tool is made very rigid and its moving parts are made as light as possible. The design of the structure is such that its natural frequency is well above that of the normally encountered disturbances to avoid resonance. In order to reduce friction, the machine tool guide way liners made of non-metallic material like Poly-Tetra Fluoro Ethylene (PTFE). In some cases, direct metal to metal contact of the moving parts is eliminated by using antifriction bearings, roller pads, hydrostatic, aerostatic (air bearing) and magnetic bearings. For very heavy machines, hydrostatic bearings are used. For machines which require very high accuracy and the load on it is fairly constant, aerostatic bearings can be used. An added advantage of magnetic bearing is that it can also provide levitation in addition to eliminating direct contact between the surfaces. CNC machines with these bearings (also linear motors) can cut at 60 to 120 m/min. However, magnetic bearings used for CNC machines are still very costly although they have been successfully used in magnetic levitation trains and lifts.

Lead screws are used in small and medium size machines to convert the rotary motion of the shaft into the sliding motion of the table. The same is achieved by rack and pinion arrangement on larger machines. Conventional lead screws have ACME (trapezoidal) thread. These have sliding contact between the screw and nut surfaces. In the lead screws of CNC machines and accurate conventional machines such as jig borers and grinders, there is no direct contact of the sliding surfaces since balls are interspersed between them. Since these balls re-circulate and always a pair of nuts are used to reduce backlash, these lead screws are called **"preloaded double nut re-circulating ball lead screws"**. The gap between the two nuts is optimally adjusted since higher gap will lead to more backlash whereas too narrow a gap will increase power consumption and wear. The adjustment of this gap is called 'preloading'. Fig. 3 shows the details of a double-nut ball lead screw. Due to the rolling contact and continuous lubrication, these ball lead screws have considerably low friction. Furthermore, the difference between the static and dynamic friction coefficients is also very less and hence they also offer negligible 'stick-slip' behaviour (erratic motion at low speed).

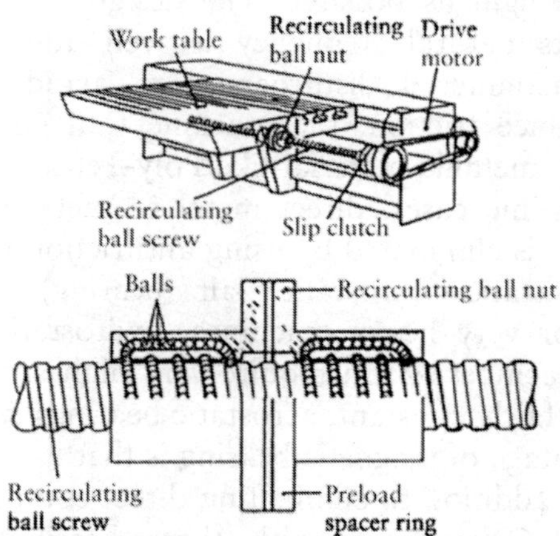

Fig. 3: Preloaded double-nut ball lead screw

5. Electrical and Electronic Hardware

The most common source of power for CNC machine is obtained from electric motors although pneumatic and hydraulic cylinders and motors are used in some specific applications. In the beginning, only stepper motors were used in CNC machines. A stepper motor moves by a fixed angle, say 7.5 degrees, for every pulse it receives. This movement is very accurate below a limiting speed and torque and hence stepper motors are generally used in open loop systems. However, when the cutter is overloaded beyond the limiting torque or when the commanded speed exceeds the limiting speed, it may fail to cope up leading to some lost motion. Therefore, today stepper motors are restricted to applications where the speed and torque are low and fairly constant such as floppy disc operations, plotter drives, etc. Stepper motors are no longer used in CNC machines. They were initially replaced by DC servo motors and subsequently AC servo motors which work in closed loop control. The DC servo motor becomes noisy and its movements become jerky when the carbon brush wears depositing the powder on the commutator. Since the commutation is electronic rather than mechanical in AC servo motor, it is free from such maintenance problems. Both these servo motors work in conjunction with a velocity feed back element in a closed loop. The velocity feed back is obtained from a tacho-generator or an encoder mounted at the rear end of the motor shaft.

One is interested in controlling the target position, the path the cutter follows from the initial position and the rate of movement in CNC machines, transducers to measure the position and velocity are required. For incorporating adaptive control of machining, transducers for measuring force, torque, temperature, etc. also will be required. Transducers may be rotary (tacho-generators, resolvers, etc.) or linear (optical scale, inductosyn, magnascale, etc.). When a rotary transducer is used to measure linear motion and vice versa, it is set to be an indirect transducer for that application. For example, the resolver mounted on the spindle motor itself can be used to calculate the table movements. Such an indirect measurement is less accurate that the direct measurement since it cannot reflect errors like pitch error, backlash, etc. The signal produced by the transducer may be analog, pulses or digital. Fig. 4 shows the optical scale used to measure linear and rotary positions.

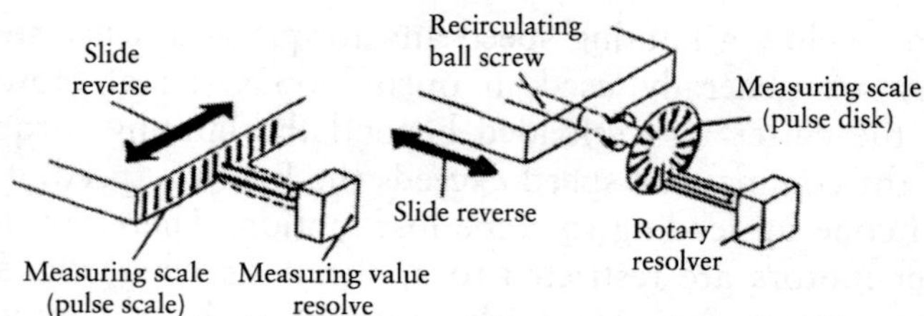

Fig. 4: Linear and rotary encoders used as feedback devices

6. Programming CNC Machines

For each part, the user has to prepare a NC programme which is often called a part programme. A programmer who has very good understanding of the drawing requirements, process planning, tooling and machining is responsible for preparing the part programme. The activity of part programming includes

- Reading the part print
- Deciding the process steps (sequence of operations)
- Deciding the tooling requirements
- Preparation of NC programmes
- Proving the programmes graphically, on soft material and finally on the actual material

The generation of NC programmes is done in the following three ways:

- Manual part programming
- Computer assisted part programming
- Automatic generation of NC programmes.

In manual part programming, the programmer calculates the path followed by the cutter center from the part geometry. He identifies the offset profiles in terms of lines and arcs and writes the NC blocks manually. An NC programme follows a word-address format in which an alphabet or a special character is followed by a number. This format is unique for each CNC controller. G and M are very commonly used codes. For example, G01 will move the cutter along a straight line path and G02 and G03 are used for circular motions. M03 and

M04 are for "spindle on" and M05 is for "spindle off". Other words used are N, U, X, Y, Z, A, B, C, T, S, F, O, etc. The programmer has to write the NC programme using these codes.

The manual part programming described earlier can be done for simple 2D shapes. When the shape is more complicated, it becomes difficult to perform the calculations manually. Therefore, the aid of computer is used. It is of two types: The first type is language based. The language used is called Automatic Programming Tool (APT) which is available in a variety of forms such as APT-AC (from IBM), APT-140 (from British Aerospace), EXAPT, TC-APT, etc. In the APT language, higher level commands like GOTO, TANTO (tangent to), PS (part surface), DS (drive surface), CS (check surface), etc. are available using which the programmer can write the machining steps and other control commands for spindle, coolant, etc. with greater ease. The latest and more sophisticated form of computer assisted part programming is an interactive environment which is based on a geometric modeller (usually a surface modeller and of late solid modeller). The programmer can interactively specify the machining steps by selecting from the menus and the machining operation are depicted on the computer screen in the form of cutter path. This is the most popular method of programming today.

In future, it will be possible to generate NC programmes directly from the solid model of the part automatically, i.e. without any user interaction. Already this has been achieved for turned parts and limited success has been reported for milled parts also. In a CAD/CAM environment, the part geometry may be already available in the database as a solid model created by the designer. If this solid model is conventional B-Rep. or CSG type, then a feature extractor converts this model into a feature-based model. Alternatively, the designer himself can create the part geometry in terms of the manufacturing features like hole, slot, boss, etc. An intelligent process planning system scans this modeller, determines the clamping schemes and cutter requirements for each set-up and then for each set-up, it also determines the cutter path for each cutter.

Important benefits that accrue out of using NC machines as compared to conventional manufacture are:

- Accuracy
- Flexibility
- High Speed tooling
- Reduction/Elimination of set-up time
- Reduced machining time
- Ease of machining complicated contours
- Less skilled operators
- Less fatigue to the operators

NC machines are the basic building blocks for Flexible Manufacturing Systems (FMS) and Computer-Integrated Manufacturing Systems (CIMS). It is the computer control of the manufacturing equipment that led to the integration of these equipment using networks. As a whole, NC plays a vital role in the conversion of Art (design) to Part (manufactured product).

OBJECTIVE TYPE QUESTIONS

Choose the correct alternative

1. Norton type gear box is an example of
   (a) Speed gear box
   (b) Feed gear box
   (c) Infinitely variable drive
   (d) Mechanical friction drive

2. The chief advantage of stepless drives in machine tools is
   (a) the drive is positive without slip
   (b) the speed change from one speed to another is very smooth
   (c) limited speed variation
   (d) they are very cheap
3. For lathes, boring machines and milling machines the desirable speed range ratio is _______.
   (a) between 5-10 (b) between 10-30
   (c) between 50-100 (d) between 100-200
4. Gear boxes used in turret lathes provide _______.
   (a) no variation of speed
   (b) stepped variation of speed
   (c) stepless variation of speed
   (d) infinite variation of speed
5. The synchronous speed of a lathe motor having 2 pairs of poles in the stator at a supply frequency of 50 Hz is _______.
   (a) 500 r.p.m. (b) 1000 r.p.m.
   (c) 1500 r.p.m. (d) 2000 r.p.m.
   (e) 2500 r.p.m.
6. Stepless regulation of speed for machine tools is usually achieved by _______.
   (a) cone pulley drive
   (b) change gear drive
   (c) gear boxes provided with clutches
   (d) mechanical friction drive
7. Ward-Leonard system of speed regulation used in machine tools surface grinding machine is
   (a) basically an hydraulic system
   (b) basically an electronic system
   (c) basically an electrical system
   (d) essentially a mechanical friction drive system.

8. The spindle speeds available in machine tools are usually arranged in ______.
   (a) Arithmetic progression
   (b) Geometric progression
   (c) Logarithmic progression
   (d) Harmonic progression
9. Very large speed ranges are required for ______.
   (a) Shaping machines
   (b) Planing machines
   (c) Semi-automatic and automatic turrets
   (d) Grinding machines
   (e) Drilling machines
10. The phenomenon of stick-slip is more predominant when the sliding speed is
    (a) zero
    (b) low
    (c) high
    (d) equal to the cutting speed
11. The range of speed regulation in a machine tool is defined as
    (a) max. speed/min. speed
    (b) min. speed/max. speed
    (c) max. speed – min. speed
    (d) (max. speed/min. speed) – 1
    (e) No. of speeds obtainable between min. and max. speed
12. The use of elastic rubber supports in machine tool foundations is mainly
    (a) to increase the speed of the machine
    (b) to increase the capacity of the machine
    (c) to prevent electrical shocks
    (d) to suppress the vibrations
13. An important disadvantage of broaching process is
    (a) the broaching tools are always longer in length
    (b) that only simple shapes can be obtained

(c) the return stroke is always idle
(d) the process is suitable only for high volume production

14. Broaching tools are usually made of
    (a) high carbon steel
    (b) High Speed Steel (H.S.S.)
    (c) ceramics
    (d) tungsten carbide
15. The person who is responsible for listing the sequence of operations or machine movements needed to complete the machining of a part on a numerically controlled machine tool (NCMT) is called a
    (a) machine operator (b) set-up man
    (c) tooling assistant (d) programmer
    (e) job analyser
16. A knowledge of ______ is essential for any one wanting to learn how to service or maintain numerically controlled machine tools (NCMT)
    (a) Chemistry (b) Thermodynamics
    (c) Psychology (d) Electronics
    (e) Automobile Engineering
17. The full load speed of a 12 pole synchronous lathe motor operating at 50 Hz is
    (a) just less than 500 r.p.m.
    (b) exactly 500 r.p.m.
    (c) just less than 600 r.p.m.
    (d) exactly 600 r.p.m.
    (e) just less than 1000 r.p.m.
18. The rotor speed of a 12 pole induction lathe motor operating at 50 cps and at full rated load is
    (a) just less than 1000 r.p.m.
    (b) exactly 1000 r.p.m.
    (c) just less than 500 r.p.m.

(d) exactly 500 r.p.m.
(e) exactly 600 r.p.m.

19. The machining operation of cutting a key-way inside a drilled hole is known as
    (a) reaming (b) broaching
    (c) boring (d) tapping

20. The machine tool in which the point-to-point numerical control system is applied is the
    (a) Drilling machine (b) Grinding machine
    (c) Milling machine (d) Lathe
    (e) Shaping machine

21. The machine tool in which the system of continuous path numerical control is applied is the
    (a) Shaping machine (b) Grinding machine
    (c) Milling machine (d) Drilling machine

22. The numerical control system which is applicable to a milling machine is called the
    (a) Point-to-point system
    (b) Continuous path system
    (c) Zig-Zag machining system
    (d) Straight-cut-system

23. The person who prepares a programme which shows the sequence of operations to be performed by a NCMT (Numerically Controlled Machine Tool) is called a
    (a) machine tool setter (b) part programmer
    (c) design-assistant (d) tool designer
    (e) skilled machinist

24. The point-to-point system of numerical control can be applied only to
    (a) conventional drilling machine or jig boring operations
    (b) conventional milling operations
    (c) conventional shaper operations
    (d) conventional lathes

25. The vertical movement of the work table on a conventional milling machine is represented in the Cartesian Coordinate System as a movement parallel to

(a) the X-axis (b) the Y-axis
(c) the O-axis (d) the Z-axis

26. The cross feed movement of a conventional milling machine tool is represented in the Cartesian coordinate system as a movement parallel to the

(a) X-axis (b) Y-axis
(c) Z-axis (d) O-axis

27. In the Cartesian Coordinate System, the longitudinal table movement of a conventional milling machine can be considered as a movement in the

(a) O-axis (b) Z-axis
(c) Y-axis (d) X-axis

28. Programming a contouring cut on a milling machine is accomplished faster when a _______ is used.

(a) punched tape (b) tabulator
(c) desk calculator (d) computer
(e) punched card

29. The surface speed in metres/minute of a workpiece of diameter D mm rotating at N r.p.m. is given by

(a) π DN (b) π DN/1000
(c) 1000/π DN (d) DN/1000
(e) 1000/DN

30. Which of the following spindle speeds (r.p.m.) are most likely to be incorporated in the main gear box of a lathe according to the Preferred Number (PN) theory?

(a) 50, 100, 150, 200, 250, 300, 350, 400, 450 r.p.m.
(b) 63, 127, 254, 381, 512 r.p.m.
(c) 14, 20, 28, 40, 56, 80, 112, 160, 224, 320, 448 r.p.m.
(d) 17, 51, 102, 204, 408, 512 r.p.m.
(e) 36, 72, 288, 1728 r.p.m.

31. The phenomenon of stick-slip is predominant
    (a) at low speeds
    (b) at high speeds
    (c) when the sliding speed is zero
    (d) when the sliding speed exceeds cutting speed
    (e) when the sliding speed is less than the cutting speed
32. The hardening of machine tool guideways is usually done by
    (a) Induction hardening
    (b) Vacuum hardening
    (c) Salt bath furnaces
    (d) Flame hardening process
33. The main advantage of the plastic guide ways is
    (a) their low coefficient of friction
    (b) their high coefficient of friction
    (c) their better appearance
    (d) the reduction in the self weight of the machine tool
34. Spark erosion machining method can be used for the machining of
    (a) Conducting materials only
    (b) Non-conducting materials only
    (c) Semi-conductors only
    (d) Both conducting and non-conducting materials
35. In spark erosion machining, for drilling deep holes of small diameters, the tool material should preferably be made of
    (a) Copper wire
    (b) Brass wire
    (c) Tungsten Carbide wire
    (d) Tungsten wire
36. For machining to take place in spark erosion
    (a) the tool must be immersed in the dielectric fluid
    (b) the work must be immersed in the dielectric fluid

(c) both tool and work must be immersed in the dielectric fluid
(d) no dielectric fluid is to be used

37. In spark erosion machining process, removal of metal takes place during
(a) charging of the capacitor
(b) discharging of the capacitor
(c) all times
(d) alternate cycles only

38. In spark erosion machining if the dielectric is air
(a) the workpiece particles will adhere to the tool
(b) the tool particles will adhere to the work piece
(c) the adherence of particles do not occur
(d) no machining takes place

39. In spark machining, erosion takes place
(a) on the job (b) on the tool
(c) on both job and tool (d) on the dielectric itself

40. Electro-discharge machining and electro-chemical machining processes _______ be used for machining ceramics and plastics.
(a) can (b) cannot

41. Surface integrity of finished workpiece is considerably _______ for electro-discharge machining process when compared to that of electro-chemical machining process.
(a) poor (b) good
(c) very good (d) excellent

42. In electro-chemical machining process metal is removed by _______.
(a) highly accelerated electron flow owing to ionization of the fluid medium
(b) anodic dissolution
(c) etchant solution
(d) stimulated emission of radiation

43. Tooling cost is ______ for electro-chemical machining process compared to that of electro-discharge machining process.
    (a) very high (b) very low
    (c) medium (d) practically nil

44. In electro-chemical machining metal removal rate depends upon
    (a) The hardness of tool material
    (b) The hardness of job material
    (c) Independent of the hardness of tool and work material
    (d) The difference between the hardness of tool and work material

45. In electro-chemical machining, the metal is removed
    (a) by the mechanical pressure of tool on job
    (b) by the mechanical pressure of job on tool
    (c) by atom by atom-dissolving of metal from the job
    (d) by the stimulated emission of radiation

46. In electro-chemical machining, best surface finish is obtained
    (a) with low current density
    (b) with high current density
    (c) with slow rate of metal removal
    (d) with high rate of metal removal

47. Electrolyte used in electro-chemical machining process is
    (a) Brine solution (b) Kerosene
    (c) Transformer oil (d) Water
    (e) Air

48. For the machining of tungsten carbide by ultrasonic machining which abrasive is used for maximum machining rate?
    (a) Silicon carbide (b) Boron carbide
    (c) Aluminium oxide (d) Glass

49. In Ultrasonic drilling process, the tool is usually given
    (a) the Rotary motion
    (b) the Reciprocating motion
    (c) the Linear motion
    (d) Both the rotary motion and the reciprocating motion
50. Diamond powder is mostly used in ultrasonic machining for the cutting of
    (a) Glass (b) Germanium
    (c) Boron carbide (d) Diamond
51. In ultrasonic machining process which part of the machine converts electric impluses to mechanical vibrations?
    (a) The trunk (b) The oscillator
    (c) The transducer (d) The abrasives
    (e) The workpiece
52. Which is the most suitable work material that can be machined by Ultrasonic Machining?
    (a) Mild steel (b) Brass
    (c) Cast Iron (d) Glass
53. The work material that can be successfully machined by all the three processes viz. Spark erosion machining, Ultrasonic machining and Electrochemical Machining is ______.
    (a) Glass (b) Tungsten carbide
    (c) Ceramic (d) Diamond
    (e) High Speed Steel (HSS)
54. The electrodes used in the Electro-Chemical Machining process must be made of ______.
    (a) Semi-conductor
    (b) an anodic material
    (c) a dielectric
    (d) an insulating material
    (e) an electrically conducting material

55. Both the electrode and the workpiece must be submerged in a dielectric fluid when machining the workpiece by the
    (a) USM process
    (b) ECM process
    (c) EDM process
    (d) AJM process
    (e) LBM process

    (*Note*: USM refers to Ultrasonic Machining
    ECM refers to Electro-Chemical Machining
    EDM refers to Electro-Discharge Machining
    AJM refers to Abrasive Jet Machining
    LBM refers to Laser Beam Machining)

56. The electrodes used in the electro-chemical machining process differ from those used in the EDM process in that the ECM electrodes
    (a) are hollow
    (b) are made of conducting materials
    (c) are made of insulating materials
    (d) are insulated at the sides

57. In the ECM process, machining is accomplished by feeding through the gap ________ between the electrically charged electrode and the workpiece.
    (a) a dielectric fluid
    (b) an electrolytic solution
    (c) Kerosene
    (d) Water
    (e) plasma effluent

58. The part of an ultrasonic machine tool which causes the tool to vibrate at a rate of 20,000 cycles/second is called the
    (a) Workpiece holder
    (b) Abrasive slurry pump
    (c) Water Jacket
    (d) Transducer
    (e) Feeding Device

59. In spark erosion machining process, the gap between tool and workpiece is filled with
    (a) a photo etchant
    (b) brine solution

(c) acid solution
(d) a liquid dielectric
(e) an electrolytic solution

60. In the electro-discharge machining process, the workpiece and the electrode are submerged in
    (a) a dielectric fluid
    (b) an abrasive slurry
    (c) an electrolytic solution
    (d) vacuum
    (e) chemical reagents

61. The cutting tool used in the spark erosion machining process is called
    (a) an Arc (b) a Capacitor
    (c) an Electrode (d) a Dielectric
    (e) a Servo

62. The machining process where the metal of a workpiece is dissolved into an electrolyte solution is called ______.
    (a) electro-discharge machining
    (b) ultrasonic machining
    (c) electro-chemical machining
    (d) chemical machining
    (e) laser machining

63. The machining method in which an abrasive slurry is used is called
    (a) Electro-discharge machining
    (b) Laser machining
    (c) Plasma Arc machining
    (d) Ultrasonic machining
    (e) Chemical machining

64. In the electrolytic grinding process generally ______ is used.
    (a) an aluminium oxide grinding wheel
    (b) a silicon carbide grinding wheel

(c) a tungsten carbide grinding wheel
(d) a diamond grinding wheel
(e) a buffing wheel

65. Chemical reagents and etchants are used in ______ machining method.
(a) Electrochemical (b) Plasma Arc
(c) Ultrasonic (d) Chemical
(e) Laser

66. In spark erosion machining process which is used for diesinking, the usual tool material is
(a) High Speed Steel (b) Brass
(c) Tungsten Carbide (d) Diamond

67. The modern machining process working on Faraday's laws of electrolysis is known as
(a) EDM (b) EBM
(c) ECM (d) LBM

68. In ECM, the current density in the discharge of channel is of the order of
(a) 10,000 Amperes per cm^2
(b) 1000 Amperes per cm^2
(c) 100 Amperes per cm^2
(d) 10 Amperes per/cm^2

69. A complicated contour is to be made exactly in a carbide plate. Which process will be used?
(a) ECM (b) EDM
(c) LBM (d) USM

70. Dielectric is a must in
(a) EDM process (b) ECM process
(c) USM process (d) LBM process

71. In EDM process the workpiece is generally connected to
(a) positive terminal (b) negative terminal
(c) earth (d) any of the above

72. Metal removal rate (MRR) in USM is more for
    (a) larger grain size of abrasives
    (b) smaller grain size of abrasives
    (c) medium grain size of abrasives
    (d) none of the above

ANSWER KEY

1. (d)	2. (b)	3. (b)	4. (b)	5. (c)
6. (d)	7. (c)	8. (b)	9. (d)	10. (b)
11. (a)	12. (d)	13. (d)	14. (b)	15. (d)
16. (d)	17. (b)	18. (c)	19. (b)	20. (a)
21. (c)	22. (b)	23. (b)	24. (a)	25. (d)
26. (b)	27. (d)	28. (d)	29. (b)	30. (c)
31. (a)	32. (d)	33. (a)	34. (a)	35. (d)
36. (c)	37. (b)	38. (b)	39. (c)	40. (b)
41. (a)	42. (b)	43. (a)	44. (c)	45. (c)
46. (d)	47. (a)	48. (b)	49. (b)	50. (d)
51. (c)	52. (d)	53. (b)	54. (e)	55. (c)
56. (d)	57. (b)	58. (d)	59. (d)	60. (a)
61. (c)	62. (c)	63. (d)	64. (d)	65. (d)
66. (b)	67. (c)	68. (c)	69. (b)	70. (a)
71. (a)	72. (a)			

15

Robotics

1. Introduction

Robots are re-programmable multi-functional mechanical manipulators. They can be programmed to perform a variety of jobs, which usually involve moving or handling objects. The following are the advantages of using robots:

- Flexibility
- High productivity
- Better quality of parts manufactured
- Less risk to human life (as robots can be deployed in the hazardous jobs).

2. Parts of a Robot

For a machine to qualify as a robot it must have the following live parts:

i. Controller
ii. Arm
iii. Drive
iv. End effector
v. Sensors

(i) *Controller*

Every robot is connected to a computer, which keeps the various parts of the arm working together and coordinates the various motions. This computer is known as the controller (Fig. 1). The controller also allows the machine to be networked to other systems so that it may work together with other machines, processes or robots. Robot control unit is the brain of the robot.

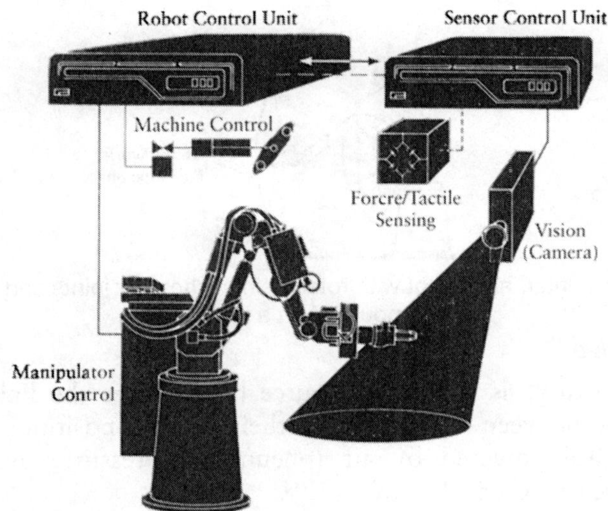

Fig. 1: Components of a robot system

Robots today have controllers that are run by programmes. Almost all robots are entirely programmed by operators, so they can do only what they are programmed for. In future, controllers with artificial intelligence (AI) could allow robots to think on their own, even programme themselves. This could make the robots more self reliant and independent.

(ii) *Arm*

Robot arms come in all shapes and sizes. The arm is the part of the robot that positions, the end effectors and the sensors to do their preprogrammed task (Fig. 2). Many but not all resemble human arms and have shoulders, elbows, wrists even fingers. This gives the robot a number of ways to position itself in its environment. Each joint is said to give the robot one degree of freedom. So simple robotic arm with 3 degrees of freedom could move in three ways, viz., up and down, left and right, forward and backward. Most working robots have six degrees of freedom.

In order to reach any position in space within its work envelope, a robot needs a total of six degrees of freedom. As a result many robots of today are designed to move in at least six ways.

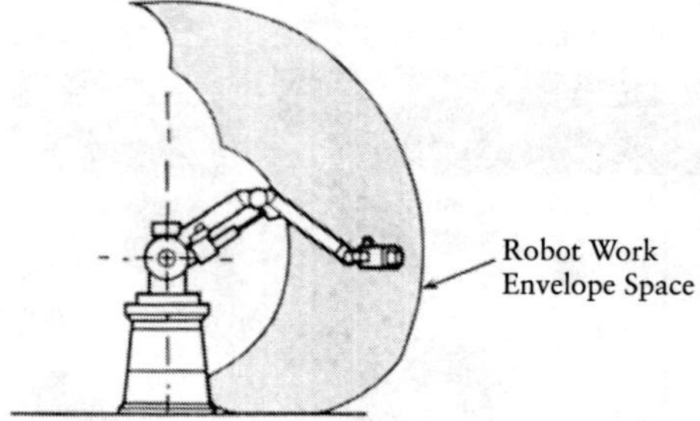

Fig. 2: Jointed arm robot with rotary joints (shoulder joints and elbow joints) mounted on a rotary base

(iii) *Drive*

The drive is the power source that drives the links (the sections between the joints) to their desired positions. Most drives are powered by air (pneumatic) pressure, hydraulic pressure or electrical power.

(iv) *End Effector*

The end effectors are the hands connected to the robots arm (Fig. 3). It is often different from a human hand; it could be a tool such as gripper, a vacuum pump, tweezers, scalpel—just about any thing that helps in doing the final job. Some robots can change the end effectors and be reprogrammed for a different set of tasks.

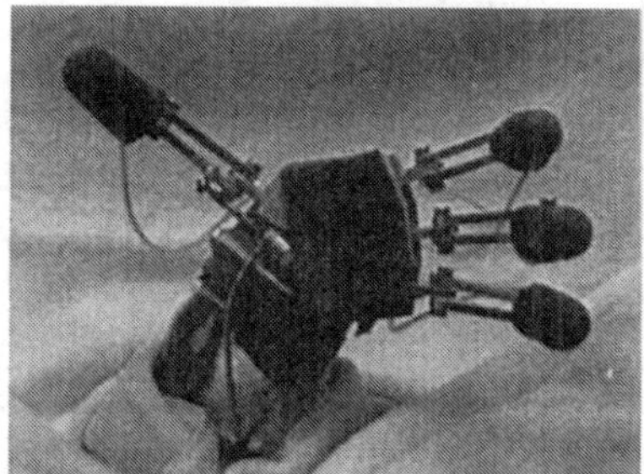

Fig. 3: End effectors of a robot

(v) *Sensors*

Most of the robots today are nearly deaf and blind. Sensors can provide some limited feedback to the robot so it can do its job. Compared to the senses and abilities of even the simplest living things robots have a very long way to go.

The sensor reads and signals back to the controller. Sensors also give the robot the controller information about its surroundings and let it know of exact position of the world around it. Sight, sound, touch, taste and smell are the other kinds of information we get from our world. Robots can be designed and programmed to get specific information that is beyond what our five senses can tell us; for instance, a robot sensor might see in the dark, detect tiny amount of movement that is too small for or too fast for the human eyes to see.

(iv) *Robot Kinematics*

The robot configurations are based on either prismatic or revolute joints whose position is described by a translation or rotation matrix respectively, measured by a feedback element such as a potentiometer or optical encoder. The servo control systems drive each joint to eliminate the error signal formed by subtracting the measured position from the commanded position. The state of the robot is defined in joint space by the set of values representing each joint position. This may be different from actuator space if gear ratios vary or some drives are coupled. It is more convenient for the user to specify positions in natural coordinate systems such as world coordinates fixed with respect to the ground, or tool coordinates fixed w.r.f. the gripper flange. In a Cartesian system, 3 translations specify the gripper position and also 3 rotations are needed to describe the gripper orientation, making 6 degrees of freedom in all. More information may be needed to specify the state of the gripper itself. Fewer degrees of freedom are possible when some motions are excluded. Other standard names for frames include the base frame, station frame, wrist frame, tool frame and goal frame.

The geometrical relationships between these frames are a matter of kinematics, based on rigid links and close fitting joints. Whether the desired positions are in fact achieved will depend also on the dynamics taking into account drive torques, components inertia and speed.

3. Classification of Robots

Robots can be classified on the basis of the following parameters:

- Arm configuration
- Shape of the work space
- Operating method
- Type of motion
- Type of power
- Size
- Type and number of joints
- Type of technology
- Configuration
- Type of motion

(i) *Classification by Arm Configuration*

The following are some of the kinematic configurations of the robot arm:

- Cartesian robot
- Cylindrical robot
- Spherical robot
- Jointed-arm or Articulated robot
- Hexapod robot

Cartesian robots move along an x, y, and z-axes in straight lines. Cylindrical robots manipulator can rotate about its base and can move linearly in horizontal and vertical direction. SCARA (Selective Compliance Articulated Robotic Arm) robot is similar to the cylindrical robot but it has a rotational reach axis rather than a linear one. Polar or Spherical manipulator can rotate about its head as well as its base. It can also move in and out. Jointed arm or articulated robot arms resemble the design of a human arm and have largest work cell of amount of required floor space. Hexapod robots have very high stiffness. For the same reason, this configuration is used even for CNC machines.

(ii) *Classification by Type of Motion*

- *Limited Sequence*: It uses mechanical stops to limit the movement. A timer may be used to activate movement in axis.
- *Point to Point*: This type of controllers have sensors for feed-back and memory to hold the coordinates of each axis. The actuators all run simultaneously at one speed, usually maximum. Some joints finish before others. The tool path is predictable, but not readily visualized.
- *Continuous Path*: These types of sensors have greater memories than point to point and can record many coordinates per second. The controller generates a sequence of interpolated via-points. This is one way of achieving joint-interpolated movement, but if the computer's arithmetic is fast enough the points can be calculated to lie on a straight line or circular arc. A large number of inverse transformations must be performed. This method is also used after auto-recording a sequence of points during lead-through teaching.
- *Joint-interpolated Motion*: All joints start and finish together, being driven at speeds calculated from that of the slowest joint (or the one with farthest to travel). I gives smoother movement, which is often closer, to what is expected, especially if intermediate via-points are included to constrain the departure from the desired track. There can be considerable variation from straight-line motion, except where the robot's geometry permits natural movement in one coordinate plane or axis.

(iii) *Classification by Type of Power Supply*

The power sources for robot actuation could be

- Electrical
- Pneumatic
- Hydraulic

Electric robots are compact, efficient, require little maintenance, and are not very noisy, but primarily meant for light duty. Pneumatic robots use compressed air and come in a wide variety of sizes. A pneumatic robot requires another source of energy such as electricity, to provide the compressed air. Hydraulic robots use oil under pressure and generally perform heavy-duty jobs. This power type is noisy, large and heavier than the other power sources. A hydraulic robot also needs another source of energy to move the fluids through its components. Pneumatic and hydraulic robots require maintenance of the tubes, fittings and hoses that connect the components and distribute the energy.

Two important sources of electric power for mobile robots are solar cells and batteries. There are lots of types of batteries like carbon-zinc, lithium-ion, lead-acid, nickel-cadmium, nickel-hydrogen, silver-zinc and alkaline to name a few. Battery power is measured in amp-hours which is the current (amp.) multiplied by the time in hours that current is flowing from the battery. For example, a two-amp hour battery can supply 2 amps of current for one hour. Solar cells make electrical power from sunlight. If you hook enough solar cells together in a solar panel you can generate enough power to run a robot. Solar cells are also used as a power source to recharge batteries.

Deep space outer space probes must use alternate power sources because beyond Mars existing solar arrays would have to be so large as to make them not feasible. The lifespan of batteries is exceeded at these distances also. Power for deep space probes is traditionally generated by radioisotope thermoelectric generators (RTG), which use heat from the natural decay of plutonium to generate direct current electricity. RTGs have been used on 25 space missions including Cassini, Galileo, and Ulysses.

(iv) *Classification by the Level of Technology*

- *Low Level Technology*: These robots are used for material handling, loading, and unloading, simple assembly operations. These machines have 2-4 axes.

- *Medium Level Technology*: These robots are used for picking and placing, loading and unloading have 4-6 axes of movement.
- *High Level Technology*: These robots can be used for manufacturing tasks as well as for other applications. These have 6-9 axes of movement and possess higher level of sensory perception and accuracy.

(v) ***Classification by Design***

- *First Generation*: These robots use fixed sequence of programmes. Reprogramming had to take place before another task could take place. They did not posses sensors and all control was through an open loop controller system.
- *1.5th Generation*: These robots can adapt to varying situations by the use of sensors and a closed loop servo controller. Although the robot may "know" when something is wrong in the system it may not be capable of doing anything about it. It requests human help. These are today's robots.
- *Second Generation*: These robots are in near future. Vision system in conjunction with better mobility systems will make the robots much more efficient and flexible. More intelligent microprocessors coupled with voice recognition and hand-to-hand coordination will expand the usefulness of the robot systems. These robots will be able to make minor decisions and will be able to make minor corrections or adjustments to manipulators.
- *2.5th Generation*: Perceptual motor functions will be the highlight of these robots. Response to sensory stimuli will produce control motion. If initial sequence of steps are provided for robot task the robot will be able to decide what to do next to accomplish the specific task.
- *Third Generation*: These are really smart robots. If the robots have generalized information about what

tasks need to be done they will be able to decide the best way to accomplish the task.

(vi) *LERT Classification System*

LERT classification system uses the following four motions as the basis for the classification, LERT being an acronym obtained by the first letters of these motions:

- *Linear Motion*: This type of motion usually occurs when one part moves along the outside of another part such as in a rack and pinion system.
- *Extensional Motion*: This is a type of telescoping motion produced when one part moves within another part. Kinematically, this is same as linear type although constructionally different.
- *Rotational Motion*: This type motion is caused by a part, moving about something other than its center.
- *Twisting Motion*: Twisting motion is produced by a part turning about its center.

LERT number for some of the robot configurations are as follows:

- Cartesian (only in front of itself) : L3
- Cylindrical (around itself) : L2R1
- Polar (above, below and around) : L1R2
- SCARA (around itself or around an obstacle) : R3

4. Programming of Robots

Robots and CNC machines require one control programme for each job they perform. Whenever they switch from one job to the other, this control programme needs to be changed, except autonomous robots which can take their own decisions based on the changing situation as they possess sensors and are intelligent. These control programmes are a series of codes that indicate the motions and switching functions they have to perform.

There are two ways in which these control programmes can be generated:

(i) Online programming; (ii) Offline programming.

Online programming is done using the robot or CNC machine itself. This is the easiest to do when the geometry and kinematics involved are simple. However, this employs the costly equipment unnecessarily and requires careful testing before actual use.

On the other hand, offline programming is carried out on another computer which is invariably cheaper than the robot. Furthermore, it requires very less input from the programmer and does all the necessary calculations by itself. In the past, programming languages which were mostly dialects of Automatically Programmable Tool (APT) language were used for this purpose. Nowadays, geometric model based interactive programming is used instead of these languages which is more elegant. There are efforts to generate these programmes automatically, i.e., without any human intervention.

Different types of techniques used for online programming robots are:

1. *Walk through or teach pendant programming*: The robot is moved using a hand held pendant that has buttons and/or joysticks for various motions. The axes values at different positions of the robot obtained using the pendent are stored which can be played back. This is the cheapest way of programming for robots. However, this is not a very elegant and easy way of programming. It takes a lot of time to learn this method of programming.

2. *Lead through programming*: In this method, the operator simply holds the arm of the robot and takes it along the required path. During this motion, the positions are recorded at regular time intervals. Typically used for paint spraying. This is the most preferred and popular method of programming industrial robot.

3. *Tele-operation*: This is based on master-slave control. The actual robot (slave) and its replica (master) are kept in two different places and these communicate using cable or wireless means. Whenever the master is moved by the operator, the slave will faithfully follow it. The operator always needs to look at the slave robot through a glass window or with the

help of video cameras. The limitation of this method is the need of the operator as long as the robot is in operation. This is used in hazardous environments such as nuclear installations, underwater, space, etc. where the conditions are not predictable and hence a standard programme cannot be used. For non-engineering applications, such as robots for handicapped persons, robots tele-operated using voice are possible.

The programme generated using above (except tele-operation where there is no programme at all) need to be validated to ensure that they produce parts that conform to design requirements and they are safe to use. This gain can be done online or offline. Solid modelling based offline simulations are becoming popular for validating the programmes of both robot and CNC machines.

5. Sensors

Robots were initially expected to become humanoids. However, non-availability of suitable sensors and the adequate hardware and software to process the sensory information at the required speed were the bottlenecks. Therefore, robots with less sensory perception could be used only for industrial applications. Developments in electronics and computer technology are overcoming these bottlenecks and more smarter and intelligent robots will be available in the near future.

The sensors for robotics can be classified as

i. Internal sensors

ii. External sensors.

(i) *Internal Sensors*

These sensors provide feedback signals for positional control of joints, except when stepping motors are used. Different types of internal sensors are

1. *Potentiometers*: These provide analogue signals, are unreliable because of track wear and are inaccurate, e.g. 0.1% linearity error represents 0.3 in 300, but this gives 2.6 mm at 0.5 m radius.

2. *Optical Encoders*: These provide digital signals easily, interfaced to a microprocessor. The incremental form uses a photocell to detect light interrupted by a series of obscured sectors of a disk, which rotates with the joint. Using two photocells, this simple counter can be used to detect direction too. An absolute encoder uses multiple tracks and photocells to provide (say) 15 bits of positional information. The sectors are patterned to provide a Gray code signal: this is more reliable than a binary code because only one bit changes from one sector to the next, eliminating spurious intermediate readings. High accuracy is possible with no contact. Typical resolution is $360/2^{15} = 0.66$ minutes of arc. Using a hybrid system with both a potentiometer and an encoder geared at 4:1 can increase resolution to 10 seconds of arc. Such accuracy demands precision manufacture, using anti backlash gearing. Only light springs are needed, as the drive torque for the encoder is low.

3. *Linear Variable Differential Transformer (LVDT)*: This is a device that works on the principle of flux linkage. It can measure the position of a body at any time and so it can be used as a sensor. If used with differential operational amplifier (opamp) it can also give the velocity and acceleration of a body.

(ii) *External Sensors— Vision*

1. *TV Camera*: This requires fast analog-to-digital converter (ADC) and frame store to capture a digitized image. 6 or 8 bit ADC gives 64 or 256 grey levels for each pixel. It collects data at a very high rate (e.g. 500 points, 625 lines, 25 frames/sec, 8 bits/pixel = 62.5 MHz bit rate) and has large storage requirements (e.g. 625 lines, 500 points = 300 kbytes per frame store).

2. *CCD Array*: Solid-state charge-coupled device (CCD) cameras are smaller and lighter, suitable for mounting on the robot. Although the pixels are captured in an array (perhaps 256 × 400 or 512 × 512 resolution), parallel ADC channels would be needed to store a digital representation directly. Instead, shift registers transfer pixels out as a standard video signal, to be captured as before using an expensive frame grabber.

3. *DRAM Solid-state Cameras*: It uses the optical properties of dynamic RAM (Random Access Memory) chips to capture binary images. The discharge rate of the individual capacitors forming the memory array is light sensitive. Using standard microprocessor memory interfacing chips allows the data to be accessed conveniently using cheap components. Different black/white threshold levels can be set by varying the discharge time after a refresh pulse. It permits image processing of 256 × 256 pixels on a computer.

Typically, we wish to identify a part and measure its position and orientation. Given isolated images of a number of non-overlapping components, the vision system might first construct a simple binary representation by coalescing similar grey levels, contrast enhancement, and by eliminating isolated pixels of one colour, noise reduction. Edge detection uses sharp gradients to identify component boundaries.

Do simple operations on pixels of one colour, then calculate recognition parameters such as

- Connectivity (number of holes)
- Area and perimeter(s)
- Centroid (x and y values)
- Second moment of area
- Principal axes (lengths and directions)

Each detected object can be analyzed and its features compared with a standard set describing each component to be recognized. Once identified, the centroid fixes the location and the angle of the major axis fixes the orientation, after which the robot can be programmed to approach the part and grasp it correctly.

(iii) *External Sensors— Tactile Sensing*

It uses rubber membrane above acrylic sheet with air gap in between. Sidelight normally contained by total internal reflection within the acrylic, but a CCD array beneath detects light scattered where the rubber is pushed down. The CCD provides an array of 'tactels', allowing tactile fingers to 'read' surface texture or detail.

Touch probe, a single-point probe connected to a variable resistance can determine component shape, or departure from a standard shape.

6. Applications of Robots

The following are the applications of robotics:

Industrial Applications

- Welding (arc welding, spot welding, seam welding and soldering)
- Material handling (loading and unloading between machines and stores)
- Drilling and deburring
- Assembly
- Inspection
- Painting
- Packaging/palletizing
- Underwater repairs
- Applications in nuclear installations
- Space applications

Applications in Other Areas

- Biorobotics
- Medical applications
- Assistant to handicapped persons
- Cutting wool from sheep
- Removal of bone from meat
- Unmanned petrol bunks

Some of these applications are discussed in the following paragraphs.

Welding

One of the common uses for industrial robots is welding. Robot welded car bodies have better safety. A welding robot (Fig. 4) never misses a welding spot and performs equally all through the day. Nearly, 25% of the all-industrial robots are used in different welding operations. Robot controls different welding parameters like wire feed and power. Seam welding robots have 5-6 axes since two and sometimes three axes are necessary for proper entry and continued operation of welds head to seam. These robots position a welding head in relation to the seam by either a trajectory method or seam sensing method. Trajectory method is simply programming a trajectory path that the robot will follow repeatedly, regardless of seam locations. Seam sensing is accomplished by either tracking the seam with a mechanical finger or by significantly costly vision system method. It is useful to consider robotic seam welding in situation where manual approach requires frequent operator movement, frequent workpiece positioning or both to complete the welding operation.

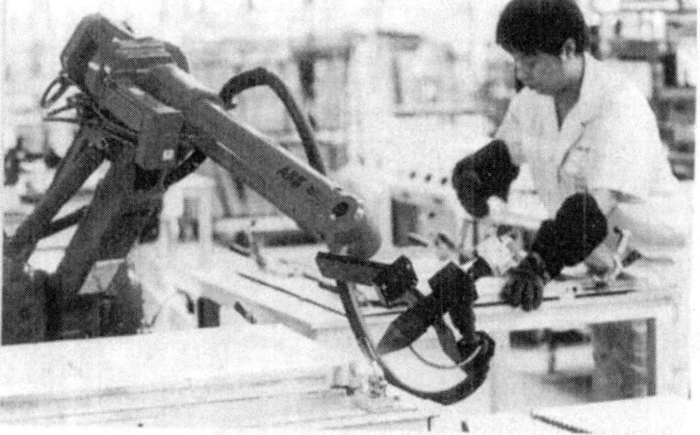

Fig. 4: Robotic arm welding an automobile door panel

Assembly

Assembly accounts for nearly 30% of the industrial robots. Many of these robots (Fig. 5) can be found in automotive and electronic industries. In addition to the robot, robotic assembly cell may require other equipment such as vision system. From a cycle time stand point a manual assembly is faster than the robotic assembly, however, robotic assembly is often a good alternative in clean rooms and other environments where human presence is not desirable.

Fig. 5: A robot assembling car parts

Painting

These robots (also known as finishing robots) are five or six axis units as shown in Fig. 6. It has the capability of painting over a complex surface at a constant velocity relative to the surface.

Fig. 6: Painting robot

Biorobotics

Recently a system of micro propagation of plants has been developed. The process involves harvesting each arriving plant, cutting its stem into segments near each node and then replanting the segments so that they can grow into new plants. The whole process is undertaken by the help of robots (Fig. 7).

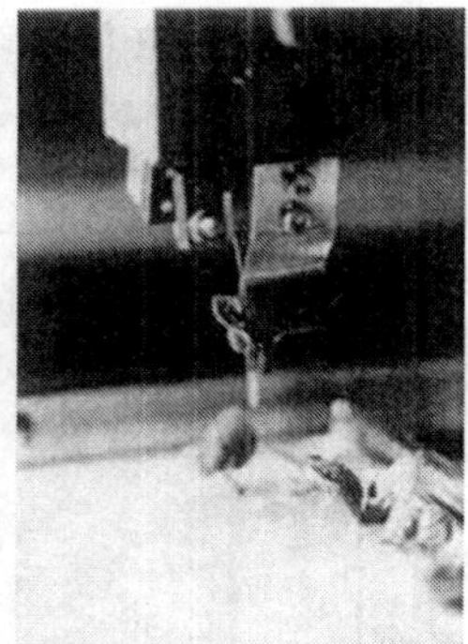

Fig. 7: A robot cutting the plant stems into segments near each node of the plant

Medical Applications

Minimally Invasive Surgery (MIS) is a revolutionary approach in surgery. In MIS, an incision is made in the human body and a robot (Fig. 8) is inserted into the body in fragments. The different parts of robots align themselves inside the human body and perform the required operations. Unfortunately, there are disadvantages improper alignment due to and sensory input to the surgeon, which is only available through a single image.

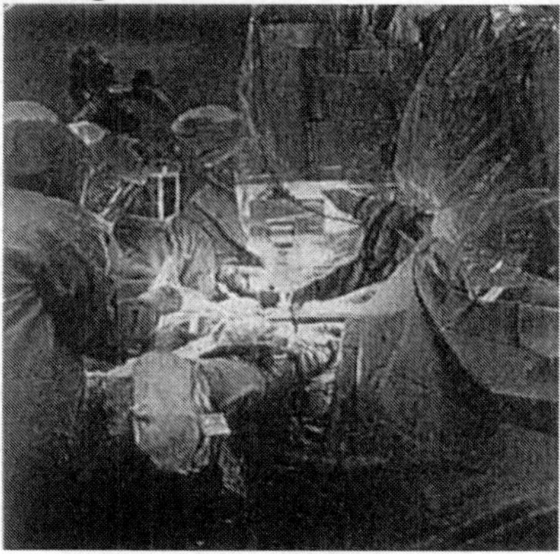

Fig. 8: Minimally invasive surgery

Underwater Applications

Conventional unmanned, submersible robots are used in ship and industry throughout the oceans of the world. To get around, automated underwater vehicles (AUV's) use propellers and rudders to control their direction of travel. An underwater robot (Fig. 9) could propel itself as a fish does using its natural undulatory motion. It is thought that robots that move like fish would be quieter, more manoeuvrable and more energy efficient.

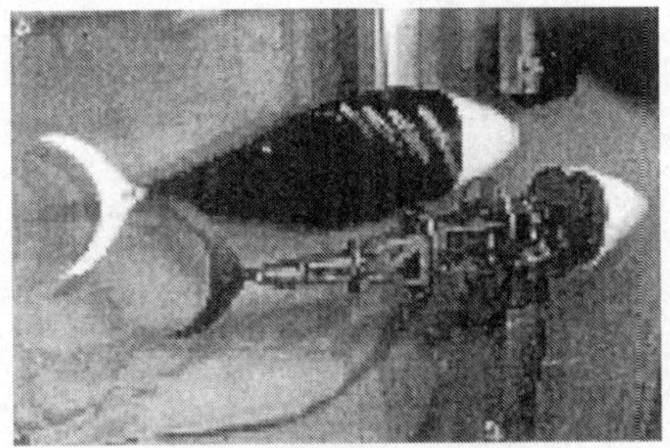

Fig. 9: An automated underwater vehicle (robot)

Space Expeditions

Space-based robotic technology developed by NASA National Aeronautical and Space Agency, U.S.A.) falls within three specific mission areas: (i) exploration robotics, (ii) science payload maintenance, and (iii) on-orbit service. Related elements are terrestrial commercial applications, which transfer technologies generated from space telerobotics to the commercial sector, and component technology, which encompasses the development of joint designs, muscle wire, exoskeletons, and sensor technology.

Today, two important devices exist which are proven space robots. One is the Remotely Operated Vehicle (ROV) and the other is the Remote Manipulator System (RMS). An ROV can be an unmanned spacecraft that remains in flight, a lander that makes contact with an extraterrestrial body and operates from a stationary position, or a rover (land rover on Mars) that can move over terrain once it has landed. It is difficult to say exactly when early spacecraft evolved from simple automatons to robot explorers or ROVs. Even the earliest and simplest spacecraft operated with some preprogrammed functions monitored closely from Earth. One of the best known ROV's is the Sojourner (Fig. 10) rover that was deployed by the Mars Pathfinder spacecraft. Several NASA Centers are involved in developing planetary explorers, robotic technology, and space-based robots. To date, the NASA RMS robot arm has performed a number of tasks on many space missions, providing a firm grip through a remote assembly device, and also as a positioning and anchoring device for astronauts working in space.

Fig. 10: ROV (Remotely Operated Vehicle): MARS PATH FINDER

Packaging/ Palletizing

This industry is still a minor application area and is expected to develop in coming years (Fig. 11). The food industry is a major area where robots are expected to play a major role in years to come (Fig. 12).

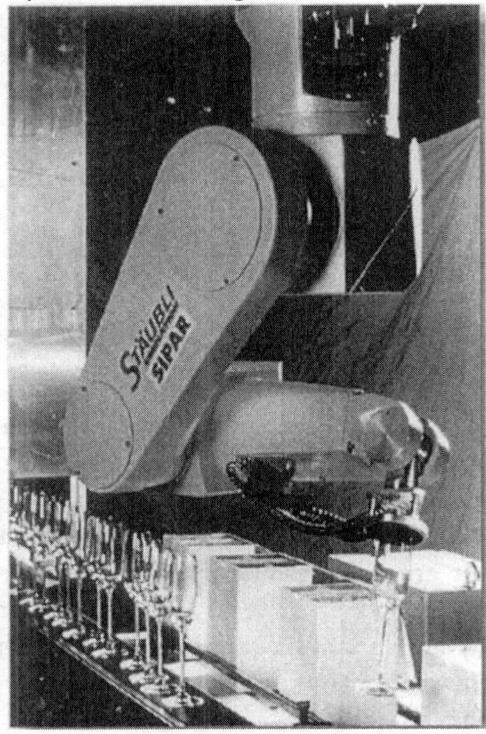

Fig. 11: Robots in packaging

Fig. 12: Robots in food industry

Although robots and CNC machines are similar in many ways, robots are less popular till recently in the industries than CNC machines. However, recent developments in sensors and signal and image processing have enhanced their capabilities substantially. Robots are now seen extensively in industries and soon they will be seen as an integral part of every household. These machines, will be widely used in future because they are:

- Adaptable to different environments.
- Capable of sensing a wide range of sensory inputs, with pattern recognition.
- Able to make decisions, set priorities and define goals.
- Able to investigate new experimental techniques.
- Easy to make programmes.

But, at the same time there remain certain social issues that need to be answered today since there is a fear that the use of robot will lead to unemployment. There has been a lot of speculation about future of human race after the development of robots that are able to take decisions of their own. It is believed that if robots are able to take decision on their own then they may develop into a race superior to the human race. But as of now these are just ideas confined to the area of science fiction.

At the beginning of new millennium, our society has several big problems to be solved to maintain good quality of life. Those are relating to saving natural resources, increase of aged population in the society, preservation of global environment and so on. We suppose that robotics and intelligent system technologies can be a powerful way and play a more important role to solve the problems our future society will have. Actually, there are strong expectations to robotics and intelligent system technologies in our society. There are new technological trends like computer network technologies, micro-nano-technologies and so on, which make us expect to develop powerful robotic and intelligent system technologies applicable to new industrial and commercial applications.

Although robots have certain social issues that go against them, like every new technology, these issues are overshadowed by the benefits of using robots.

16

Computer Aided Process Planning (CAPP)

1. Introduction

With the rising importance of competitiveness, technological improvements in the manufacturing arena are necessary. Many of these improvements will be in the field of automation. Manufacturing automation can speed up product turnaround, reduce the need for retooling, and lead to a more efficient allocation of resources. Automation can be effective for small batch manufacturing and in spare parts production. While this is a desirable goal, many small shops cannot afford to fully automate. By using clearly defined interfaces, a shop can support both manual and automated operations. In reality a factory will have manual, semi-automated and fully automated facilities.

There are a number of obstacles to the implementation of a fully integrated automated manufacturing facility. One major gap is the lack of smooth information flow between Computer-Aided Design (CAD) systems and Computer-Aided Manufacturing (CAM) systems. Traditionally, these two functions have been treated as completely separate activities. There is no feedback from CAM to CAD to reflect the manufacturability of a particular design. There are very few commercial/production systems, which actually integrate CAD with CAM.

Extending these ideas to general part families is a much more difficult task. A brief example of the information flow illustrates a number of problem areas. A client requests some product to be manufactured and provides a loose set of requirements. This request is translated into representation, usually a part drawing. This representation includes some simple translation of functional attributes into specific tolerance information. The information is loosely organized as notes or text on the part drawing. This step can be done manually or through CAD. The step is complete when the client and designer agree that the drawing adequately represents the client's needs. The problem with this approach is that the information is not represented in a computer database form. This implies that a human will have to interpret the notes sometime later in the process, leading to ambiguities. More importantly, this approach does not allow any feedback to the designer as to the manufacturability of the design. The next step in the process is to bridge the link to the CAM systems. This is called process planning.

Process planning transforms the design information into process specification structure used by the manufacturing organization. This step includes defining a group of machinable features and their associated processing steps, selecting appropriate machine tools to be used to process the part, generating tool and fixturing orders, and any other information needed to actually produce the part. The CAM system then expands each process step into more detailed instructions including robot or NC machine tool programme tool, offsets, etc. It is at this point that important information is generated which should be communicated back to the designer. The important point is to produce a product at minimum cost while retaining the desired quality and functionality. The above described information flow is illustrated in Fig. 1.

Thus, the step called process planning is the transformation of information from the CAD representation to the CAM representation. The transformation rules that humans apply are not well understood even by those who use them. Clearly this makes it difficult to encode those rules in process planning systems. It is only when these rules can be represented in automatic systems that any feedback can be given during the design process. To accomplish this, a more powerful product representation is needed. This representation must serve the needs of the designer who is striving for functionality, as well as the manufacturing engineer who wants high quality at low cost. As many computer-aided tools support the design process, computer-aided process planning (CAPP) has evolved to simplify and improve process planning and achieve more effective use of manufacturing resources. In essence, CAPP can be considered as a bridge between CAD and CAM.

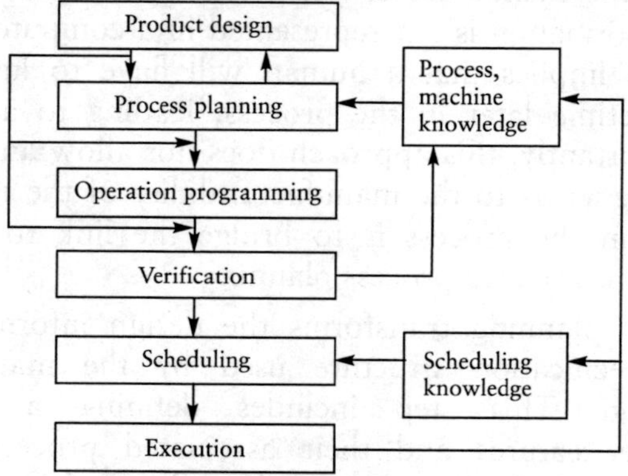

Fig. 1: Product realization

2. Methodology in Conventional Process Planning

Because of the interrelationship between the product design, material selection and process planning, some manufacturing engineering decisions need to be taken early in the design process, even while the design is in the conceptual stage.

In order to prepare a process plan, a process planner needs to acquire the following knowledge:

- Ability to interpret an engineering drawing
- Familiarity with manufacturing processes and practice
- Familiarity with tooling and fixtures
- Knowledge about the resources available in the shop:
    - Knowledge how to use reference books, such as Tool Engineers/Machinability data handbook

    - Ability to do computations on machining times and cost
    - Familiarity with raw material, properties and specifications.
- Knowledge of the relative costs of processes, tooling and raw materials.

To prepare a process plan, the following steps needed to be taken:

- Study the overall shape/design of the part. Use this information to classify the part and determine the type of manufacturing processes/workstation needed.
- Thoroughly study the drawing. Try to identify special manufacturing features and notes.
- If raw stock is not given, determine the best raw material shape to use.
- Identify datum surfaces. Use information on datum surfaces to determine the setups.
- Select machines for each setup.
- For each setup, determine the rough sequence of operations necessary to create all the features.
- Sequence the operations determined in the previous step. Check whether there is any interference between operations or dependency between operations. Use this information to modify the sequence of operations.
- Select tools for each operation. Try to use the same tool for several operations if it is possible. Keep in mind the trade off on tool change time and estimated machining time.
- Select or design fixtures for each setup.
- Evaluate the process plan generated thus far and make necessary modifications.
- Select cutting parameters for each operation.
- Prepare the final process plan document.

Process planning encompasses the above-mentioned activities and functions to prepare a detailed set of plans and instructions to produce a part. The planning begins with engineering drawings, specifications, parts or material lists and a forecast of demand. The results of the planning are:

- Route sheets, which specify operations, operation sequences, work centers, standards, tooling and fixtures. This routing becomes a major input to the manufacturing resource planning system to define operations for production activity control purposes and define required resources for capacity requirements planning purposes.
- Process plans, which typically provide more detailed, step-by-step work instructions including dimensions, related to individual operations, machining parameters, set-up instructions, and quality assurance checkpoints.
- Fabrication and assembly drawings to support manufacture (as opposed to engineering drawings to define the part).

3. Manual Process Planning

Manual process planning is based on a manufacturing engineer's experience and knowledge of production facilities, equipment, their capabilities, processes, and tooling. Process planning is very time-consuming and the results vary based on the experience of the person doing the planning. This methodology of process planning is aptly illustrated in Fig. 2.

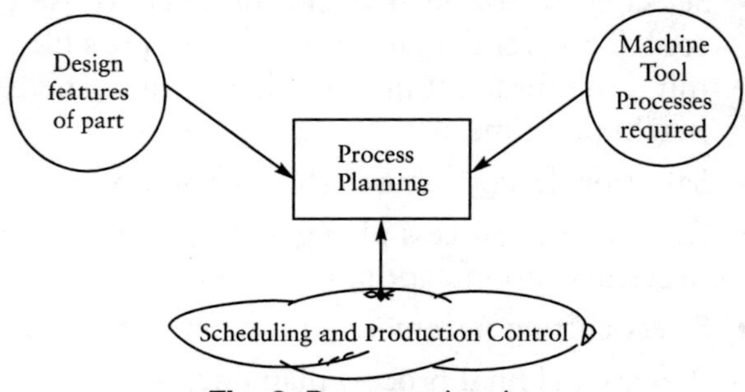

Fig. 2: Process planning

4. Evolution of CAPP

Manufacturers have been pursuing an evolutionary path to improve and computerize process planning in the following five stages:

Stage I – Manual classification; standardized process plans

Stage II – Computer maintained process plans

Stage III – Variant CAPP

Stage IV – Generative CAPP

Stage V – Dynamic, generative CAPP

Prior to CAPP, manufacturers attempted to overcome the problems of manual process planning by basic classification of parts into families and developing somewhat standardized process plans for parts families (Stage I). When a new part was introduced, the process plan for that family would be manually retrieved, marked-up and retyped. While this improved productivity, it did not improve the quality of the planning of processes and it neither took into account the differences between parts in a family nor improvements in production processes.

Computer-aided process planning initially evolved as a means to electronically store a process plan once it was created, retrieve it, modify it for a new part and print the plan (Stage II). Other capabilities of this stage are cost and standard time estimating systems.

This initial computer-aided approach evolved into what is now known as "variant" CAPP. However, variant CAPP is based on a Group Technology (GT) coding and classification approach to identify a larger number of part attributes or parameters. These attributes allow the system to select a baseline process plan for the part family and accomplish about ninety per cent of the planning work. The planner will add the remaining ten per cent of the effort modifying or fine-tuning the process plan. The baseline process plans stored in the computer are manually entered using a expertmaster planner concept, that is, developing standardized plans based on the accumulated experience and knowledge of multiple planners and manufacturing engineers (Stage III).

The next stage of evolution is toward generative CAPP (Stage IV). At this stage, process planning decision rules are built into the system. These decision rules will operate based on a part's group technology or features technology coding to produce a process plan that will require minimal manual interaction and modification (e.g., entry of dimensions).

While CAPP systems are moving more and more towards being generative, a pure generative system that can produce a complete process plan from part classification and other design data is a goal of the future. This type of purely generative system (in Stage V) will involve the use of artificial intelligence type capabilities to produce process plans as well as be fully integrated in a CIM environment. A further step in this stage is dynamic, generative CAPP, which would consider plant and machine capacities, tooling availability, work center and equipment loads, and equipment status (e.g., maintenance downtime) in developing process plans.

The process plan developed with a CAPP system at Stage V would vary over time depending on the resources and workload in the factory. For example, if a primary work center for an operation(s) was overloaded, the generative planning process would evaluate work to be released involving that work center, alternate processes and the related routings. The decision rules would result in process plans that would reduce the overloading on the primary work center by using an alternate routing that would have the least cost impact. Since finite scheduling systems are still in their infancy, this additional dimension to production scheduling is still a long way off.

Dynamic, generative CAPP also implies the need for online display of the process plan on a work order oriented basis to insure that the appropriate process plan was provided to the shop floor. Tight integration with a manufacturing resource planning system is needed to track shop floor status and load data and assess alternate routings vis-á-vis the schedule. Finally, this stage of CAPP would directly feed shop floor equipment controllers or, in a less automated environment, display assembly drawings online in conjunction with process plans.

The development of CAPP through the years is illustrated in Fig. 3.

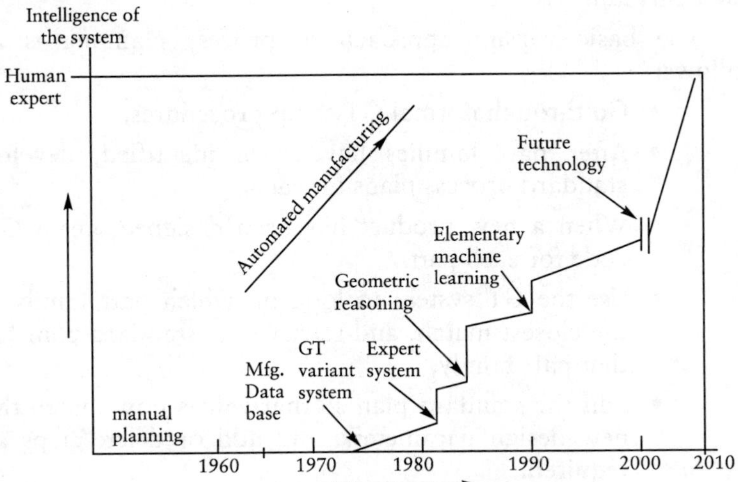

Fig. 3: Evolution of CAPP systems

5. Two Important Types of CAPP Systems

There are two main types of CAPP systems:

i. Variant type

ii. Generative type

A number of CAPP systems combine both methods, so that a third category is now recognized, the semi-generative method. It is an interim method to the time of complex development generative method.

Variant CAPP

The variant approach to CAPP was the first approach used to computerize process planning. Variant CAPP is based on the concept that similar parts have similar process plans. The computer is used as a tool to assist in identifying similar process plans, as well as in retrieving and editing the plans to suit the requirements for specific parts. Variant CAPP is related to part classification and GT coding. In these approaches, parts are classified and coded based upon several characteristics or attributes. A GT code can be used for the retrieval of process plans for similar parts. The initial challenge is in developing the GT classification and coding structure for the part families and in manually developing a standard baseline process plan for each part family.

The basic variant approach to process planning is as follows:

- Go through normal GT setup procedures.
- After part families have been identified, develop standard process plans for each.
- When a new product has been designed, get a GT code for each part.
- Use the GT system to look up which part family is the closest match, and retrieve the standard plan for that part family.
- Edit the standard plan so that values now match the new design parameters, and add or delete steps as required.

Some benefits of the variant CAPP system are:

- It is well suited to medium to low product mixes
- It can be developed quickly for most companies
- Can be used with CIM
- One programme can be used in different type of industries

Its disadvantages are:

- GT codes can become obsolete quickly
- While it is fast to setup, it is slower for planning than generative systems
- More prone to error than generative systems

This variant approach is illustrated in Fig. 4.

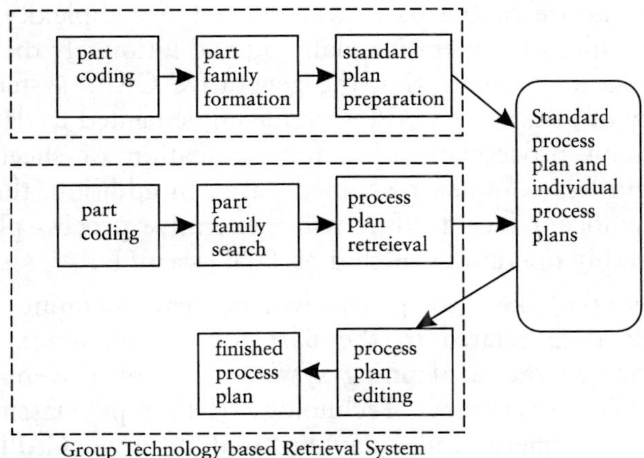

Fig. 4: Architecture of Variant CAPP system

Generative CAPP

Generative CAPP has come into development in the late seventies. It aims at the automatic generation of process plans, starting from scratch for every new workpiece description. Often, the workpiece description is a CAD solid model, as this is an unambiguous product model. The model will contain not only shape and dimensioning information, but also tolerances, which are special features. The process plan can be routed directly to the production planning system and production control system. Time estimates and resource requirements can be sent to the production planning system for scheduling. The NC part programme and material handling control programme can also be sent to the control system. A manufacturing database, decision-making logic and algorithms are the main ingredients of a generative CAPP system. In the early eighties, knowledge based CAPP made its introduction using AI techniques.

The first key to implementing a generative CAPP system is the development of decision rules appropriate for the items to be processed. These decision rules are specified using decision trees, computer languages involving logical "if-then" type statements, or artificial intelligence approaches with object-oriented programming.

The nature of the parts will affect the complexity of the decision rules for generative planning and ultimately the degree of success in implementing the generative CAPP system. The majority of generative CAPP systems implemented to date have focused on process planning for fabrication of sheet metal parts and less complex machined parts. In addition, there has been significant recent effort with generative process planning for assembly operations, including PCB assembly.

A second key to generative process planning is the available data related to the part for the planning. Simple forms of generative planning systems may be driven by GT codes. GT or Features Technology (FT) type classification without a numeric code may be used to drive CAPP. This approach would involve a user responding to a series of questions about a part that in essence capture the same information as in a GT or FT code. Eventually when features-oriented data is captured in a CAD system during the design process, this data can directly drive CAPP.

Possible sources of input vary from system to system, but they are essentially:

- Interpret designs from CAD directly
- User defines features then answers questions about them
- The user designs directly on the CAPP system
- The users creates a special product description file

The advantages of generative CAPP system are:

- It has the potential to become fully automatic
- Runs faster when planning

Its disadvantages are:

- It is still elusive due to its complex nature and its interaction with the vast manufacturing information. Only for simple shapes like axi-symmetric and prismatic parts, considerable success has been reported.
- Heavy computing resources are required.

Some of the approaches for generative CAPP are:

- Decision trees
- Decision tables
- Axiomatic approach
- Artificial intelligence-based approach and expert systems.

Decision Trees

A decision tree is comprised of a root and a set of branches originating from the root. In this way paths between alternate courses of action are established. Branches are connected to each other by nodes, which contain a logic operation such as an "and" or "or" statement. When a branch is true, travelling along the branch is allowed until the next node is reached, where another operation is assigned or an action is executed. Decision trees can call one another recursively. Decision trees can either be used as computer or represented as data. As a computer code, the tree is converted to a flowchart. The starting node is root and every branch represents a decision statement, which is either false or true. Decision trees have certain definite benefits over decision tables:

- Trees can be updated and maintained more easily than a decision tables.
- Selected branches of the decision tree may be extended to a considerable depth if necessary, while other branches may be quite short, which is more difficult to do with decision tables.
- Some branches of the decision tree may be used to define TYPE, and others ATTRIBUTES, which results in relatively small trees.
- Trees are easy to customize, visualize, develop and debug.

Decision Tables

Decision tables organize conditions, actions and decision rules in tabular form. Conditions and actions are placed in rows while decision rules are identified in columns. The upper part of the table includes the conditions that must be met in order for the actions (represented in the lower part of the table) to be taken. Decision table should contain the actual rules conditions specified in the design. According to the rule representation, decision tables can be classified as follows:

- Limited entry decision tables that represent the exact conditions (input values) as true or false entries.
- Extended entry decision tables that specify the condition but not the value.
- Mixed entry decision tables whereby sequenced and unsequenced actions can be entered. Sequenced actions rate a sequence number while unsequenced actions do not rate one.

Axiomatic Approach

Its intention is to provide a logical framework for designing products and processes. A set of desired characteristics of the design, known as Functional Requirements (FRs) must be defined to establish a design range. A set of Design Parameters (DPs) must also be identified so that the system range can be defined. Axiomatic approach to process planning is used to define and simplify the relation between the FRs and DPs through a set of axioms. The entire framework is based on two axioms:

Axiom 1: The independence axiom—Maintain the independence of functional requirements.

Axiom 2: The information axiom—Minimize the information content.

To implement axiomatic design in process planning, the following steps are applied:

- List all the design (or production) parameters to be evaluated.
- Divide the surfaces to be produced into surface groups, each of which is to be machined by a single machine.
- List candidate machines for each surface group.
- Evaluate all alternatives for the production and machine parameters.

- Obtain the total information content and select the best machine combination based on the information content.

Artificial Intelligence (AI) Technique and Expert Systems (ES)

Artificial intelligence techniques like formal logic, describing components and expert systems for codifying human processing knowledge are also applicable to process planning problems.

Due to the rapid development of artificial intelligence techniques, generative process planning systems are built as expert systems with technological knowledge representation, which specifies general and detailed rules of process planning. They have involved object-orientated programming techniques and virtual single manufacturing database techniques. Some new generation systems have employed fuzzy logic and neural network techniques and machine learning approach.

The major areas of activities in generative approach to process planning are:

- determination of part characteristics
- partial product designing
- selection of basic machining processes
- machine selection
- technological instrumentation selection for dimensional measurements
- technological operation sequencing
- tool selection
- machining parameters selection
- time and cost calculation

Semi-Generative Method

Semi-generative method is an interim approach used when problems have occurred in building purely generative process planning system. The semi-generative method can be characterized as an advanced application of variant technology employing generative type features. Systems utilized semi-generative method to process planning should co-operate with planner, who possess technology knowledge. The planner's responsibility is the interpretation of decision data and/or working drawing. The following are the methods used for this purpose:

- Variant method can be used to develop the general process plan, then the generative method can be used to modify it.
- Generative method can be used to create as much of the process plan as possible then the variant method can be used to fill in the details.
- Process planner can select either generative mode for complicated part features or variant mode for fast process plan generation.

6. CAD/CAM Integration and CAPP Features

A frequently overlooked step in the integration of CAD/CAM is the process planning that must occur. CAD systems generate graphically oriented data and may go so far as graphically identifying material to be removed during processing. In order to produce such things as NC instructions for CAM equipment, basic decisions regarding equipment to be used, tooling and operation sequence need to be made. This is the function of CAPP. Without some element of CAPP, there would not be such a thing as CAD/CAM integration. Thus, CAD/CAM systems that generate tool paths and NC programmes include limited CAPP capabilities or imply a certain approach to processing.

CAD systems also provide graphically oriented data to CAPP systems to produce assembly drawings, etc. Further, this graphically oriented data can then be provided to manufacturing in the form of hardcopy drawings or work instruction displays. This type of system uses work instruction displays at factory workstations to display process plans graphically and guide employees through assembly step by step. The assembly is shown on the screen and as an employee steps through the assembly process with a footswitch, the components to be inserted or assembled are shown on the CRT graphically along with text instructions and warnings for each step. Some of the approaches to computer aided process planning are shown in Fig. 5.

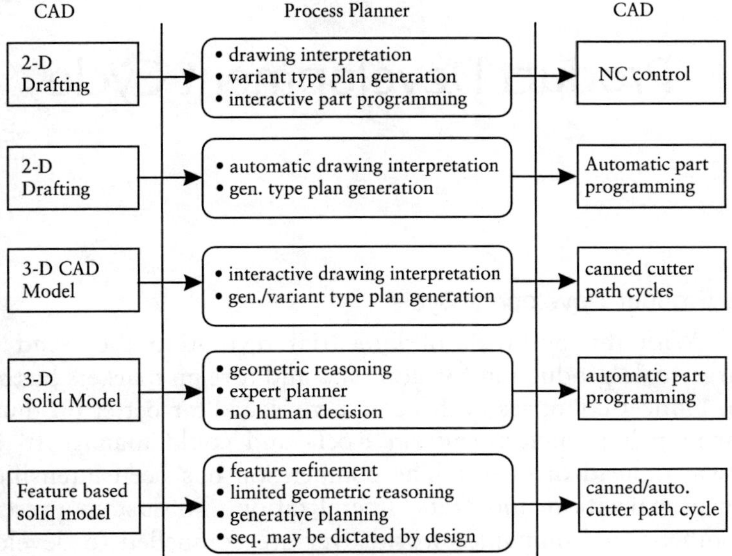

Fig. 5: Approaches in CAPP

The advantage of computer aided process planning is that the entire procedure is automated via computers. The specific advantages are: reduced clerical effort, fewer calculations, fewer oversights in logic, immediate access to up-to-date information, consistent information, faster response to engineering or production changes, use of latest revisions, more detailed and uniform planning, more efficient use of resources.

CAPP is a highly effective technology for discrete manufacturers with a significant number of products and process steps. Rapid strides are being made to develop generative planning capabilities and incorporate CAPP into a Computer-Integrated Manufacturing (CIM) architecture.

The complete success in automatic CAPP systems for realistic applications is yet to be realized in spite of decades of research efforts. This led to the development of alternate routes from CAD to CAM, via Rapid Prototyping (RP).

17

Product Development Cycle

1. Product Developent Cycle

With the outbreak of industrial revolution the trend of large-scale production for domestic and foreign markets has set in. Pioneer countries in this trend envisaged far better products through huge investments in R&D and could manage to be always ahead of others. The competition has been intensified by globalization and trade liberalization. To sustain present competition, companies world over are compelled to develop new and better products. It is also important that the manufacturing is economical and the product reaches the market ahead of the competitors.

The huge investments required for new product development is associated with a very high risk as a product's success or failure in the market depends on its acceptance by the user. For any progressive company it becomes essential to take this risk to retain the market position and to survive in the present day competition, which makes it even more important to ensure commercial success of the product. New products have very little chance of survival without proper Unique Selling Propositions (USPs) attached to it. With growing awareness, consumer's choices are more governed by the price-features offered with the product.

Definitions of Design

The various definitions of design are given below:

- Finding the right components of a physical structure/system.

- Engineering design is the use of specific principles, technical information and imagination in the definition of a mechanical structure, machine or system to perform specified functions with the maximum efficiency.
- A creative activity that involves a process of developing something new and useful that has not existed previously.

Most of these descriptions refer to the various aspects of Design Process, but not of designing itself (Fig. 1). We prefer to think of design as a process that initiates change in the man-made objects which leads to the new experiences most appropriate for the situation, brings out the real meaning to the users and at the same time offers rewards to the companies. When such experiences become part of culture, they create a foundation for the significant real world change. The act of designing, as of now is more concerned not with the end product, but with the changes that manufacturers, distributors, users and the society as a whole have to make, in order to adapt and to get benefit from the new design (Fig. 2).

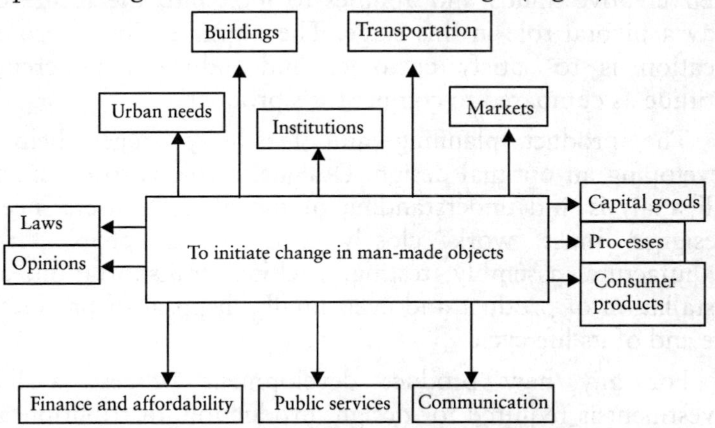

Fig. 1: Factors influencing product design

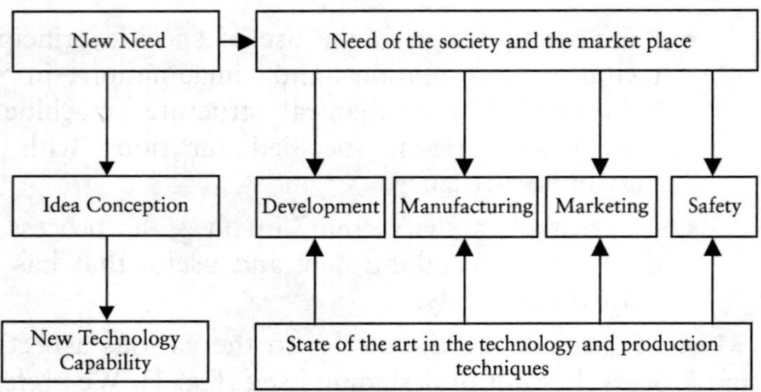

Fig. 2: Interactive model of innovation

Identification of various stages of Product Development Cycle, that starts at the initiating stage to the last phase of the product cycle and effective management of every stage is crucial for the successful implementation of product. These stages are shown diagrammatically in Fig. 3.

The basic product idea or idea generation may be initiated from the various sources such as customer's need, from shop floor or new research and new technologies. Designers with their creative minds and abilities to look into the future can play a pivotal role in this stage. The ultimate aim of product creation is to satisfy customer and induce a preferential attitude as compared to competitor's product.

The product planning and feasibility stages help in developing an optimal design. Designer's role is to ensure the full analysis and understanding of the ultimate users' needs. Designer must work closely with other people from manufacture, assembly, testing, packing, transportation, and installation of product and even finally disposal of product at the end of its life cycle.

For any new product development process a huge investment is required for design, production, distribution and sales promotion without any confirmed commitment of buyers. For retaining the market position and to survive in today's competitive market, this risk activity of new product development is essential for progressive companies. Therefore, it has become even more important to ensure commercial success of the product at the market place.

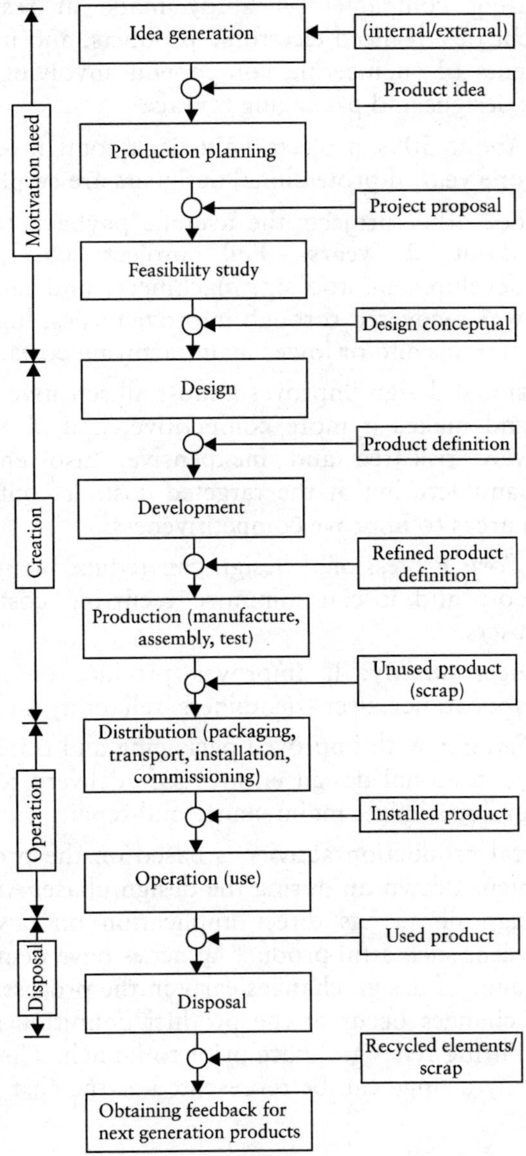

Fig. 3: Stages in the product development cycle

Interestingly the recent research also indicates that investment in design yields the highest returns for manufacturing companies. A study made in respect total development of advanced electronic products, and incremental improvements of engineering components involving graphics, ergonomic designs and packaging reveals:

- About 50% projects paid back total investment in one year, if professional designers are employed.
- For other projects the average payback period was about 2 years. Full project cost including development, tooling, machinery, and launch costs was recovered through improved sales, higher profit margins and/or lower manufacturing costs.

Professional design improves almost all sensitive aspects of products and makes it more competitive, and also makes it comparatively risk-free and inexpensive, also ensures the product manufacturing at the targeted cost. It contributes in three main areas to improve competitiveness:

(i) *Cost*: Professional design can reduce manufacturing cost and it can minimize recurring costs for the users.

(ii) *Acceptability*: It improves product uniqueness in appearance, user friendliness, reliability and safety.

(iii) *Service*: With improved packaging and transportation professional design ensures safe delivery to the user and minimizes maintenance and repair.

The total production activity is based on the product and assembly plans drawn up during the design phase. Any change during design phase has direct implication on development process. Recent successful product launches have demonstrated the importance of design changes early in the process. In Japan over 60% changes occur at the product definition stage and only 10% during redesign phase prior to launch. The potential impact of any change can be represented as the cost of change at that stage.

Product Development Cycle	Cost increase
Concept Stage	1x (nil)
Design Stage	10x
Tooling Stage	100x
Testing Stage	1000x
After Release of product	10000x

2. Rapid Prototyping

During last decade many new technologies were developed for Rapid Prototyping (RP) by which a physical three-dimensional plastic models are created, using CAD file data through special sintering, layering or deposition techniques, without building actual mould. This technology also known as Solid Freeform Fabrication (SFF) or Layered Manufacturing (LM) offers tremendous advantages in terms of saving the cost of prototyping, shorter lead times and improving total product quality.

Though currently most of these processes are very expensive, the use of RP will become much more acceptable with better, stronger and accurate models being built and with novelty. These processes are bridging the gap between design conceptualization and manufacturing.

Rapid Tooling

An important phase of product development cycle is manufacture of production tooling. Characteristically all tools require long time for production and demand high investments. Many efforts have been put into improve this phase to achieve savings in time and cost. Making these tools using conventional machine tools, CNC machines, EDM and so on, give very high dimensional accuracies but these machining operations are time consuming and costly. Rapid Tooling (RT) is an economical alternative to traditional tool manufacture.

Professional designer can play a very important role at the conceptualization stage of the product development that is critical for the success of a product in the competitive market. Product design is used as an effective tool in the competitive market. Hence, design is now being discussed by all industries and business houses. Designer's prime commitments to satisfy the users requirements, his needs, aspirations and safety considerations, become a pivotal activity for the design process. Designer's spatial creative abilities help him giving shape to the idea by effective 3D space utilization.

Feedback from those involved with the product after it leaves the factory—purchasers, operators, service engineers, etc.—is vital for the designers providing essential information for the design of the next generation products.

Finally, and increasingly important as environmental factors influence profits more and more, designers need to be concerned about how the product is to be disposed of in order to simplify recycling or remanufacture and avoid pollution hazards. Rapid product development facilitates "green manufacturing".

During the course of product development, various decisions are taken that influence the cost and performance of the product. Any incorrect decision taken at one stage of the design process will get amplified by the subsequent stages. Therefore, early detection and correction of the design flaws cost less. Design has to take into account the features of subsequent activities in order to arrive at optimal decisions. Design under Concurrent Engineering environment helps achieve this goal.

OBJECTIVE TYPE QUESTIONS

Choose the correct alternative

1. Of the following behaviour characteristics of a supervisor, the one that is most likely to lower the morale of the men he supervises is
   (a) enthusiasm (b) favouritism
   (c) punctuality (d) thoroughness
2. The ability of an employee to take the first step and follow through on a job is known as
   (a) manners (b) laziness
   (c) initiative (d) individuality
3. The best method of getting an employee who is not working upto his capacity to produce more work is to
   (a) have another employee criticize his production
   (b) privately criticize his production but encourage him to produce more
   (c) criticize his production before his associates
   (d) criticise his production and threaten to dismiss him
4. The best thing for a supervisor to do when a subordinate has done a very good job is to
   (a) tell him to take it easy
   (b) praise his work
   (c) reduce his work load
   (d) say nothing because he may feel proud
5. If a design staff cannot possibly complete a drawing on time, then the best action for him to take is
   (a) work during lunch time
   (b) work overtime
   (c) ask an employee to assist him
   (d) notify the supervisor
6. The best way to correct a mistake made by your subordinate is to

(a) correct the mistake yourself and privately explain the correction to subordinate
(b) correct the mistake yourself and say nothing to subordinate
(c) give it to another subordinate to correct
(d) insult him and then have him correct the mistake

7. The best method of allocating the job to the technicians would be to allocate them according to the technician's
(a) seniority
(b) desire to do the work
(c) ability to do the work
(d) attitude towards other employees

8. The most important factor that an individual must fulfill in order to ensure his own safety in a plant or on a construction job is to
(a) work slowly
(b) be familiar with the specifications
(c) wear clothing to suit climatic conditions
(d) be alert

9. Two subordinates under your supervision dislike each other to the extent that production is cut down. Your best action as a supervisor is to
(a) ignore the matter and hope for the best
(b) transfer the more aggressive man
(c) cut down on the work-load
(d) talk to them together about the matter

10. A design staff doing drawing work complains that the lighting at his desk is poor. Of the following, the best action to take is to
(a) exchange seats with another staff
(b) tell him to have his eyes checked
(c) do nothing about it
(d) check the light intensity with a meter, and improve the lighting, if necessary

11. A subordinate who continuously by-passes his immediate supervisor for technical information should be
    (a) reprimanded by his immediate supervisor
    (b) ignored by his immediate supervisor
    (c) given more difficult work to do
    (d) given less difficult work to do
12. Of the following, the best way to handle an over talkative subordinate is to
    (a) have your superior talk to him about it
    (b) have a subordinate talk to him about it
    (c) talk to him about it in a group conference
    (d) talk to him about it in private
13. A supervisor from another shop asks you to assist him on a job which would require a full day of your time. Of the following, the best immediate action for you to take is to
    (a) refuse to assist him
    (b) ask for compensation before doing it
    (c) assist him promptly
    (d) notify his department head
14. Of the following, the best way to develop a subordinate's potential is to
    (a) give him a fair chance to learn by doing
    (b) give him more than his share of work
    (c) criticize only his work
    (d) urge him to do his work rapidly
15. For best execution of work, the work should be assigned according to a person's
    (a) attitude towards the work
    (b) ability to do the work
    (c) salary
    (d) seniority

16. One of the following characteristics which a supervisor should not display while explaining a job to a subordinate is

(a) enthusiasm (b) confidence
(c) self-pity (d) determination

17. When a worker feels he has been unfairly treated, or handicapped or is dissatisfied with general working conditions, he is likely to submit a

(a) suggestion (b) argument
(c) complaint (d) grievance

18. Complicated instructions should not be written

(a) accurately (b) clearly
(c) factually (d) verbosely

19. The purpose of "earthing" in electrical devices and equipment is

(a) to eliminate interference from adjacent electrical equipment
(b) to complete the necessary electrical circuit
(c) to secure the mains cable to the chassis (metal framework) of the apparatus or equipment
(d) to provide a means of making the apparatus or equipment safe in case a fault occurs.

20. The following component(s) is/are required to initiate and to sustain a fire

(a) a source of heat
(b) a source of any combustible material
(c) a supply of Oxygen or air
(d) all the above components

21. Which of the following can minimize the occurrence of accidents in factories?

(a) Proper and thorough training schemes
(b) A proper attitude towards safety, health and welfare

(c) A strict adherence to safety rules and procedures
(d) Wearing of the appropriate safety clothing
(e) All the above factors

22. ______ is your mental, emotional, physical, and behavioural response to anxiety-producing events or situations.
(a) Technology (b) Stress
(c) Environment (d) Information
(e) Sensitivity

23. ______ is the science addressing the compatibility of people and machines.
(a) Ergonomics
(b) CNC programming
(c) HRD (Human Resource Development)
(d) Robotics

24. ______ views motivation as dependent on the strength of an individual's desire for a set of goals and the likelihood that a specific type of behaviour will lead to the achievement of the individual's goals.
(a) Differentiation model of motivation
(b) Needs model of motivation
(c) Expectancy model of motivation
(d) All of the above

25. The theory of ______ explains the forces and behaviour that occur within a group.
(a) information systems (b) barriers
(c) change (d) group dynamics

26. Two facets in any process of change are
(a) Technological and social
(b) Behavioral and psychological
(c) Familiar and unfamiliar
(d) Good and bad
(e) All of the above

27. The term used to justify automated systems for professional or management staff is

(a) Fail safe (b) Soft skills
(c) Value added (d) All of the above.

28. Machine-paced work does not cause stress among the workers.
    (a) True (b) False
29. Behaviour differences between individuals are produced by physical differences, mental capabilities, life experiences, culture, and perception of a situation.
    (a) True (b) False
30. According to the needs model of motivation, money remains important only to the extent that it contributes to the fulfillment of social or ego needs.
    (a) True (b) False
31. If half the potential of human resources could be used, the productivity increase would exceed that possible from technology.
    (a) True (b) False
32. The highest need in the needs hierarchy is the need for self-fulfillment.
    (a) True (b) False
33. The feasibility study implies the practicability of a proposed project and involves an analysis of the total requirements for a ______ evaluation of a proposed solution.
    (a) operator (b) economic
    (c) technical (d) environment
    (e) All of the above
34. A feasibility study can result in
    (a) a recommendation to do the system analysis
    (b) management asking for additional information
    (c) the project's being dropped or shelved
    (d) All of the above

ANSWER KEY

1. (b)	2. (c)	3. (b)	4. (b)	5. (d)
6. (a)	7. (c)	8. (d)	9. (d)	10. (d)
11. (a)	12. (d)	13. (c)	14. (a)	15. (b)
16. (c)	17. (d)	18. (d)	19. (d)	20. (d)
21. (e)	22. (b)	23. (a)	24. (c)	25. (d)
26. (a)	27. (c)	28. (b)	29. (a)	30 (a)
31. (a)	32. (a)	33. (e)	34. (d)	

18

Impact of Innovations on Product Development

Impact of Innovations on National Economy

International scenario is changing rapidly. Globalization and integration wave has swept the world trade and investment scene. These changes place number of challenges to all developing countries. Technological developments in the field of information, communication, production and marketing have been contributing to rapid growth in world trade and investment. Trade and financial flows have been increasing much faster than before, overtaking the growth of world output. Tremendous growth in trade and investments resulted in the changes in the attitude and philosophy of world trade. Competitive spirit is growing among countries, each one is striving to be a leader in this race.

Until recently General Agreement on Trade and Tariff (GATT) was incorporating trade in goods but now World Trade Organization (WTO) with its enlarged scope tries to liberate trade in services and Trade Related Intellectual Property Rights (TRIPS) and Trade Related Investment Measures (TRIMS) considering their impact on world trade and production.

No doubt that the globalization of trade and other changes are advantageous in the long run, but they have also added to the difficulties of developing countries. If developing countries want to be successful in this race, they have to make vigorous and consistent efforts. Being one of the developing countries,

India is no exception to this. If we look at the performance of some of recently developed countries in total world export, emerging pattern is a clear indication how innovation is playing pivotal role in developing economy of these countries (Table 1).

Table 1: Asian and Other Countries Leading in Exports

Country
China M.L. (Main Land)
China H.K. (Hong Kong)
India
Japan
South Korea
Malaysia
Singapore
Taiwan
Thailand
Canada
Australia

The percentage share of India in total world exports is only about 0.5-0.6% whereas the share of exports of Asian counties has increased considerably. East Asian countries' success in increasing their share in world exports in short span of time is really a commendable task. This led to naming them as Newly Industrialised Countries or Asian Tigers.

Trade Policy: Outward Oriented and Inward Oriented

According to World Development Report Classification, trade policy can be divided into outward oriented and inward oriented. An outward oriented policy refers to trade and industrial policies that do not discriminate between production for domestic consumption and for export, whereas inward oriented policy favours production for domestic market than for the export market. Outward oriented policy promotes exports and thereby industrialization and economic development, but inward oriented policy encourages import substitution, protect domestic industries and encourages them to develop rapidly.

Objectives such as increasing market share, popularizing Indian brands in the world market, creating reputation of Indian industries over the world did not appear in Indian trade policy. In contrast to this Japan, Taiwan, Korea and other East Asian countries followed outward oriented policy, chose export led growth and took aggressive approach. These countries since 1950s tried to create their position in the world market by increasing their share in world export.

Export Competitiveness

Export growth depends upon competitiveness of exports in the world market. Competitiveness is influenced by price and non-price factors. Indian goods lag behind in both ways, i.e., the prices of goods are not competitive to their rivals and quality of Indian goods is also not up to the mark. However, competitiveness score of India has improved to a certain extent now.

The position of successful countries indicate that their policies gave enough attention to the quality of production, design aspects to make it more user friendly, cost structure and better marketing efforts.

Product Design as a Strategy to Win International Market

The strategy used by Asian Tigers is product design and innovation to win international markets. As per this strategy, market can be captured if qualitative user-friendly products are innovated. These countries tried to find market niche and develop new products for exports.

In the process of industrialization, these countries initially imitated the products of leading brands then improved them and finally developed new products indigenously and popularized their brands in the world market. According to this design strategy a comprehensive approach is adopted to incorporate all the good features in a product like:

- Better quality,
- Reasonable pricing,
- Attractive packaging, and
- User-friendliness.

These countries have concentrated their efforts in innovating new products for export markets. They also invested heavily in R&D. Many innovations are based on a deeper level of customer and market understanding and go beyond what current customers say they need. These innovations create needs and continuously improve their performance.

The number of patents filed by these Asian countries has also increased considerably.

The product design approach has been appreciated by Indian industries to succeed in the world competition in the coming years. Unless India takes up this product design approach comprehensively, it will not be able to sustain foreign competition.

Innovation Process

Innovation Process calls for an intimate understanding of currrent customers and markets, potential new customers and markets, team and organization competencies and improvement opportunities, vision, values, and mission of the company. Effective innovation depends on disciplined management systems and processes, that encourages a search for creative ways to do things better, different, or more effectively. Beyond the management tools of surveys and establishing focus groups, innovation leaders find a multitude of ways to understand and satisfy their customers. Many new products and services come from a deep understanding of the intended customers' problems and aspirations.

Asking the customers question like "what if..." and using these answers as a wishful thinking and as a source of inspiration for developing new products and services are essential for product innovation.

Innovation process can be used to develop end products in a way, they have never been thought of previously. These creative engineering products and devices are often derived by combination of known components, principles in a unique, novel and non-conventional ways. Driving curiosity, will to explore unknown frontiers and wide imagination are playing important roles in this act of creativity. Albert Einstein has rightly said "Imagination is more important than knowledge, for knowledge is limited, whereas imagination embraces the entire world—stimulating process, giving birth to evolution".

Almost every one of us is creative enough in early childhood, but very few can remain highly creative (about 2% in average at any given age) as they grow old. It is interesting to know, what goes wrong in the process of growing. The creativity of most of the individuals is limited or inhibited by three different mental blocks, viz.:

Mental Blocks for Creativity

i. Perceptual blocks,
ii. Cultural blocks, and
iii. Emotional blocks

(i) Perceptual Blocks

Perceptual blocks are caused by not seeing the real problems or ignorance about what may be actually wrong. These blocks tempt us to begin our problem-solving activity without proper goal in mind. This tendency at the end, brings lot of frustration after looking at the most obvious solution for the problem.

These perceptual blocks may be classified as merely having a mental set or predisposition (bias) towards seeing the situation in a particular way only, no matter how closely or thoroughly we examine the problem. This attitude makes it difficult to define the problem in the beginning and the way the problem gets defined gives a specific direction to solution. This imposed blindness prevents looking at the solution from different areas and its application to the given problem.

(ii) Cultural Blocks

Being members of a society, we are governed by the rules of behaviour, and our thoughts and actions in a particular situation are reflections of customs and traditions. The society requires that an individual behaves in a certain way only because it is customary. While the creativity, on the other hand, calls for changes in the present ways, investigation of alternatives and if necessary proposes radical solutions. These blocks, first start from home and in pre-school years and are consolidated further by remarks like "Good Behaviour, Good Grades, etc.".

People too curious and asking many questions are considered troublesome. By avoiding questioning, we are attacking the very heart of creativity, i.e., killing curiosity. Only inquisitive minds can take new ways to make better things. Most of the times feasibility of solution and economic considerations stop our imagination.

(iii) Emotional Blocks

These blocks to creativity are within ourselves and we are afraid of making mistakes or making a fool of oneself. These over powering emotions like fear, love, hatred and anger can cause of these emotional blocks.

Under the pretext of time pressure, we normally take the first idea occurring to our mind, as a good solution for the given problem and stop exploring further. Many a times the "really good idea" comes quite later as we warm up in this mental exercise. There is no magic formula to become more creative. Positive attitudes, self-analysis and the desire to improve are the most essential instruments for creativity.

The creativity should become like a personal trait by which, an individual sticks to the difficult problem, with an obsession and does not leave it until a good solution is struck. Most of the times creative discoveries happen when one

- lets his imagination soar to the sky,
- select the alternative and then engineer it (develop it) in the best possible ways to bring it to reality.

An American aluminum company coined a very precise word for this act "Imagineering". Only after getting and off the traditional thinking, building immunity towards negative thinking and accepting boldly the challenges to achieve "impossible" one can succeed in the act of "Imagineering".

A larger pool of less significant ideas allows us to select the most effective idea, as the mathematical probability of finding a really creative solution increases manifold. The process through which one can make a long, exhaustive list of alternatives is called ideation.

The important source of building a bank of ideas is through the phenomenon of association of ideas. In this process, one imagines something else when seeing, hearing, tasting, feeling or smelling the existent and builds one idea based on another. The skill of coming up with new ideas can be increased with constant practice and close awareness of nature, man made objects and day-to-day occurrences of the events. One can analyze the problem most critically by asking some of the following questions:

- What is wrong with the present product? An exhaustive list of all the lacuna (short comings) should made.
- How can this be improved? All those bad features listed above are to be improved without giving any concern for feasibility.
- Can this be used for other applications? List all other possible users for the present product and also, those applications with some modifications in the present product.
- How an appearance of product be improved? Various options like colour, form, textures, graphics, packaging, the material for the casing (or enclosure) are to be listed without worrying about economics of change.
- How can the present product be made more convenient to use and safer to operate? Make it easier to work with, easier to carry around, easier to dispose, faster in operation, automatic, silent, safer to use, safer to children, safer to old people non-toxic, non-flammable, stain proof and so on.

Ideation Process

Another approach can be taken in Ideation is by transformation of existing product:

- *Modify*—changes trims, color, form, material
- *Magnify*— make it heavier, stronger, larger, longer
- *"Minify"*—make it smaller, shorter, lighter, narrower
- *Reverse*—try to make it up side down, around, operating in opposite way.,
- *Rearrange*—interchange components, put it in new order
- *Replace*—plastics for metal, light color instead of dark, round form instead of square
- *Combine* components, principles of working, methods of assembly, various functions in one simplify
- *Adapt*—New application for old product, association of something else, copying old principles in newer context.

Various techniques of innovation described so far and many other like brain storming, idea matrix, role playing, synectics, bionics and so on are to be used for making an exhaustive list of alternatives, which enables one for "seeing it different ways" the existing situation, so that many new avenues are thrown open for new product development.

19

Computer Integrated Manufacturing (CIM)

1. Introduction

Computer Integrated Manufacturing (CIM) involves the linking together of all business functions in a manufacturing organization to form an efficient unified enterprise. CIM is not confined to the shop floor, but embraces the total business. CIM systems establish the communication links to bring together the numerous computer tools into a cohesive network, which in turn provides prompt up-to-date information for decision-makers and the control functions to implement their decisions.

CIM should not be implemented by using merely the existing customized links within the organization; it rather calls for reengineering of the entire organization. The process should start with a basic appraisal of the total business before moving on to looking at the long-term plan for the provision of facilities such as equipment and network selection. The planning and implementation of a CIM project is through a top-down and bottom-up process. An approach that is taken from a software engineering point of view is like more of a waterfall model (i.e. to top-down approach).

The components required for CIM are available and in use; however, there is no one supplier who provides the complete CIM system. This leaves users with the problem of combining hardware and software from heterogeneous suppliers into an integrated system. Many organizations, therefore, spend a lot of time and money in investing in expensive interfaces and customization. Over the years standards have developed and integrated within manufacturing sectors.

CIM through the use of technology enhances competitiveness. Indicative of this are the following benefits directly attributed to CIM:

- 5 to 20 per cent reduction in personnel labour costs.
- 15 to 30 per cent reduction in engineering design costs.
- 30 to 60 per cent reduction in overall manufacturing lead times.
- 30 to 60 per cent reduction in work-in-process.
- 40 to 70 per cent gain in overall production.
- 200 to 300 per cent gain in capital equipment utilization.
- 200 to 500 per cent gain in product quality.
- 300 to 3500 per cent gain in engineering productivity.

2. Definitions of CIM

CIM is a tool to be used to achieve or facilitate defined corporate goals or strategies. It is a way of doing business that emphasizes an automated coordination of information and effort throughout all the functional areas of a corporation. CIM enhances particular strengths and capabilities of the systems or services the companies provide. CIM also can be considered as a new important trend in the field of using information technology to improve manufacturing industry. CIM require the total management system in which it can combine the development of product and its related business activities through the utilization of networks. The objective of CIM can also be aimed at lead-time reduction, productivity improvement, and "quality and reliability" control for future cost reduction. CIM is becoming very popular in automated assembly lines mainly because it can introduce dynamic management through the direct connection of production and sales.

The definition of CIM immediately points out three areas for consideration:

i. CIM requires management to clearly express its corporate goals and manufacturing strategies. If these goals or strategies are not established and expressed, there can be no rational basis for committing resources to CIM or allocating resources for initiating CIM in areas of no long-term interest to the corporation. Since CIM is a way of doing business, there must be a total commitment to CIM by management at all levels.

ii. CIM is expensive in terms of hardware costs, software development, and requires dedicated personnel for the planning and implementation phases. A typical corporate CIM plan may take as long as five to ten years to fully implement.

iii. CIM facilitates both access to and exchange of information. In many cases, CIM cuts across historical and traditional departmental lines of responsibility.

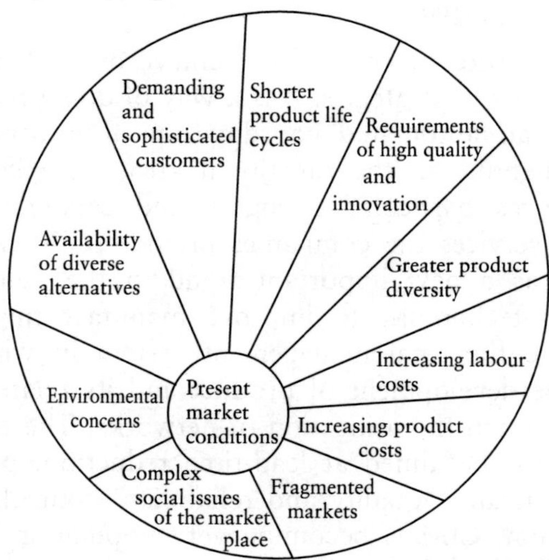

Fig. 1: Present market conditions which demand CIM

Fig. 1 depicts various market conditions which demand CIM. These keep changing. Because of these frequent changes in a buyers market, the life of products keeps coming down as illustrated in Fig. 2. It can be seen that the product is almost obsolete even as it enters the market. Therefore, shortening of time-to-market is vital for the survival of organizations. In other words, organizations have to react quickly to changes in market demands (i.e. should become agile), more so in the global economy. More and more manufacturing organizations are integrating computer technology into their operations to achieve this. CIM is concerned with providing computer assistance, control and integration to all levels of the manufacturing industries, by linking automation into a distributed process system. The technology that is applied in the use of CIM makes intensive use of distributed network of computers, data processing and an effective database management system.

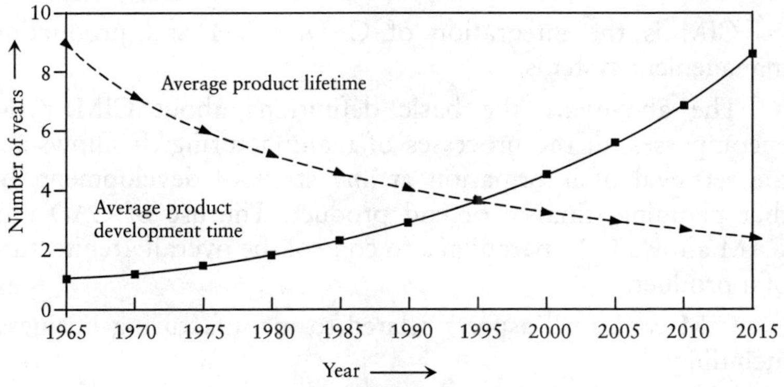

Fig. 2: Product Development Time vs. Product Life Time

CIM concerns itself with information retrieval at all levels of the production factory and its management. The term CIM has widely become accepted for linking together different applications through the use of computers in manufacturing in such a way that information can be shared across and transferred among these applications. CIM will lead to "paperless factory" of the future.

CIM is a unified network of computer systems performing and/or controlling the total integration of a business function.

CIM integrates automated systems into the processes of producing consistent products; it can be seen as the logical organization of individual engineering processes, production and market support functions into a computer-integrated system.

CIM is a way of bringing together computer power to influence the four main functions of manufacturing:

i. Product design
ii. Process control
iii. Planning
iv. Control systems

The concept of CIM encompasses an integrated database that flows through the entire process from design to financial control, with all transactions—such as production, process, automatic process controls, storage, maintenance—all being driven from the same database (common data base).

CIM is the integration of CAD, CAM and production management systems.

The above are the basic definitions about CIM. CIM encompasses all the processes of manufacturing. It allows for the retrieval of information at any stage of development, be that planning, finance or end product. The use of CAD and CAM allows CIM paradigms to control the overall architecture of a product.

CIM covers all aspects related to manufacturing business, including:

- Evaluation and development of different processes.
- Market analysis and future forecasts.
- Generation of future systems through analysis of product and market characteristics.
- Design and analysis for the processes relating to the manufacture of the required components (i.e., welding, cutting, painting, etc.).

- Determination of the capacity of production relating to a particular project.
- Economics analysis of the total system and its disturbances.

The list above indicates that CIM aims to let information processing technology to penetrate into all areas of manufacture in order to:

- Make the entire manufacturing process more productive and efficient.
- Increase product reliability.
- Reduce maintenance cost related to manufacturing system and the product.

As more and more manufacturing businesses begin to integrate computers into their operations, then a paperless innovation culture will emerge. The new culture will allow information to be retrieved, sent and viewed for a product at any given stage of its life cycle. The use of CIM systems will provide a heavy reliance on organizations in using an active and efficient database system that allows the free flow of information quickly and with ease.

CIM is revitalizing manufacturing processes and their industries. As CIM concentrates on data information flow and retrieval, its overall business philosophy is geared towards integrated manufacture. This forces out any unnecessary waste by reducing complexity and cost whilst improving quality to become more competitive.

There are three prerequisites for CIM implementation in an organization:

i. All activities must happen on computers.

ii. There must be reliable connectivity among these computers through LAN/Internet/Intranet.

iii. There must be well-established communication protocols among the computer systems. Furthermore, there shall also be a reliable security system to monitor and control the data that may be distributed across the network.

3. Components of CIM

The following are the components of a CIM system:

i. *Computer-Aided Design* (*CAD*): CAD is a core module in the CIM system and has been widespread for many years. CAD enables computer assistance while the human operator converts his/her ideas into a graphical or a mathematical model. The key to CAD is its conversion from a model represented in the human brain into the machines in the form of digital data.

ii. *Computer-Aided Manufacture* (*CAM*): CAM is seen as a system, which supports production engineering by defining operating sequences, establishing requirements for fixtures, tooling and simulating operations prior to "first-off" machining. CAM takes a focus towards Flexible Manufacturing Systems (FMS). These consist of two or more machine tools linked by an integrated part handling system and operating under the control of a computer. The control system can route parts between machines so as to achieve maximum utilization of the manufacture process.

iii. *Automated Materials Handling*: In order to move materials efficiently during manufacture, full integration must be allowed. Automated handling systems include cranes, Automated Guided Vehicles (AGVs) and robots.

iv. *Group Technology* (*GT*): This is classification and coding of parts in order to ensure that manufacture of similar products is done in the same way. The benefits of GT:
   - Design Retrieval: The rapid identification and retrieval of information.
   - Cellular Manufacture (CM): The creation of cells to manufacture identical parts.
   - Purchase Support: Related parts, being procured together.

- Repair and Maintenance Service Streamlining: Identification of alternate spare parts and reduction of variety.

v. *Manufacture Resource Planning (MRP)*: This deals with the inventory control, ordering, bills of materials, considering the overall requirements of the business.

The elements listed above are the main building blocks of CIM. The main aim of a CIM is to enable a manufacture organization to retrieve information about its manufacture process efficiently with speed and accuracy. In any organization information is retrieved through an effective database system that operates through a network.

4. Barriers to CIM Implementation

There are several major obstacles to the acceptance of CIM as a practice management solution. The following are five of the most frequently encountered difficulties in implementing CIM:

i. *CIM is resource intensive and an expensive undertaking*: Most managers see CIM in terms of its technology. This technology is inevitably expensive, not easy to implement and requires huge resources for testing and debugging the systems. In addition, these investments are seen as ones that must be made upfront. For many firm these costs represent a significant barrier.

ii. *The implementation of CIM is a long-term undertaking*: When most managers look at the experiences of firms that have been successful with CIM, they see endeavours that took many years to complete. For a company that is faced now by dropping sales, falling margins, and rapidly eroding profits, there is little attraction in examining a system that may require four or five years before it yields any benefits.

iii. *There are cheaper alternatives to CIM available*: For many managers who are interested in improving manufacturing performance and enhancing corporate competitiveness, there are other alternatives available such as system simplification, time-based competition, competing on quality, and just-in-time manufacturing. What makes these alternatives so appealing is that they appear to require less investment.

iv. *There are relatively few success stories*: The concept of CIM has been around since the early 1970s. During this period, many firms across the country have attempted to take advantage of the potential benefits of CIM. Firms have tried to implement technology such as FMSs, robotics, MRP, NC machines, AGVS, and CAD systems. However, very few of these attempts have succeeded. The success stories in CIM are both few and are less known.

v. *The fear of loss of management control*: For many managers, CIM represents a potential loss of management control. These managers tend to understand the business of building and delivering their firm's products. They tend to have "hands-on" experience. They have been with their companies for a long time. Often, their authority in their firms is based on their knowledge of the firm. CIM represents a major challenge to their position and authority within the firm. By introducing automation and technology, CIM often redefines the production process. It changes how the firm plans, schedules, executes, and tracks production. Much of the skills and knowledge possessed by the current management may no longer be pertinent to this new CIM-run environment. These managers may feel that they no longer understand operations in this climate and gradually may lose control.

Move to CIM

A move to CIM involves some strategic steps:

1. The problems, if any, should be clearly identified and studied.

2. Any justification for CIM proposal should specify anticipated cost savings from improved throughput (daily output) and reduced inventory.
3. In approving the proposal, top management should recognize that the project may be a long-term project. The required investment is to be made for acquiring new technology.

Guidelines to be followed in implementing CIM

To improve the chances of success, certain guidelines are be followed in the implementation of CIM:

i. *Training and education*: Every one in the company, from top management to the labour, should be exposed to the CIM system. Training and education are very important in understanding CIM.
ii. *There should not be any resistance to change*: Education and training helps in implementing CIM successfully.
iii. *The CIM implementation should be focussed (i.e. it should be specific)*: The changes demanded by CIM could be confined to the manufacturing system initially and later to other systems.
iv. *Manufacturing basics to be improved*: Attention should be devoted to the areas of master production scheduling, production planning, materials planning, and shop floor control. These areas that are currently being done less efficiently should be improved first.
v. *The manufacturing system is to be protected as much as possible from external changes*: External changes are viewed as very disruptive. The system is to be protected in three ways. The first is through the master production schedule. Time limits are to be introduced. An order review and release mechanism to be introduced to manage the flow of work from the planning system to the shop floor. Finally, extra capacity is to be added on the shop floor. This reduces the potential impact of any bottlenecks. It

also simplifies the entire process of capacity planning.

vi. *System simplification is to precede system automation*: The difficulties with work flow and shop layout, should be eliminated by changing the layout of the floor and streamlining the materials flow.

vii. *Manufacturing facilities are to be consolidated*: All production facilities are to be concentrated at one location/plant as far as possible.

viii. *Use "plain" systems wherever possible*: The systems are to be as much as possible "as is." People have to change their practices to conform to the requirements of the new systems. This would reduce the implementation lead-time. Given the limited availability of certain critical management personnel, any activity that lengthened the lead-time is to be avoided.

ix. *Get the right people involved and get them committed*: The new, CIM system has to be a system that everyone believed in and used. There has to be ownership by the users of this new system. To achieve this goal, systems and people are treated as support.

These guidelines provide the general structure within which the implementation can take place. From the perspective offered by hindsight, they also significantly simplify the implementation process by identifying in advance many of the potential pitfalls in the implementation process.

Tangible and Intangible Results (Benefits)

The signs of success effectiveness of the new CIM system could result in

- Reduction in lead times (by 40 per cent).
- Fall in inventories (by 33 per cent to 50 per cent).
- Orders delivered on time (90 per cent).

- Data accuracy increases significantly, bills of materials routings and production figures are more accurate (95 per cent).
- The lead time on spare parts drops form eight to nine months to about three to four days.
- Higher profits competitiveness in the marketplace and reputation for service improves.

In addition to the above tangible benefits, there could be other equally important intangible benefits:

- People involvement increases significantly. People now make decisions and take responsibility for their actions.
- Cross-functional communication improves.
- The communication with suppliers improves resulting in better schedules.
- Confidence in the CIM system and in the overall ability of the company improves significantly. The planning becomes better. Cost information, cost data from the CIM system becomes accurate and in time.
- Better data leading to better actions.
- The overall operation environment of the system improves greatly.

In this way most of the goals initially set for the CIM project could be achieved.

5. Developments in CIM

Today, we are at the threshold of rapid changes in information processing and communication technology. While the mass production is still needed for certain sectors of manufacturing, it is no longer necessary in a rapidly changing global economy. The challenge in reviving the manufacturing industry to face the competitive threat depends on the ability to respond to all the requirements of the global market. However, many of the manufacturing industries in the developed world are still having islands of automation. The manufacturing industries have to face the growing competition in the global market. Today many companies are relocating offshore to reduce labour costs. This phenomenon can be avoided if manufacturing companies invest in CIM and related technologies and become globally competitive. The efficient application of CIM in a fully integrated manner will help the industries face the global and local competition with a high degree of confidence. With the advent of new networking technology and Internet, the application of CIM has extended beyond geographical boundaries and embraced the global application of CIM. To facilitate the investment in CIM and to help manufacturing industries, many developments taking place in application of CIM. Some of the developmental trends which are evolving today in CIM can be categorized as follows:

- Evolving effective management strategies for CIM.
- Enterprise integration for CIM within geographical boundaries and beyond.
- Better network communication for the implementation of CIM.
- Advanced tools and technologies for the application of CIM.
- Manufacturing system modelling.
- Application of Artificial Intelligence (AI) such as fuzzy logic, neutral network, genetic algorithms and intelligent systems for evolving integrated Intelligent Manufacturing Systems (IMS).

Each of these developments diverse in nature and consists of many sub-groups and technologies.

Justification and management strategies for CIM consist of multitude of research directions. The research trends focus on providing managers with principles and guidelines for the application of CIM in their working environment and approaches to overcome the resistance to change towards integrated manufacturing. Further, this development addresses the organizational and social issues in an enterprise, issues involving benchmarking the features of the manufacturing supply chain and manufacturing strategy implementation through real time systems, optimized enterprise integration and Decision Support Systems (DSS) and Expert Systems (ES) for manufacturing managers. DSS and ES may further include ES for quality control, ES for production planning and selecting manufacturing parameters, DSS to evaluate design decisions, financial decisions, and education, training and information systems.

Enterprise integration for CIM within geographical boundaries and beyond consists of a developments which include architectures and modelling for enterprise integration, evaluation methodologies for enterprise integration, and international CAD/CAM collaboration for CE and CIM implementation through integration of subsystems.

Network communications for the implementation of CIM include application of wide-area networks and Internet for CIM, information enhancement by data integration, issues related to integration of client and server for manufacturing shop floor automation, application of multimedia and hypermedia in CIM environment, and data management in CIM system.

The advanced tools and technologies for the application of CIM is one of the important areas in CIM-related research. It consists of robotics and automation for a CIM framework, vision-based manufacturing systems and intelligent AGVs and intelligent and knowledge-based enabling systems for CIM. Manufacturing system-modelling consists of integration of information models with functional models for CIM, object-oriented resource modelling, integrated simulation modelling approaches for CIM, and modelling approaches and methodologies for designing CIM systems.

Application of AI for fully integrated IMS consists of research in the application of neural network on manufacturing automation and technologies, intelligent scheduling systems with genetic algorithms, fuzzy scheduling systems, hybrid systems of CIM components, adaptive models for CIM architectures and embedding AI in CIM software.

6. Future Direction of CIM

In today's competitive global market survival of any industry depends on its ability to communicate and transfer the right information at the right time to the right people. Manufacturing cannot escape from this present requirement. Having an ability to communicate for effective management and manufacturing activities across the geographical boundaries among the globally distributed resources will significantly benefit manufacturing industries. Today a number of global conglomerates are formed in many facets of industry. A virtual enterprise is a network of interconnected global conglomerates. Today's competitive and agility requirements of the global market can be only met by virtual enterprises. To provide a better future in the present market requirements, research in virtual CIM and the application of it in worldwide manufacturing industries are beginning to emerge (Fig. 3). Application of virtual CIM is a step towards the future in manufacturing to face competitive challenges. However, many development works need to be carried out to face challenges faced by virtual enterprises. Virtual CIM will satisfy the globalized and distributed manufacturing enterprises of future in order to meet the competitive and agility requirements of market conditions.

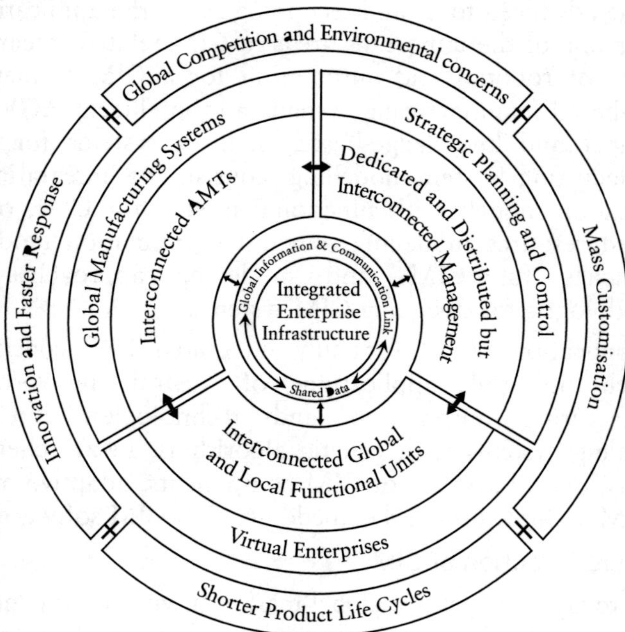

Fig. 3: Virtual CIM wheel representing the global market conditions

In a virtual enterprise the integration of information is important, as only through information can a virtual organization become meaningful, and only by adopting a new generation of information technology can this vision be realized.

In the justification of CIM and management strategies for CIM, recent developments in management information systems have strengthened the application of a DSS in a group decision environment. In addition, rapidly changing technology and the increasing reliance on knowledge and communication create significant challenges to static models of an enterprise. Therefore, in order to justify and optimize investments in a virtual enterprise, a Group Decision Support System (GDSS), is required which has a capacity to consider the appropriate Advanced Manufacturing Technologies (AMTs) in a global framework. A GDSS can improve decision outcomes by using structured approaches for the unstructured problems in a group decision-making endeavour. A GDSS will help decision-makers in a team environment, analyze the investment decisions of a company in an interactive manner, either in geographically distributed or in a localized manufacturing enterprise with diverse functional units.

The rapid developments in technology provide innovative approaches to manufacturing and the recent trend towards global manufacturing has increased the number of geographically distributed and multi-location enterprises. In the research direction of enterprise integration for CIM beyond and within geographical boundaries, a collaborative framework for remote machining and planning and control systems have been developed. The prime technology, which propels the advances in manufacturing system and international networking, is the communication technology. Hence, the research on network communications for the implementation of CIM has produced many developments. In addition, the use of Internet and the World Wide Web in manufacturing applications is gaining momentum. However, research should be further strengthened towards providing integrated architectures and communication protocols for virtual CIM framework to enable real-time machining, processing, and decision-making facilities in geographically dispersed enterprises.

Research in robotics and AGVs is predominant in the development of advanced tools and technologies for the application of CIM. In today's technologies the remote manipulation of these machines has already been achieved. Mobile robots and global positioning satellites have played a significant role in remote manipulation of robots. Collaborative robots are being developed to machine complex surfaces and large components. In addition, vision-based systems have been developed towards Rapid Prototyping, part identification and processing. These developments need to be enhanced towards providing practical solutions for a globally dispersed manufacturing enterprise and be tested in a virtual CIM framework.

Today, many manufacturing industries are under pressure to achieve their objectives within limited capital available for new investments. Optimization of resource allocation for strategic investments could provide a mechanism to achieve the goal of the enterprise. The future of CIM depends on optimized integrated systems. Traditionally, to improve the effectiveness and competitiveness of an organization, all the efforts are focused on achieving a single goal or achieving the objective within a functional unit. However, the traditional method of single goal seeking mechanism or optimizing the results of an individual unit does not satisfy the overall objective of a manufacturing company. Therefore, a multiple goal seeking or multicriteria optimization which considers all the relevant factors and overall benefits of a company as a single unit, has been proposed to make CIM an important application in the present global economic circumstances. The existing multicriteria optimization mechanisms should be further enhanced or new approaches be developed for application in the globally distributed virtual CIM enterprises, so that the benefits of virtual CIM is fully realized. In addition, global integration approaches, which consider a globally distributed company as a single entity, should be developed and utilized to make virtual CIM a possibility. Intelligence manufacturing is the path for the future. Therefore, manufacturing technologies should be blended with intelligence, which will help manufacturing enterprises to produce better quality products and manufacturing equipment to solve problems encountered during the normal course of operation itself.

There are various ways of developing a product at the least possible time. Rapid Product Development (RPD) is one such way.

RPD is a synergic integration of virtual and physical technologies. Synergy of the activities involved is essential to carry forward the benefits achieved at any one activity till the end. For instance, when a human operator with a variable cycle time and a robot with a constant cycle time are in series production, the variance of the production line is much more than that of the human operator as the robot may be idling at times waiting for the operator to finish. The situation can be corrected by replacing the operator by an automated system or introducing a buffer station in between them.

In today's economy driven by global competition, RPD makes more sense than ever before. Companies no longer can afford to remain complacent with their current performance. They have to continuously improve to stay in business. For this, adopting new technologies is essential. Often, they invest in new technologies such as CAD/CAM but do not train the employees adequately. There is a misconception that value addition to the employees by way of education and training is not value addition to the organization as he may leave any time, with the less skilled engineers, the new technologies cannot be exploited to the fullest extent. Continuous upgradation of the employees and keeping them satisfied with their working conditions and social life are very essential for the success of the organizations.

Glossary of Terms

1. Cutting Tools, Jigs and Fixtures

abrasives

Garnet, emery, carborundum, aluminium oxide, silicon carbide, diamond, cubic boron nitride, or other hard materials in various grit sizes used for grinding, lapping, polishing, honing and pressure blasting. Each abrasive particle acts like a tiny, single point tool that cuts a small chip providing a good surface finish.

abrasive belt

Abrasive-coated belt used for production finishing, deburring, and similar functions.

bonded abrasives

Abrasive grains mixed with a bonding agent. The mixture is pressed to shape and then fired in a kiln or cured. They are in the form of wheels, segments, cup wheels, etc. Bond types include oxychloride, vitrified, silicate, metal, resinoid, plastic, rubber, and shellac. Another type of bond is electroplated, wherein the abrasive grains are attached to a backing by a thick layer of electroplated material.

boring bar

Essentially a cantilever bar that holds one or more cutting tools in position during a boring operation. Can be held stationary and moved axially while the workpiece revolves around it, or revolved and moved axially while the workpiece is held stationary or a combination of these actions. Boring bars are installed on milling, drilling, and boring machines, as well as lathes and machining centers.

boring cutter, boring tool

Cutting tool mounted in a boring bar (the holder) that enlarges a cored or drilled hole. May be a single-point or multiple-cutting-edge tool. Can be adjustable.

broach

Tapered tool, with a series of teeth of increasing length, that is pushed or pulled into a workpiece, successively removing small amounts of metal to enlarge a hole, slot, or other opening to final size.

dressing

Removal of undesirable materials from 'loaded' grinding wheels using a diamond tip, single-or multi-point. The process also exposes unused, sharp abrasive points.

grinding ratio (G)

Ratio of work material removed to grinding wheel material lost.

grinding wheel

Wheel formed from abrasive materials mixed in a suitable matrix. Takes a variety of shapes, but falls into two basic categories, those that cut on their periphery, as in reciprocating grinding, and those that cut on their side or face, as in tool-and-cutter grinding.

grit size (in μm)

Specified size of the abrasive particles in grinding wheels and other abrasive tools. Determines metal-removal capability and quality of finish. Higher Grit no. indicates smaller grit size and vice-versa.

gundrill

Self-guided drill for producing deep, long holes with good accuracy and fine surface finish; has coolant passages that deliver coolant to the tool/workpiece interface at high pressure.

fixture

A device, that holds a specific workpiece.

honing tool

Abrasive segments fixed to the periphery of a tool head and used to bring internal bores to an accurate, fine finish. Most often used for precision sizing and finishing of bores, but can be used to hone other shapes and to impart thin, ultrasharp cutting edges. For certain applications, honing tools may be hand-held.

Jig

Jigs assist in the assembly or manufacture of a part or device. It holds the workpiece while guiding the cutting tool with a bushing. A jig used in subassembly or final assembly might provide assembly aids such as alignments and adjustments.

magnetic chuck

Workholding device used on surface grinders and milling machines for holding ferrous parts with large, flat sides. Holding power may be provided by permanent magnets or by electromagnetic system.

mandrel

Workholder for turning that fits inside hollow workpieces. Types available include expanding, pin, and threaded.

truing

Using a diamond or other dressing tool to ensure that a grinding wheel is round and concentric and will not vibrate at required speeds. Also performed to impart a contour to the wheel's face.

vise

Workholding device that is mounted on various machine tables. Designs vary from plain to flanged to swiveling. Multi-angle vises, such as the toolmaker's universal vise, accurately hold work to allow machining at virtually any angle.

wheel-balancing stand

Used to ensure that a grinding wheel is balanced before mounting it on the grinding machine.

chuck

Workholding device that is fixed to a milling machine lathe, or drill-press spindle. It holds a tool or workpiece by one end allowing it to be rotated. May also be fitted to the machine table to hold a workpiece. Two or more adjustable jaws actually hold the tool or part. May be actuated manually, pneumatically, hydraulically, or electrically.

collet

A device that secures a tool or workpiece. Similar in function to a chuck, but can accommodate only a narrow size range. Typically provides greater gripping force and precision than chucks.

drill jig

Accessory that holds a workpiece securely while guiding a drill or other tool into the workpiece; ensures accurate, repeatable location.

2. Machining Operations

abrasive machining

Various grinding, honing, lapping, and polishing operations that utilize abrasive particles to impart new shapes, improve finishes, and part stock by removing metal or material.

boring

Enlarging a hole that already has been drilled. Generally, it is an operation of truing the previously drilled hole with a single-point, lathe-type tool. Boring is essentially internal turning, in which usually a single point cutting tool forms the internal shape. Some boring tools are available with two cutting edges to balance cutting forces.

broaching

An operation in which a cutter progressively enlarges a slot or hole, or shapes a workpiece externally. Small teeth start the cut, intermediate teeth remove the majority of the material, and large teeth finish the task. Broaching can be a one-step operation, as opposed to milling and slotting, which require repeated passes.

brushing

Use of rapidly rotating metallic wires or fibers to effectively and economically remove burrs, scratches, and similar mechanical imperfections. The greatest application has been made in the finishing of gears and bearing races for removal of sharp edges and stress raisers.

buffing

Smoothing and brightening a surface by pressing an abrasive compound, embedded in a soft wheel or belt, against the workpiece.

centerless grinding

The workpiece rests on a knife-edge support, rotates through contact with a regulating or feed wheel, and is ground by a grinding wheel. This method allows grinding long, thin parts without steady rests; also provides good finish and dimensional accuracy.

creep-feed grinding

Grinding operation in which the grinding wheel is slowly fed into the work at sufficient DOC (Depth of Cut) to accomplish in one pass which otherwise would require repeated passes.

disc grinding

Operation in which the workpiece is placed against the side of a wheel rather than the wheel's periphery.

laser cutting

A process related to laser-beam cutting utilizing beam-spot intensities several orders of magnitude greater, so complete vaporization occurs along the beam's path of travel.

electron-beam machining (EBM)

Use of high energy electron beams to generate thermal energy for removing material. A pulsating stream of high-velocity electrons produced from a filament source is focused by electrostatic and electromagnetic fields to concentrate energy on a very small area of work. High-power beams are used with electron velocities exceeding half the speed of light. As the electrons impinge on the work, their kinetic energy is transformed into thermal energy and melts or evaporates the material locally.

grinding

Machining operation in which material is removed from the workpiece by a powered abrasive wheel, stone, belt, paste, sheet, compound, slurry, etc. various grinding operations are: precision surface grinding (creates flat and/or squared surfaces); cylindrical grinding (for external cylindrical and tapered shapes, fillets, undercuts, etc.); centerless grinding; chamfering; thread and form grinding; tool-and-cutter grinding; offhand grinding; lapping and polishing (grinding with extremely fine grits to create ultra-smooth surfaces); honing; and disc grinding.

honing

A low-velocity abrading process, material removal is accomplished at lower cutting speeds than in grinding. Therefore, heat and pressure are minimized, resulting in excellent size and geometry control. The most common application of honing is on internal cylindrical surfaces. The cutting action is obtained using abrasive sticks (aluminum oxide and silicon carbide) mounted on a metal mandrel. In honing there is no distortion. Also used to give cutting tools ultrafine edges.

jig boring

High-precision machining (a sophisticated form of milling) that was originally used in jig and fixture manufacturing. Basic jig-boring processes include centering, drilling, reaming, through and step boring, counterboring, and contouring. The increasing demands for accuracy in metal-working operations has extended the application of jig-boring machines.

Non-traditional machining

Variety of chemical, electrical, mechanical, and thermal processes for machining workpieces; such as EDM (spark erosion machining), ultrasonic machining, laser beam machining, electron beam machining, ECM, etc.

lapping

Finishing operation in which a loose, fine-grained abrasives in a liquid medium abrades material. Extremely accurate process that corrects minor shape imperfections, refines surface finishes, and produces a close fit between mating surfaces.

point-to-point system

Numerical control system normally used for drilling and other operations where center-point location is readily determined. Tool is rapidly moved to a position, then drills, taps, reams, bores, counterbores, countersinks, or performs other machining operations.

polishing

Abrasive process that improves surface finish and blends contours. Abrasive particles attached to a flexible backing abrade the workpiece.

power brushing

Any process that uses a power-driven, rotary industrial brush to deburr, clean, or finish a metal part. Depending on application, the brush fibers, collectively known as brush fill material, may be metal wires; fiberglass-coated, abrasive-filled plastics; synthetics such as nylon and polypropylene; natural animal hairs such as horsehair; or vegetable fibers can be used as brush material.

shaping

Using a shaper primarily to produce flat surfaces in horizontal, vertical, or angular planes. It can also include the machining of curved surfaces, helixes, serrations, and special work involving odd and irregular shapes. Often used for prototype or short-run manufacturing to eliminate the need for expensive special tooling or processes.

shear forming (metal spinning)

Also known as flow turning, is similar to spinning, but during flow turning the metal is intentionally thinned by shear forces. This is also known as power spinning. The process is done on a lathe.

slicing, slitting

Cutting along single lines. A gang of circular blades cuts strips from a sheet. This operation also is used to cut along lines of given length of contour in a sheet or in products.

slotting

Machining, normally milling, that creates slots, grooves, and similar recesses in workpieces, including T-slots and dovetails. Also refers to a punching operation in which elliptical and rectangular holes are cut.

boring machine

Cutting tool (single-point or multiple-cutting-edge), rotates to perform internal cuts. Boring can also be accomplished by holding the tool stationary and turning the workpiece. The machine may be vertical, slanted, or horizontal, and has one or more spindles. Typically a large, powerful machine, it can give fine finishes.

broaching machine

Machine designed to use broaching tools; typically designated by operating characteristics (pull, push, rotary, continuous, spline), type of power used (hydraulic/mechanical), and tonnage ratings. Broaching is also performed on arbor presses (manual and powered).

computer numerical control (CNC)

Microprocessor-based controller dedicated to a machine tool that permits the production of parts. Programmed numerical control activates the machine's servos and spindle drives, and controls the various machining operations.

direct numerical control (DNC)

Actions of multiple machine tools controlled by a single computer, located centrally.

drilling machine, drill press

Machine designed to rotate end-cutting tools. Can also be used for reaming, tapping, countersinking, counterboring, spotfacing, and boring.

grinding machines

Powers a grinding wheel or other abrasive tool for the purpose of removing metal and finishing workpieces to close tolerances. Provides smooth, square, parallel, and accurate workpiece surfaces. When ultra-smooth surfaces and finishes of the order of microns are required, lapping and honing machines (precision grinders that run abrasives with extremely fine, uniform grits) are used. In its "finishing" role, the grinder is perhaps the most widely used machine tool. Various types are available; bench and pedestal grinders for sharpening lathe tool bits and drills; surface grinders for producing square, parallel, smooth, and accurate workpieces; cylindrical and centerless grinders; center-hole grinders; form grinders; facemill and endmill grinders; gear-cutting grinders; jig grinders; abrasive belt (belt-roll) grinders; tool-and-cutter grinders for sharpening and resharpening cutting tools; carbide grinders; hand-held die grinders; and abrasive cutoff saws.

jig bore

Precision boring machine that resembles a milling machine. Originally designed to make precision jigs, fixtures, dies, and other tooling, this machine is now used for production machining of precision parts, extremely accurate hole location, and similar tasks. Employs a precision spindle that drives the cutting tool and an accurate, stable workholding table. Basic types include fixed-bridge, open-side or "C-frame" machines, and adjustable-rail. Often used under climate-controlled conditions (i.e. A/c rooms).

machining centre

A computer-controlled machine tool capable of drilling, reaming, tapping, milling, and boring. Normally comes with an automatic tool changer (ATC); available as Vertical Machining Centre (VMC) and Horizontal Machining Centre (HMC).

numerical control (NC)

Any controlled equipment that allows an operator to programme its movements by punching in a series of coded numbers, symbols, etc.

3. Nontraditional Machining (Modern Machining Methods)

abrasive flow machining (AFM)

Finishing operation for holes, inaccessible areas or restricted passages. Done by clamping the part in a fixture, then passing semisolid abrasive media through the passage. Often, multiple parts are loaded into a single fixture and finished simultaneously.

abrasive water jet machining (AWJM)

A process that uses high-pressure waterjets in combination with a slurry of fine abrasive grains to machine materials.

chemical machining (CHM)

Removal of base material by a carefully controlled dissolution. Material not to be removed is usually protected by a strippable mask.

electrochemical machining (ECM)

Operation in which electrical current flows between a workpiece and conductive tool through an electrolyte. Initiates a chemical reaction that dissolves metal from the workpiece at a controlled rate. Unlike traditional cutting methods, workpiece hardness is not a factor, making ECM suitable for machining hard and tough materials. Variations of ECM are electrochemical grinding, electrochemical honing, and electrochemical turning.

electrical-discharge grinding

A process utilizing EDM in which a grinding-wheel type of electrode is used.

electrical-discharge machining (EDM)

Process that vaporizes conductive materials by controlled application of pulsed electrical current that flows between a workpiece and electrode (tool) in a dielectric fluid. Permits machining shapes to closer accuracies without the internal stresses conventional machining often generates, useful in press tool-diemaking and mould making.

Electrodes are made from graphite or copper. The electrode is separated by a dielectric liquid and maintained at a close distance of 25 μm. A DC voltage is pulsed to the electrode and jumps to the conductive workpiece. The resulting sparks erode the workpiece and generate a cavity in the reverse shape of the electrode, or a through hole in the case of a plain electrode. Also known as "die sinker" or "die-sinking machine".

EDM— standard electrodes with CNC

Similar to the electrical-discharge-machining process, but uses a CNC control technique to generate shapes with standard electrodes. Standard electrodes along with CNC controls produce the required shapes.

EDM— CNC wire cut machine

A process similar to electrical-discharge machining; a small-diameter copper or brass wire is used as a travelling electrode. The process is usually used along with a CNC control and will only work when a part is to be cut completely through. It is an ultraprecise, electrical, contour sawing operation. Used widely in press tool/die manufacturing.

electrical-discharge sawing

A process similar to the conventional electrical-discharge machining; a continuous-blade type tool is employed as an electrode. The process is similar to wire electrical-discharge machining but generally used for rough cutting of thick steels.

electrochemical deburring (ECD)

A variation in electrochemical machining designed to remove burrs and impart small radii to corners. The process normally uses a specially shaped electrode to carefully control the process to a specific product. The process will work on materials regardless of hardness.

electrochemical discharge grinding (ECDG)

A combination of electrochemical grinding and electrical-discharge machining. Material is removed by both processes. The workpiece and the grinding wheel never come into contact as in any other electrical-discharge machining process. Electrolyte flows between job and wheel.

electrochemical grinding (ECG)

A variation of electrochemical machining that uses a conductive rotating abrasive wheel. The chemical solution (electrolyte) is forced between the wheel and the workpiece. The shape of the wheel determines the final shape. The rotating wheel ensures a constant supply of fresh chemical solution (electrolyte). Upto 10% of the material removal is via abrasive action between the workpiece and the wheel and 90% of material is removed by electrochemical machining.

electrochemical honing (ECH)

A process very similar to electrochemical grinding; a honing tool is used. The process works well on hardened materials and has a material-removal rate faster than that of conventional honing.

electrochemical polishing (ECP)

A variation of electrochemical machining that usually involves a stationary electrode with a larger gap and lower current densities. The process is designed to enhance surface finish without removing significant amounts of material.

electrochemical turning

A variation of electrochemical machining that uses a simple electrode in conjunction with a rotating workpiece.

electron-beam machining (EBM)

Removal of material with a concentrated beam of electrons that melts and vaporizes the material.

electropolishing

A specialized form of chemical machining that uses a combination of chemical cutting action and a deplating process. An electrical potential is maintained between the workpiece and the cathode. The process is normally used to remove very small amounts of material and improve the surface.

hydrodynamic machining (HDM)

General term for various forms of waterjet and abrasive-waterjet machining. In all cases, a fine, highly pressurized jet of water cuts and removes the material.

laser-beam machining (LBM)

A process that removes material by focusing a concentrated laser beam on to the workpiece. The material is melted and vaporized.

laser machining

Intense, pulsed beams of light generated by lasers—typically carbon dioxide or neodymium-doped yttrium aluminum garnet (Nd:YAG)—that drills welds, engraves, marks, slits and casehardens the job. Usually under computer numerical control with high cutting rates and high power.

electromagnetic forming

Cold-forming technique wherein intense magnetic pulses at a high frequency reshape the workpiece surface.

photochemical machining

A variation of chemical machining that uses a chemically resistant mask sensitive to light. Light activates the mask only in the areas to be protected. The remaining mask is washed away. The process is typically used to produce parts such as circuit boards and other delicate items.

plasma arc machining (PAM)

Process in which an electric arc between a torch and a workpiece heats and ionizes a gas such as nitrogen, argon, hydrogen or air, which is forced through the torch in a swirling action. Plasma temperatures about 30,000°C maximum, melt and vaporize workpiece material. Secondary, gases or water flow are often used to clean the kerf of molten metal during cutting. Plasma arc machining used for cutting or planing of thick workpieces. Also used to clad or heat-treat surfaces.

plasma beam cutting

Material is removed by focussing a beam of superheated electrically ionized gas (plasma). The material is melted and vaporized and must be electrically conductive.

4. Materials and Matallurgy

aluminium oxide

Abrasive material for grinding wheels; Al_2O_3; also called as oxide ceramics and is used to coat tools.

aluminizing

Formation of an aluminum or aluminum-alloy coating on a metal by hot dipping, hot spraying, or diffusion.

amorphous material

Material not having a crystal structure. Glass is an amorphous material.

anode

The electrode of an electrolyte cell at which oxidation occurs. Electrons flow away from the anode in the external circuit. It is usually at the electrode that corrosion occurs and metal ions enter the electrolyte solution.

alloy

A material having metallic properties and composes of two or more chemical elements of which at least one is a metal.

alloy steel

Steel containing specified quantities of alloying elements (other than carbon, such as manganese, copper, silicon, sulfur and phosphorous) within the specified limits recognized for formation or alloy steels. Alloy steels have improved physical properties.

anodizing

Forming a conversion coating on a metal surface by anodic oxidation: most frequently applied to aluminium to form a protective, non-corrosive surface, often in various colours.

carbon steel

Steel combined with varying amounts of carbon. Alloying elements in small percentages of manganese, silicon and copper, sulfur and phosphorus are used.

cubic boron nitride (CBN)

Crystal manufactured from boron nitride under high pressure and temperature; of cubic structure. Used to cut hard-to-machine ferrous and nickel-based materials up to R_c 70. Second in hardness only to diamond.

cold working

Deforming metal plastically under conditions of temperature and strain rate that induce strain hardening. Working below the recrystallization temperature.

corrosion

The chemical or electrochemical reaction between a material, usually a metal, and its environment that results in deterioration of the material and its properties.

ceramic

Ceramic tools made from finely powdered aluminum oxide sintered into the desired form. Ceramic tools operate at higher speeds than carbides, they wear less, provide smoother finishes, and can machine harder materials. They are, however, less resistant to shock. Typically used for high-speed turning.

cast iron (CI)

A generic term for a large family of cast ferrous alloys in which the carbon content exceeds the solubility of carbon in austenite at the eutectic temperature. Most cast irons contain at least 2% carbon, plus silicon and sulfur and may contain other alloying elements. Various forms of grey cast iron are: white cast iron, malleable cast iron and ductile cast iron.

cathode

The electrode of an electrolytic cell at which reduction is the principal reaction, (electrons flow toward the cathode in the external circuit). Typical cathodic processes are cations taking up electrons and being discharged, oxygen being reduced and the reduction of an element or group of elements from a higher to a lower valence state.

electroplating

Electrodepositing a metal or alloy in an adherent form on a component serving as a cathode. Improves corrosion resistance and appearance of the parts.

elongation

In tensile testing, the increase in the gauge length, measured after fracture of the specimen within the gauge length, usually expressed as a percentage of the original gauge length. Higher elongation represents higher ductility of the material.

embrittlement

Reduction in the normal ductility of a metal due to a physical or chemical change. Hydrogen embrittlement, and temper brittleness are common types of embrittlement.

erosion

Destruction of metals or other materials by the abrasive action of moving fluids, usually accelerated by the presence of solid particles or matter in suspension. When corrosion occurs simultaneously, the term "erosion-corrosion" often is used.

etching

(1) Subjecting the surface of a metal to preferential chemical or electrolytic reaction in order to reveal structural details for metallographic examination.

(2) Chemically or electrochemically removing the outside layers or films from a metal surface to condition the surface for a subsequent treatment, such as painting or electroplating.

fatigue endurance limit

The maximum stress below which a material can be subjected to an infinite number of stress cycles.

galling

A condition whereby excessive friction between high spots results in localized welding with subsequent spalling and further roughening of the rubbing surface(s) of one or both of two mating parts.

galvanic cell

A cell in which chemical change is the source of electrical energy. It usually consists of two dissimilar conductors in contact with each other and with an electrolyte, or of two similar conductors in contact with each other and with dissimilar electrolytes.

galvanic corrosion

Corrosion associated with the electrical current of a galvanic cell consisting of two dissimilar conductors in an electrolyte or two similar conductors in dissimilar electrolytes. Corrosion occurs at the anode of the cell.

hardenability

The relative ability of a ferrous alloy to form martensite when quenched from a temperature above the upper critical temperature. Hardenability is commonly measured as the distance below a quenched surface at which the metal exhibits a specific hardness (50 R_c for example) or a specific percentage of martensite in the microstructure.

hardness

Resistance of metal to plastic deformation, usually by surface indentation. However, the term may also refer to stiffness or temper, or to resistance to scratching, abrasion or cutting. Indentation hardness may be measured by various hardness tests, such as Brinell, Rockwell and Vickers.

heat-affected zone (HAZ)

That portion of the base metal that was not melted during brazing, cutting or welding, but whose microstructure and mechanical properties were altered by the heat.

hot working

Deforming a metal plastically at a temperature above the recrystallization temperature. Recrystallization also occurs without any strain hardening.

machinability, machinability rating

Determines acceptability of a tool for the workpiece to be machined. Indicates workpiece's hardness, chemical composition and qualities, microstructure, tendency to workharden.

pickling

Removing surface oxides from metals by chemical or electrochemical reaction.

pitting

Localized corrosion of a metal surface, confined to point or small area, that takes the form of cavities.

plasma arc spraying

The most versatile of the thermal spraying processes. In plasma arc spraying, coating material in wire or powder form is fed into the tip of a plasma arc torch, where it is melted and propelled against the substrate (base material). The torch is moved to provide controlled coverage of the target area.

plastic deformation

The permanent (inelastic) distortion of metals under applied stresses that strain the material beyond its elastic limit.

precipitation hardening

Hardening caused by precipitation of a constituent from a supersaturated solid solution.

recrystallization

(1) The formation of a new, strain-free grain structure from that existing in cold-worked metal, usually accomplished by heating. (2) The change from one crystal structure to another, as occurs on heating or cooling through a critical temperature.

recrystallization temperature

The minimum temperature at which complete recrystallization of a cold worked metal occurs within a specified time.

red hardness

Ability of a cutting tool material to withstand high temperatures at the point of cutting without softening and deformation.

5. Manufacturing Systems

agile manufacturing

Agile manufacturing improves competitiveness of manufacturing enterprises. An agile enterprise could reconfigure design operations, processes and business relationships quickly bringing about continuous improvements and changes to meet changing market requirements, reducing time to market.

automation

All or part of a machining or manufacturing process is accomplished by setting in motion a sequence that completes the process without further human intervention. May be mechanical (controlled by stops, cams, etc.) electrical (controlled by relays, contact switches, etc.) or electronic (computer-or microprocessor-controlled). "Fully automated" implies computer-integrated manufacturing.

CAD, computer-aided design

Product-design functions performed with the help of computers and special software.

CAE, computer-aided engineering

Engineering functions performed with the help of computers and special software. Includes functions such as determining a material's ability to withstand stresses.

CAM, computer-aided manufacturing

Use of computers to control machining and manufacturing.

CIM, computer-integrated manufacturing

All phases of production from management to sales to order processing, design, quality control, and machining, are controlled or monitored by interconnected computers.

FMS, flexible manufacturing system

Automated manufacturing system designed to machine a variety of similar parts. System is designed to minimize production changeover time. Computers link machine tools with the workhandling system and peripherals. Also associated with machine tools grouped in "cells" for efficient production.

lean production

Lean production refers to producing more with less. Lean production is the result of the use of several newer manufacturing developments that have taken place worldwide. Waste elimination in the use of all resources including people and time is the essence of lean production. Inventory reduction and zero rejection are other dimensions of lean production.

machining or metal cutting system

Any machining process used to produce a component or a given workpiece to produce a new part. Conventionally applies to machining operations in which a cutting tool mechanically removes material in the form of chips; applies to any process in which metal or material is removed to create new shapes.